AF506288

DESIGNING CMOS CIRCUITS FOR LOW POWER

Designing CMOS Circuits for Low Power

Edited by

Dimitrios Soudris
Democritus University of Thrace

Christian Piguet
CSEM, Neuchâtel

and

Costas Goutis
University of Patras

KLUWER ACADEMIC PUBLISHERS

BOSTON / DORDRECHT / LONDON

A C.I.P. Catalogue record for this book is available from the Library of Congress.

ISBN 1-4020-7234-1

Published by Kluwer Academic Publishers,
P.O. Box 17, 3300 AA Dordrecht, The Netherlands.

Sold and distributed in North, Central and South America
by Kluwer Academic Publishers,
101 Philip Drive, Norwell, MA 02061, U.S.A.

In all other countries, sold and distributed
by Kluwer Academic Publishers,
P.O. Box 322, 3300 AH Dordrecht, The Netherlands.

Printed on acid-free paper

Printed in the Netherlands.

Contents

6
Computer Arithmetic Techniques for Low-Power Systems   97
Vassilis Paliouras and *Thanos Stouraitis*

7
Reducing Power Consumption in Memories   117
Alexander Chatzigeorgiou and *Spiridon Nikolaidis*

8
Low-Power Clock, Interconnect and Layout Designs   141
Christian Piguet

9
Logic Level Power Estimation   169

George Theodoridis and *Costas Goutis*

Part II   LOW POWER DESIGN STORIES

10
Low-Power Design for Safety-Critical Applications   205

List of Figures

List of Tables

Contributing Authors

Labros Bisdounis was born in Agrinio, Greece, in 1970. He received the Diploma and the Ph.D. degrees in Electrical Engineering from the Department of Electrical and Computer Engineering, University of Patras, Greece, in 1993 and 1999, respectively. His Ph.D. thesis is in the area of timing and power modeling of submicrometer CMOS circuits. During his collaboration with the above department he participated as a researcher in several EU projects related to VLSI design. Since 2000, he is with the Development Programmes Department of INTRACOM S.A., Athens, Greece, working on the design and development of VLSI circuits and systems. His main research interests is on various aspects of VLSI circuits and systems design such as: circuit timing analysis, power dissipation modeling, design and analysis of submicrometer CMOS circuits, low-power and high-speed CMOS digital circuits and systems design and System-on-Chip design for telecom applications. Dr. Bisdounis has published 20 papers in international journals and conferences on these areas, and he has received more than 70 references. He is a member of the Technical Chamber of Greece.

Alexander Chatzigeorgiou received the Diploma in Electrical Engineering and the Ph.D. degree in Computer Science, from the Aristotle University of Thessaloniki, Thessaloniki, Greece, in 1996 and 2000, respectively. From 1997 to 1999 he was with Intracom S.A., Greece, as a telecommunications software designer. Since May 2002 he has been with the Department of Applied Informatics at the University of Macedonia, Thessaloniki, Greece, as a Lecturer in Software Engineering and Object-Oriented Design. His research interests include low-power hardware and software design, embedded systems architecture and timing and power modeling of digital CMOS circuits. He is a member of the IEEE and the Technical Chamber of Greece.

Costas Goutis was a Lecturer at the School of Physics and Mathematics at the University of Athens, Greece, from 1970 to 1972. In 1973, he was the Technical Manager in the Greek P.T.T., responsible for the installation and maintenance of the telephone exchanges in a large provincial region. He was Research Assistant and Research Fellow in the Department of Electrical and Electronic Engineering at the University of Strathclyde, U.K., from 1976 to 1979, and Lecturer in the Department of Electrical and Electronic Engineering at the University of Newcastle upon Tyne, U.K., from 1979 to 1985. Since 1985 he has been Associate Professor and Full Professor in the Department of Electrical and Computer Engineering, University of Patras, Greece. His recent research interests focus on VLSI Circuit Design, Low Power VLSI Design, Systems Design, Analysis and Design of Systems for Signal Processing and Telecommunications. He has published more than 160 papers in international journals and conferences. He has been awarded a large number of Research Contracts from ESPRIT, RACE, IST and National Programs. Prof. Goutis has close collaboration with the industry. His group has recently won an international low power design contest sponsored by Intel and IBM.

Dimitrios Gouvetas received the Diploma degree in Electrical Engineering from the Department of Electrical and Computer Engineering, University of Patras, Greece in 1997. His Diploma thesis is in the area of CMOS Technology for Low Power and High Speed Circuits. He is a member of the Technical Chamber of Greece.

Thomas Heselhaus received the Dipl.-Ing. degree in 1998 from Aachen University of Technology. Since 1998 he has been working as a research assistant at the Chair of Electrical Engineering and Computer Systems, Aachen University of Technology. His main topic are memory architectures and the modelling and optimization of basic cells.

Athanassios Kakarountas is a postgraduate student at the VLSI Laboratory of the University of Patras. He received his Diploma in Electrical and Electronics Engineering at the University of Patras. The research field his Ph. D. thesis is based on is in the area of Fault detection and fault tolerance in VLSI designing. He has the working experience of two ESPRIT European projects : i) LPGD : Contribution in the formation of a low power methodology/ flow and its application to the implementation of DCS1800 - GSM/DECT Modulator Demodulator for a DECT/GSM (DCS1800) dual mode mobile phone and ii) CoSafe : Design of a Low Power safety-critical pump for in-vein infusion in collaboration with Micrel Medical Devices Ltd.

Odysseas Koufopavlou received his Diploma and the Ph.D. degrees in Electrical Engineering in 1983 and 1990, both from University of Patras, Greece. From

1990 to 1994 he was at the IBM Thomas J. Watson Research Center, Yorktown Heights, NY, USA. He is currently an Associate Professor with the Department of Electrical and Computer Engineering, University of Patras. His research interests include VLSI, low power design, and high performance communication subsystems architecture and implementation. He served as conference General Chair of the 1999 International Conference on Electronics, Circuits and Systems, and on organizing committees and technical program committees of many major conferences. He leads several projects, the Greek government, and major companies. Dr. Koufopavlou has published 83 journal and conference papers and received patents and inventions. He is a member of IEEE, IFIP 10.5 WG and the Technical Chamber of Greece.

Spiridon Nikolaidis was born in Eleochori of Kavala, Greece in 1965. He received the Diploma and Ph.D. degrees in Electrical Engineering from Patras University, Greece, in 1988 and 1994, respectively. Since September 1996 he has been with the Department of Physics of the Aristotle University of Thessaloniki, Thessaloniki, Geece. He is now an assistant professor at this Department in the field of digital design. His research interests include CMOS gate propagation delay and power-consumption modeling, high speed and low power CMOS circuit design techniques, power estimation of DSP architectures, and design of high speed and low power DSP architectures. He is author and co-author in more than 60 papers and articles in conferences. He also participates in many European (ESPRIT, IST) and Greek government projects.

Tobias G. Noll received the Ing. (grad.) degree in Electrical Engineering from the Fachhochschule Koblenz in 1974, the Dipl.-Ing. degree in Electrical Engineering from the Technical University of Munich in 1982, and the Dr.-Ing. degree from the Ruhr-University of Bochum in 1989. From 1974 to 1976, he was with the Max-Planck-Institute of Radio Astronomy, Bonn. Since 1976 he was with the Corporate Research and Development Department of Siemens and since 1987 he headed a group of laboratories concerned with CMOS circuits for digital signal processing. In 1992, he joined the RWTH Aachen University of Technology where he is a Professor holding the Chair of Electrical Engineering and Computer Systems. His activities focus on low power deep submicron CMOS architectures, circuits and design methodologies, as well as digital signal processing for communications and medicine electronics.

Vassilis Paliouras received the Diploma in electrical engineering in 1992 and the Ph.D. degree in electrical engineering in 1999, from the Electrical and Computer Engineering Department, University of Patras, Greece. He works as a researcher at the VLSI Design Laboratory, ECE Dept., while teaching microprocessor-based system design at the Computer Engineering and Informatics Dept., both at the University of Patras, Greece. His research interests include computer arithmetic algorithms and circuits, microprocessor architecture, and VLSI signal processing, areas where he has published more than 30 conference and journal articles. Dr. Paliouras received the MEDCHIP VLSI Design Award in 1997. He is also the recipient of the 2000 IEEE Circuits and Systems Society Guillemin-Cauer best paper Award. He is a Member of ACM, SIAM, and the Technical Chamber of Greece.

Kyriakos Papadomanolakis is a postgraduate student at the VLSI Laboratory of the University of Patras. He received his Diploma in Electrical and Electronics Engineering at the University of Patras. He is currently working on his Ph. D. thesis in the area of Hardware safety in VLSI designing at the VLSI Laboratory of the University of Patras, at Rion. He has the working experience of two ESPRIT European projects: i) ASPIS: Contribution in the designing of a DECT/GSM (DCS1800) dual mode mobile phone, in INTRACOM S.A. at the department of R&D and ii) CoSafe: Design of a Low Power safety-critical pump for in-vein infusion in collaboration with Micrel Medical Devices Ltd.

Christian Piguet received the M. S. and Ph. D. degrees in electrical engineering from the EPFL, respectively in 1974 and 1981. He joined the Centre Electronique Horloger S.A., Neuchâtel, Switzerland, in 1974. He is now Head of the Ultra Low Power Circuits section at the CSEM S.A. He is presently involved in the design of low-power low-voltage integrated circuits. He is Professor at EPFL and also lectures at the University of Neuchâtel, Switzerland. He is author or co-author of more than 100 scientific papers and has contributed to numerous advanced engineering courses.

Robert Schwann received the Dipl.-Ing. degree in 1997 from Aachen University of Technology. Since 1998 he has been working as a research assistant at the Chair of Electrical Engineering and Computer Systems, Aachen University of Technology. His field of research are the signal processing and image quality of medical ultrasound systems.

Dimitrios Soudris received his Diploma in Electrical Engineering from the University of Patras, Greece, in 1987. He received the Ph.D. Degree from in Electrical Engineering, from the University of Patras in 1992. He is currently working as Ass. Professor in Dept. of Electrical and Computer Engineering, Democritus University of Thrace, Greece. His research interests include low power design, parallel architectures, embedded systems design, and VLSI signal processing. He has published more than 80 papers in international journals and conferences. He was leader and principal investigator in numerous research projects funded from the Greek Government and Industry as well as the European Commission (ESPRIT II-III-IV and 5th IST). He has served as General Chair and Program Chair for the International Workshop on Power and Timing Modelling, Optimisation, and Simulation (PATMOS). Recently, received an award from INTEL and IBM for the project results of LPGD #25256 (ESPRIT IV). He is a member of the IEEE, the VLSI Systems and Applications Technical Committee of IEEE CAS and the ACM.

Thanos Stouraitis received a B.S. in Physics and an M.S. in Electronic Automation from the University of Athens, Greece, in 1979 and 1981, respectively, an M.S. in Electrical Engineering from the University of Cincinnati in 1983 and the Ph.D. degree from the Univ. of Florida in 1986. He was awarded the Outstanding Ph.D. Dissertation award of the University of Florida and a Certificate of Appreciation by the IEEE Society of Circuits and Systems in 1997. He is a Professor of Electrical and Computer Engineering at the University of Patras, Greece. He has served on the faculty of the University of Florida and the Ohio State University. He has published 2 books, several book chapters, about 30 journal and 70 conference papers in the areas of computer architecture, computer arithmetic, VLSI signal and image processing, and low-power processing. He is currently the chair of the IEEE Circuits and Systems Society's technical committee on VLSI Systems and Applications, while he serves on the digital signal processing and the multimedia systems committees. He has served as the General Chair of the IEEE 3rd Int. Conference on Electronics, Circuits, and Systems (ICECS '96), co-Program chair for EUSIPCO '98, and regularly serves on the Program Committee of various circuits-and-systems and signal-processing conferences.

Antonios Thanailakis was born in Greece on August 5, 1940. He received B.Sc. degrees in physics and electrical engineering from the University of Thessaloniki, Greece, 1964 and 1968, respectively, and the MSc . and Ph.D. Degrees in electrical engineering and electronics from UMIST, Manchester, U.K. in 1968 and 1971, respectively. He has been a Professor of Microelectronics in Dept. of Electrical and Computer Eng., Democritus Univ. of Thrace, Xanthi, Greece, since 1977. He has been active in electronic device and VLSI system design research since 1968. His current research activities include microelectronic devices and VLSI systems design. He has published a great number of scientific and technical papers, as well as five textbooks. He was leader for carrying out research and development projects funded by Greece, EU, or other organizations on various topics of Microelectronics and VLSI Systems Design (e.g. NATO, ESPRIT, ACTS, STRIDE).

George Theodoridis received the Diploma in Electrical Engineering for the University of Patras in 1994, and the Ph.D. Degree in Electrical and Computer Eng. from the same institution in 2001. He is currently working as researcher at the VLSI Design laboratory of Electrical and Computer Eng. Department of Patras University. His working interests include several aspects of development of methodologies and techniques for power estimation and optimization, reconfigurable computing, parallel architectures and embedded DSP systems. He has published more than 20 papers in international journals and conferences. He also has participated in numerous research projects funding from the European Union and Greek Government and Industry.

Oliver Weiss received the Dipl.-Ing. degree in 1995 from Aachen University of Technology. Since 1995 he has been working as a research assistant at the Chair of Electrical Engineering and Computer Systems, Aachen University of Technology. His main research interest is the development of a datapath generator for the physically oriented design of optimized CMOS macros for digital signal processing.

Foreword

This book is the fourth in a series on novel low power design architectures, methods and design practices. It results from of a large European project started in 1997, whose goal is to promote the further development and the faster and wider industrial use of advanced design methods for reducing the power consumption of electronic systems.

Low power design became crucial with the wide spread of portable information and communication terminals, where a small battery has to last for a long period. High performance electronics, in addition, suffers from a permanent increase of the dissipated power per square millimeter of silicon, due to the increasing clock-rates, which causes cooling and reliability problems or otherwise limits the performance.

The European Union's Information Technologies Programme 'Esprit' did therefore launch a 'Pilot action for Low Power Design', which eventually grew to 19 R&D projects and one coordination project, with an overall budget of 14 million EURO. It is meanwhile known as European Low Power Initiative for Electronic System Design (ESD-LPD) and will be completed in the year 2002. It involves to develop or demonstrate new design methods for power reduction, while the coordination project takes care that the methods, experiences and results are properly documented and publicised.

The initiative addresses low power design at various levels. This includes system and algorithmic level, instruction set processor level, custom processor level, RT-level, gate level, circuit level and layout level. It covers data dominated and control dominated as well as asynchronous architectures. 10 projects deal mainly with digital, 7 with analog and mixed-signal, and 2 with software related aspects. The principal application areas are communication, medical equipment and e-commerce devices.

The following list describes the objectives of the 20 projects. It is sorted by decreasing funding budget.

CRAFT CMOS Radio Frequency Circuit Design for Wireless Application

- Advanced CMOS RF circuit design including blocks such as LNA, down converter mixers & phase shifters, oscillator and frequency synthesiser, integrated filters delta sigma conversion, power amplifier

- Development of novel models for active and passive devices as well as fine-tuning and validation based on first silicon fabricates

- Analysis and specification of sophisticated architectures to meet in particular low power single chip implementation

PAPRICA Power and Part Count Reduction Innovative Communication Architecture

- Feasibility assessment of DQIF, through physical design and characterisation of the core blocks

- Low-power RF design techniques in standard CMOS digital process

- RF design tools and framework; PAPRICA Design Kit.

- Demonstration of a practical implementation of a specific application

MELOPAS Methodology for Low Power Asic design

- To develop a methodology to evaluate the power consumption of a complex ASIC early on in the design flow

- To develop a hardware/software co-simulation tool

- To quickly achieve a drastic reduction on the power consumption of electronic equipment

TARDIS Technical Coordination and Dissemination

- To organise the communication between design experiments and to exploit their potential synergy

- To guide the capturing of methods and experiences gained in the design experiments

- To organise and promote the wider dissemination and use of the gathered design know-how and experience

LUCS Low Power Ultrasound Chip Set.

- Design methodology on low power ADC, memory and circuit design
- Prototype demonstration of a handheld medical ultrasound scanner

ALPINS Analog Low Power Design for Communications Systems

- Low-voltage voice band smoothing filters and analog-to-digital and digital-to-analog converters for an analog front-end circuit of a DECT system
- High linear transconductor-capacitor (gm-C) filter for GSM Analog Interface Circuit operating at supply voltages as low as 2.5V
- Formal verification tools, which will be implemented in the industrial partners design environment. These tools support the complete design process from system level down to transistor level

SALOMON System-level analog-digital trade-off analysis for low power

- A general top-down design flow for mixed-signal telecom ASICs
- High-level models of analog and digital blocks and power estimators for these blocks
- A prototype implementation of the design flow with particular software tools to demonstrate the general design flow

DESCALE Design Experiment on a Smart Card Application for Low Energy

- The application of highly innovative handshake technology
- Aiming at some 3 to 5 times less power and some 10 times smaller peak currents compared to synchronously operated solutions

SUPREGE A low power SUPerREGEnerative transceiver for wireless data transmission at short distances

- Design trade-offs and optimisation of the micro power receiver / transmitter as a function of various parameters (power consumption, area, bandwidth, sensitivity, etc)
- Modulation / demodulation and interface with data transmission systems
- Realisation of the integrated micro power receiver / transmitter based on the super-regeneration principle

PREST Power REduction for System Technologies

- Survey of contemporary Low Power Design techniques and commercial power analysis software tools
- Investigation of architectural and algorithmic design techniques with a power consumption comparison
- Investigation of Asynchronous design techniques and Arithmetic styles
- Set-up and assessment of a low power design flow
- Fabrication and characterisation of a Viterbi demonstrator to assess the most promising power reduction techniques

DABLP Low Power Exploration for Mapping DAB Applications to Multi-Processors

- A DAB channel decoder architecture with reduced power consumption
- Refined and extended ATOMIUM methodology and supporting tools

COSAFE Low Power Hardware-Software Co-Design for Safety-Critical Applications

- The development of strategies for power efficient assignment of safety critical mechanisms to hardware or software
- The design and implementation of a low-power, safety-critical ASIP, which realises the control unit of a portable infusion, pump system

AMIED Asynchronous Low-Power Methodology and Implementation of an Encryption/Decryption System

- Implementation of the IDEA encryption/decryption method with drastically reduced power consumption
- Advanced low power design flow with emphasis on algorithm and architecture optimisations
- Industrial demonstration of the asynchronous design methodology based on commercial tools

LPGD A Low-Power Design Methodology/Flow and its Application to the Implementation of a DCS1800-GSM/DECT Modulator/Demodulator

- To complete the development of a top-down, low power design methodology/flow for DSP applications
- To demonstrate the methods at the example of an integrated GFSK/GMSK Modulator-Demodulator (MODEM) for DCS1800-GSM/DECT applications

SOFLOPO Low Power Software Development for Embedded Applications

- Develop techniques and guidelines for mapping a specific algorithm code onto appropriate instruction subsets
- Integrate these techniques into software for the power-conscious ARM-RISC and DSP code optimisation

I-MODE Low Power RF to Base band Interface for Multi-Mode Portable Phone

- To raise the level of integration in a DECT/DCS1800 transceiver, by implementing the necessary analog base band low-pass filters and data converters in CMOS technology using low power techniques

COOL-LOGOS Power Reduction through the Use of Local don't Care Conditions and Global Gate Resizing Techniques: An Experimental Evaluation.

- To apply the developed low power design techniques to the existing 24-bit DSP, which is already fabricated
- To assess the merit of the new techniques using experimental silicon through comparisons of the projected power reduction (in simulation) and actually measured reduction of new DSP; assessment of the commercial impact

LOVO Low Output VOltage DC/DC converters for low power applications

- Development of technical solutions for the power supplies of advanced low power systems, comprising the following topics
- New methods for synchronous rectification for very low output voltage power converters

PCBIT Low Power ISDN Interface for Portable PC's

- Design of a PC-Card board that implements the PCBIT interface
- Integrate levels 1 and 2 of the communication protocol in a single ASIC
- Incorporate power management techniques in the ASIC design:
  - system level: shutdown of idle modules in the circuit
  - gate level: precomputation, gated-clock FSMs

COLOPODS Design of a Cochlear Hearing Aid Low-Power DSP System

- Selection of a future oriented low-power technology enabling future power reduction through integration of analog modules
- Design of a speech processor IC yielding a power reduction of 90% compared to the 3.3 Volt implementation

The low power design projects have achieved the following results:

- Projects, who have designed a prototype chip, can demonstrate a power reduction of 10 to 30 percent.

- New low power design libraries have been developed.

- New proven low power RF architectures are now available.

- New smaller and lighter mobile equipment is developed.

Instead of running a number of Esprit projects at the same time independently of each other, during this pilot action the projects have collaborated strongly. This is achieved mostly by the novelty of this action, which is the presence and role of the coordinator: DIMES - the Delft Institute of Microelectronics and Submicron-technology, located in Delft, the Netherlands (http://www.dimes.tudelft.nl). The task of the coordinator is to co-ordinate, facilitate, and organize:

- The information exchange between projects.

- The systematic documentation of methods and experiences.

- The publication and the wider dissemination to the public.

The most important achievements, credited to the presence of the coordinator are:

- New personnel contacts have been made, and as a consequence the resulting synergy between partners resulted in better and faster developments.

- The organization of low power design workshops, special sessions at conferences, and a low power design web site, http://www.esdlpd.dimes.tudelft.nl. At this site all public reports of the projects can be found and all kind of information about the initiative itself.

- The used design methodology, design methods and/or design experience are disclosed, are well documented and available.

  Based on the work of the projects, in cooperation with the projects, the publication of a low power design book series is planned. Written by members of the projects this series of books on low power design will disseminate novel design methodologies and design experiences, which were obtained during the runtime of the European Low Power Initiative for Electronic System Design, to the general public.

In conclusion, the major contribution of this project cluster is that, except the already mentioned technical achievements, the introduction of novel knowledge on low power design methods into the mainstream development processes is accelerated.

We would like to thank all project partners from all the different companies and organizations who make the Low Power Initiative a success.

Rene van Leuken, Reinder Nouta, Alexander de Graaf, Delft, May 2002

Introduction

Modern electronic systems have reached a significant turning point in the last decade, from low performance products such as wristwatches and calculators to high performance products such as laptops and personal digital assistants. The introduction of these devices to the consumer market raised to the surface a characteristic that had been previously omitted. This was low power dissipation. Gradually, engineers invented novel techniques, which may be included in efficient design methodologies, for designing and implementing efficient circuits not only in terms of area and performance, as they were used to, but in the term of low power consumption.

The material in this book is based on the background and the innovative results of the different partners involved in the AMIED, LPGD, PREST, COSAFE, and LUCS projects of the European Low Power Initiative for Electronic System Design under the successful coordination of DIMES, Delft. The partners have been studying for many years low-power design field introducing novel concepts and efficient techniques. Due to close collaboration of academic and industrial research groups the presented material have been influenced by the plethora of disseminations e.g. public deliverables, technical meetings, workshops, during the projects execution.

The book consists of two parts: The first part includes the low power design techniques for power optimization and estimation, while the second one provides the results from the projects COSAFE and LUCS. Starting from the description of the power consumption sources, low power optimization and estimation techniques for logic design level, circuit/transistor design level, and layout design level are provided in eight chapters (i.e. Chapters 2-9). The next two chapters describe the novel low power techniques, which were used during the implementation of the safety-critical Application Specific Instruction Processor designed in COSAFE project, and the implementation of the low power 16-channel ultrasound beamformer application specific integrated circuit (ASIC) designed for LUCS project.

A top-down approach with respect to the design level is adopted in the presentation of the low power design techniques . However,it was not possible to present in detail manner all the low power optimization and estimation techniques from the logic level to the layout level. Only the most important research contributions presented in a tutorial manner, are included. The book can also be used as a textbook for undergraduate and graduate students, VLSI design engineers,and professionals, who have had a basic knowledge of VLSI digital design.

The authors of the chapters of this book together with the editors would like to use this opportunity to thank the many people, i.e. colleagues and Ph.D. students, whose dedication and industry during the projects execution lead to the introduction of novel scientific results and realization of innovative integrated systems.

The authors of Chapter 3 and 9 would like to thank Dr. S. Theoharis for his contribution in the software development of logic optimization and estimation tools, which were necessary for making power measurements.

The authors of Chapter 6 wish to acknowledge the discussions with Dr. Karagianni, who significantly influenced the particular chapter. Also the authors appreciate the financial support from the "C. Caratheodory's" fund for the University of Patras.

The authors of Chapter 10 would like to thank V. Spiliotopoulos for his support in the designing and realization of the COSAFE ASIP. Also, many thanks to V. Kokkinos for his contribution in the implementation of the many fault-secure multiplier designs.

All authors of this book together with the editors would like to thank DIMES, Delft, for their continuous support during the running of the low power projects for the dissemination of the scientific results. This book is one of the activities of this dissemination task.

The authors of Chapter 4 would like to thank the people, who contributed to PREST, whose public deliverable reports used as an inspiration source for chapter's preparation.

Last but not least, D. Soudris would like to thank his parents for being a constant source of moral support and for firmly imbibing into him from a very young age that *perseverantia omnia vincit* - it is this perseverance that kept him going. This book is dedicated to them.

Dimitrios Soudris, Christian Piguet, Costas Goutis, May 2002

I

LOW POWER DESIGN METHODS

Chapter 1

MOTIVATION, CONTEXT AND OBJECTIVES

Dimitrios Soudris
Democritus University of Thrace
Xanthi, GR-67100, Greece
dsoudris@ee.duth.gr

Christian Piguet
CSEM: Centre Suisse d'Electronique et de Microtechnique
Neuchâtel, Switzerland
LAP-EPFL,Lausanne, Switzerland
christian.piguet@csem.ch

Costas Goutis
University of Patras
Patras, GR-26500, Greece
goutis@ee.upatras.gr

Power dissipation has emerged as a very significant design constraint, that must every designer take into consideration. Portability, as well as packaging, cooling, circuit reliability, cost and operation frequency, have steered researchers to try to find approaches in order to confront the power requirements. This is especially true in the field of personal computing devices (portable desktops, audio and video based multimedia products), wireless communication systems (personal digital assistants, personal communicators, videophones), home entertainment (consumer set-top boxes) and wearable computers, which are becoming increasingly popular. There also exists a strong pressure for designing of high-end products to reduce their power dissipation.

D. Soudris et al. (eds.), Designing CMOS Circuits for Low Power, 3–8.
© 2002 *Kluwer Academic Publishers. Printed in the Netherlands.*

A top-down ordinary VLSI design approach is illustrated in Figure 1.1(a), which summarizes the flow of steps that are required to follow from a system-level specification to the physical design. This approach aimed at performance optimization and area minimization. However, the introduction of power dissipation as the third design parameter forced the designers to update the existing design procedure, as it is shown in Figure 1.1(b). Incorporating in each design level two fundamental design steps, namely the *power optimization* and the *power estimation*, the resulting procedure is the low power design flow shown in 1.1(b). By the term "power optimization", we define the process of deriving the best design, given certain constraints, without any violation of design specifications. A set of power optimization techniques should be applied in each design level to meet the design goals. Power estimation is concerned about the calculation of the power or energy dissipation, given a percentage of accuracy, at different phases of the design process. The purpose power estimation is to increase confidence of the design ensuring that the power dissipation goals are violated. At the same time, power estimation techniques evaluate the effect of various optimizations and design modifications on power at each abstraction level.

A designer should follow a certain trajectory of design levels in a top-down approach. Although a designer performs first power optimization step and then power estimation step, within a certain design level there is no exact design procedure. Each design level includes a large collection of low power techniques, each of which may result into a significant reduction of the power dissipation. However, a certain combination of low power techniques may lead to better results than an another series of the same techniques. Actually, this issue is an open research topic, strongly-depended on the application requirements. The degree of ordering accuracy of the low power techniques is inversely-proportional to the range of the application.

The purpose of this book is twofold: first to provide in a tutorial way the fundamentals of low power design for logic, circuit, physical design level and second to present recent research results, which derived by the projects COSAFE and LUCS in the context of European Low Power Initiative for Electronic System Design (ESD-LPD). Low power design techniques and methodologies for system and architecture level are beyond the scope of this book. Significant contributions for higher design levels are presented in the book entitled "Unified low-power design flow for data dominated multi-media and telecom applications" edited by F. Catthoor, IMEC, in the context of ESD-LPD.

The objective of the first part of this book (i.e. Chapters 2-9) is to assist readers develop in-depth analytical and design capabilities for low power design CMOS circuits. Determining the sources of power dissipation, in-depth description of the main existing low power optimization and estimation techniques, and, their corresponding advantages, drawbacks and comparisons will

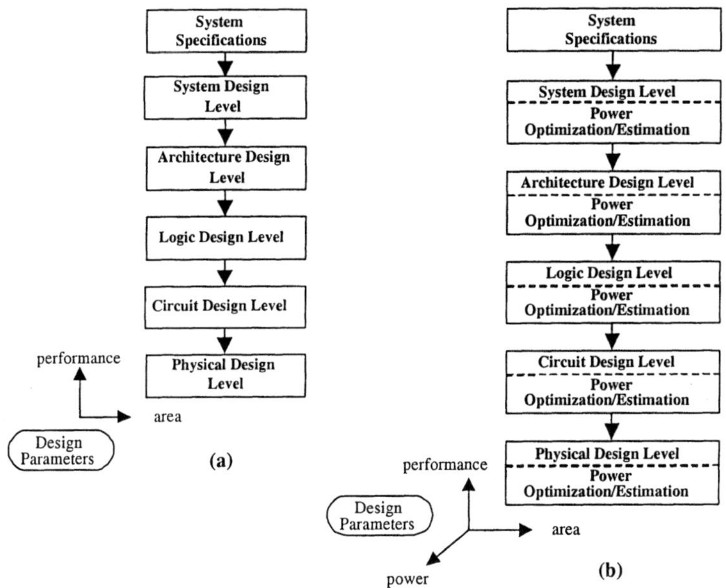

Figure 1.1. VLSI Design Flow: (a) traditional design flow and (b) low power design flow with power optimization and estimation steps.

be discussed. This material can be used as a textbook for undergraduate and graduate students, VLSI design engineers, and professionals, who have had a basic knowledge of Microelectronics and CMOS digital design. Much of the above-mentioned low power information was exploited during the design and implementation of the novel systems described in Chapter 10 and 11. Consequently, a reader can realize *how* low-power design concepts, techniques and flows can be utilized in realistic and competitive applications.

The book consists of ten contributions, where the eight of them cover the fundamentals of low power design field for logic, circuit, and physical level, while the remaining two contributions concern the "design story" of two innovative low power systems. Brief description of each contribution will be provided in the following paragraphs.

The power dissipation sources of CMOS circuits are presented in *Chapter 2*. Specifically, the main principles of dynamic, short-circuit, static, and leakage current dissipation are illustrated together with the low power strategies for reducing each power component. Due to technology advance, the chapter includes the emerging design approach with multiple supply voltages and/or multiple threshold voltages, which allow significant reduction of power components.

A typical low power design flow consists of power optimization and estimation techniques, which should be applied in each design level. The next seven contributions describe such techniques. *Chapter* 3 addresses the techniques for power optimization of logic circuits. Starting with the formulation of logic optimization problem, design techniques for combinational and sequential circuits are presented. The power optimization of a logic circuit implies the reduction of its switching activity employing techniques and algorithms in the technology-independent and technology-dependent optimization steps. Furthermore, the categorization of the existing techniques as well as the optimization design flow followed by a designer is provided.

The power characteristics of different logic styles and alternative implementations of basic digital circuits are reviewed and compared in *Chapter* 4. In particular, the critical characteristics of several logic styles, such as dynamic logic and pass transistor logic, are studied in terms of performance, area and power dissipation. Then, an overview of latches and flip-flops focusing on power characteristics are provided. Furthermore, techniques for reducing power dissipation based on transistor sizing and reordering are presented.

Efficient implementations of adder and multiplier circuits with respect to power, silicon area, delay, and power-delay product, for various topologies are presented in *Chapter* 5. Considering a several static and dynamic CMOS circuit design styles, different circuits of of 1-bit full adder are designed. The different design styles are compared by performing detailed transistor-level simulations on a benchmark full adder circuit using HSPICE, and analyzing the results in a statistical way. Furthermore, power measurements and comparisons for a number of well-known multipliers are provided.

Chapter 6 investigates techniques that reduce the power dissipated in digital circuits based on alternative arithmetic schemes. By exploiting transformations that alter both the data representation and the corresponding operators, it is found that power dissipation can be significantly reduced, for particular computationally-intensive applications. Two classes of transformations, namely the Residue Number System (RNS) and the Logarithmic Number System (LNS). Several properties of the LNS and the RNS are reviewed, the exploitation of which reduces the data activity, the strength of the operators, or even the number of actual operations required to perform certain computational tasks.

Power reduction techniques for SRAM and DRAM memories, which depend on three key issues that affect energy dissipation: static current, supply voltage and charging/discharging capacitance, are presented in *Chapter* 7. In the first half of this chapter the sources of power dissipation in SRAMs are described briefly, while several well-known methods for active power reduction in SRAMs are considered. Most of these methods concentrate on the reduction of static currents and on power optimizations of circuits, whose power dissipation relies heavily on static current, such as sense amplifiers. In DRAMs however, the dominant component of the power dissipation comes from the memory array, especially during charging/discharging of the bit-lines. Therefore, techniques that allow the reduction of this source of power are described.

Chapter 8 describes low power issues at the physical implementation of CMOS circuits. The first very important issue is the clocking scheme, as clock trees consume more and more power due to clock skew minimization. Critical techniques are presented, such as two loaded and unloaded parallel clock trees and latch-based clocking schemes including gated clocks. Larger and larger interconnect delays and crosstalk effects in deep submicron technologies are the second issue, resulting in new routing strategies with hierarchical metal layers. The third issue is layout design for low-power standard cell libraries.

Chapter 9 addresses the issue of power estimation of a logic circuit. The use of power estimators is very important, since it allows checking if the power specifications are met, avoiding the cost of the redesign process. This contribution starts with the formulation of power estimation problem and continues with basic terminology used in literature. The estimation methods are classified into the simulation-based power estimation and probabilistic methods. The main characteristics of each estimation method as well as its advantages and limitations are provided.

The next two chapters include the architecture and the design techniques used for the implementation of the safety-critical Application Specific Instruction Processor designed in COSAFE project, and the implementation of the low power 16-channel ultrasound beamformer application specific integrated circuit designed for LUCS project.

Safety critical applications are demanding redundancy to increase observability and achieve the required safety levels. An approach to achieve high levels of safe operation while system's power dissipation is maintained low is presented in *Chapter* 10. The terms related to the reliability of a system and design for testability are presented in detail. The controversial nature of low-power design and design for on-line testability is exhibited presenting unsuitable low-power techniques. Design of self-checking circuits, combinational and sequential, is presented and how they can be combined to form a safe operating system. A portable system designed to infuse medicine to patients, which is a safety demanding application, has served as a reference system.

Electronic beamforming, as being well known from radar technique, has become standard in today's medical diagnostic ultrasound scanner systems. *Chapter* 11 describes a design experiment on the implementation of the digital part of a low power ultrasound beamformer ASIC. The design is based on a physically oriented design methodology applying a datapath generator for automatic layout generation and a quantitative optimization of all building blocks for lowest possible power dissipation. State-of-the-art power reduction strategies were applied on all levels of the design process, from the algorithmic system level down to the physical level. Appropriate number representation in arithmetic building blocks, clustered supply voltage technique, customized memory blocks, just to name few of the further power reduction strategies being applied, allowed to reach unequaled power figures.

The last chapter of this book provides a brief history of microelectronics revolution underlining the major scientific contributions, for instance the transistor's invention, and the future perspectives of the microelectronics technology for the next decades. Emphasis is given on the potential of the emerging nanotechnology.

Chapter 2

SOURCES OF POWER DISSIPATION
IN CMOS CIRCUITS

Dimitrios Soudris

Antonios Thanailakis
Democritus University of Thrace, Xanthi, Greece
{ dsoudris, thanail } @ee.duth.gr

Abstract   The study of the power dissipation sources of CMOS circuits is presented. Specifically, the main principles of dynamic, short-circuit, static, and leakage power dissipation are illustrated together with the low power strategies for reducing each power component. Furthermore, we enlighten to some innovative techniques of power reduction, which are based on multiple supply voltages and multiple threshold voltages.

Keywords: CMOS inverter, dynamic power dissipation, short-circuit power dissipation, leakage power dissipation, static power consumption, multiple supply voltage, multiple threshold voltage

2.1   Introduction

Power dissipation is recognized as a critical parameter in modern VLSI design field. The development of competitive market sectors such as wireless applications, laptops, and portable medical devices, depends on the power dissipation as the most important parameter because the growth rate of the battery technologies is not so promising [1], [2]. In most cases, the low power dissipation is equally important with the remaining two design parameters, namely area and speed. Consequently, the design of an efficient integrated circuit in terms of power, area, and speed simultaneously, has become a very challenging problem.

D. Soudris et al. (eds.), Designing CMOS Circuits for Low Power, 9–22.
© 2002 *Kluwer Academic Publishers. Printed in the Netherlands.*

2.2 The components of power dissipation in CMOS circuits

The power dissipation in digital CMOS circuits can be described by:

$$P_{avg} = P_{dynamic} + P_{short-circuit} + P_{leakage} + P_{static} \qquad (2.1)$$

where P_{avg} is the average power dissipation, $P_{dynamic}$ is the dynamic power dissipation due to switching of transistors, $P_{short-circuit}$ is the short-circuit current power dissipation when there is a direct current path from the power supply down to ground, $P_{leakage}$ is the power dissipation due to leakage currents, and P_{static} is the static power dissipation.

2.2.1 Dynamic Power dissipation

The dynamic power dissipation, $P_{dynamic}$, is caused by the charging and discharging of capacitances in the circuit. We will illustrate the computation of dynamic power dissipation through the example of a CMOS inverter driving a load capacitor C_L, as it is shown in Figure 2.1. The output capacitor C_L represents the cumulative effect due to parasitic capacitances of the nMOS and pMOS transistors (source- and drain-diffusion to bulk), the capacitance associated with internal and external wires of the inverter cell, and the input capacitance (gate to bulk) of the circuits driven by the inverter.

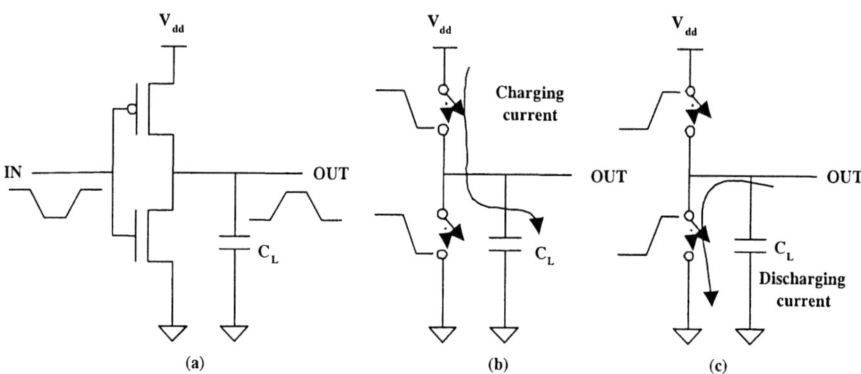

Figure 2.1. The operation of a CMOS inverter: (a) CMOS inverter, (b) Charging phase, and (c) Discharging phase.

We consider the operation of a CMOS inverter assuming that the circuit is initially in a steady state, having as input the logic value '1' and, obviously as, output the logic value '0'. In this case the output capacitor is discharged. When the input waveform undergoes a falling transition, the pMOS transistor conducts (ON) and the nMOS transistor turns off, as it is depicted in Figure

2.1(b). The current drawn from the power supply charges now the capacitor C_L up to V_{dd}. During this charging process, the energy drawn from the power supply is $C_L \cdot V_{dd}^2$, of which half is stored in the capacitor C_L and the other half is dissipated in the parasitic capacitances of pMOS transistor and the interconnect. When the input waveform undergoes a rising transition, the nMOS transistor conducts and the pMOS transistor turns off, as it shown in Figure 2.1(c). Now, there is a current path directly from the output capacitor to the ground and, thus, a discharging current flows through this path. The energy $\frac{1}{2} C_L \cdot V_{dd}^2$ stored in the load capacitor is dissipated into the nMOS transistor as well as in the interconnect. Therefore, the dynamic power dissipated by a CMOS inverter over a time interval $[0, T]$ can be computed by:

$$P_{dynamic} = C_L \cdot V_{dd}^2 \cdot N_{0 \to 1} \cdot \frac{1}{T} \tag{2.2}$$

where $N_{0 \to 1}$ is the number of rising transitions at the inverter's output, or equivalently the number of times C_L is charged, over the period $[0, T]$. Without loss of generality, we assume that the inverter operates in a clock frequency f, and that the number of rising transitions is half the total number of transitions. Thus, eq. 2.2 may be rewritten as follows [3], [4]:

$$P_{dynamic} = C_L \cdot V_{dd}^2 \cdot N \cdot f \tag{2.3}$$

where N is the average number of transitions per clock cycle at the inverter's output, and it will, henceforth, be referred to as the *switching activity*.

The dynamic power dissipation is the dominant factor compared with the other components of power dissipation in digital CMOS circuits. For technologies up to 0.35 μm, the dynamic dissipation is about 80% of a circuit's total dissipation. As the technology scales down, i.e. for submicron technologies, the contribution of dynamic power dissipation also increases because of increased functionality requirements and the clock frequencies [9]. Consequently, the majority of existing low power design and power estimation techniques focuses on this dynamic component of dissipation.

2.2.1.1   Rower Reduction Approaches of Dynamic dissipation.   Equation 2.3 derived for the average switching power dissipation of CMOS logic gates indicates that $P_{dynamic}$ is proportional to the the load capacitance, C_L, the square of V_{dd}, the switching activity and clock frequency f. Consequently, the power reduction can be achieved by various manners:

- Reduction of output capacitance, C_L,

- Reduction of power supply voltage, V_{dd},

- Reduction of the average number of transitions per clock cycle, N, (or switching activity) and

- Reduction of clock frequency.

Apparently, low power reduction can be achieved by the combination of two or more aforementioned manners. A very popular low power strategy aims at the reduction of the *effective capacitance* or *switched capacitance*, which is defined as the product of output capacitance times switching activity, i.e. $C_L \cdot N$.

Generally speaking, the two main low power reduction strategies concern the reduction of voltage supply and the effective capacitance and therefore, a designer should devise novel techniques based on these strategies. In particular, the reduction of power supply voltage is one of the most aggressive techniques because the power savings are significant due to the quadratic dependence of V_{dd} (eq. 2.3). Although such reduction is usually very effective, a designer has to deal with many important issues to avoid decrease of the system performance. Specifically, the reduction of the power supply voltage leads to an increase to the delay propagation (i.e. decrease of circuit speed), as it is shown in Figure 2.2. Since the input and output signal levels should be compatible with the peripheral circuitry, the shift of industry from a supply voltage value to a smaller one is quite expensive and slow. In contrast, the reduction of the switching activity and/or the capacitance for a certain technology depends mainly on the invention of novel design techniques. Therefore, an existing circuit can be re-designed to be a low-power circuit avoiding the significant expenses of purchasing new technology. The reduction of switching activity requires a detailed analysis of signal transition probabilities, and implementation of various circuit-level design techniques, for instance logic synthesis optimization and balanced paths. The output capacitance can be reduced, for instance, by transistor resizing and selection of the appropriate logic family. A typical example, where power reduction came from the use of both supply voltage reduction and low power design techniques, is the Pentium-I with 15 watts (5V@66MHz) versus the Pentium-II with 8 watts (3.3V@133 MHz). Plethora and detailed description of low power design techniques will be provided in the next chapters.

2.2.2 Short-circuit power dissipation

The short-circuit power consumption, $P_{short-circuit}$, is caused by the current flow through the direct path existing between the power supply and the ground during the transition phase. Consider again the CMOS inverter shown in Figure 2.1. When the input signal changes from the logic value '1' to the logic value '0', or vice versa, there exists a very small time interval during which both nMOS and pMOS transistors are ON, and hence a short-circuit current flows between the power supply and the ground. Figure 2.3 illustrates the short-circuit current effect in a CMOS inverter. More specifically, if the rising input voltage exceeds the threshold voltage, V_{thn}, the nMOS transistor of the inverter circuit starts

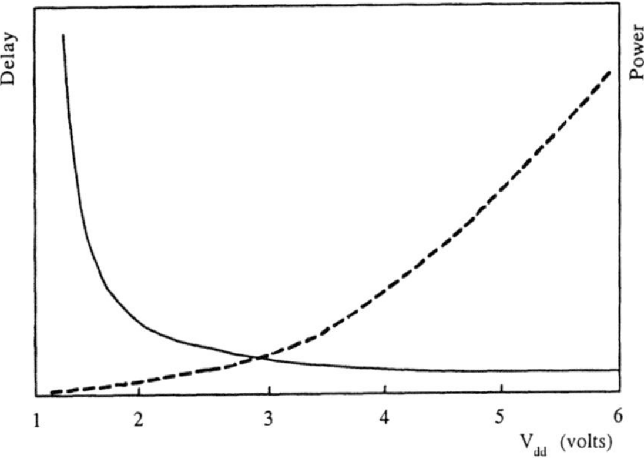

Figure 2.2. Delay Propagation and Power dissipation vs Supply Voltage V_{dd}

conducting, while the pMOS transistor conducts until the input voltage reaches to the value of $(V_{dd}- \mid V_{thp} \mid)$. Hence, there exists a time interval where both transistors are turned on. As the capacitance C_L is discharged through the nMOS transistor, the output voltage starts decreasing. The drain-to-source voltage drop of the pMOS transistor becomes nonzero, which allows the pMOS to conduct as well. The short-circuit current is ceased when the input voltage transition is completed and the pMOS is turned off. Considering a symmetrical inverter and identical rise and fall times, a similar situation occurs for the short-circuit current component deriving from the falling edge of the input signal, where the output waveform starts rising and both MOSFET transistors are ON. The average of both the short-circuit current component of the rising edge of the input signal and the corresponding current component from the falling edge determine the total amount of power drawn from the power supply.

The short-circuit current is especially dominant when the output load capacitance is small, and/or when the input signal rise and fall times are large. Assuming symmetric rise and fall delays and threshold voltages, the time-averaged short-circuit current drawn from the power supply and the short-circuit current power dissipation of a CMOS inverter can be approximated by [5]

$$I_{short-circuit} = K/V_{dd} \cdot (V_{dd} - 2V_{th})^3 \cdot \tau \cdot N \cdot f \qquad (2.4)$$

$$P_{short-circuit} = K \cdot (V_{dd} - 2V_{th})^3 \cdot \tau \cdot N \cdot f \qquad (2.5)$$

respectively, where K is a constant that depends on the transistor sizes, as well as on the technology, V_{th} is the threshold voltage of the nMOS and pMOS transistors, τ is the rise or fall time of the input signal, N is the average number of transitions in the inverter's output, and f is the clock frequency.

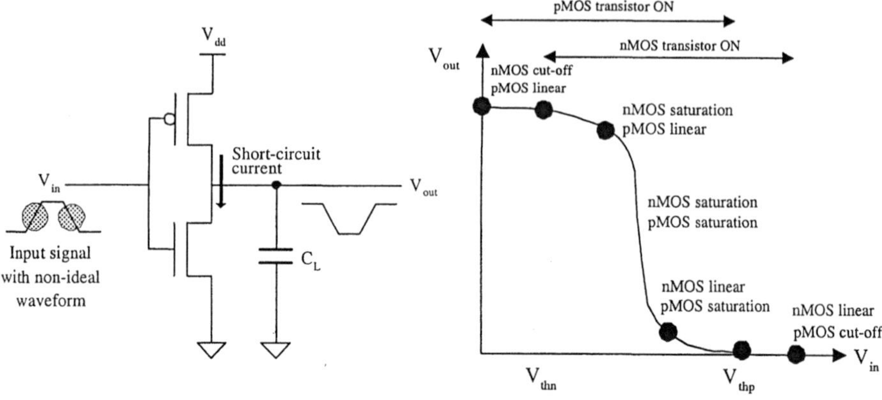

Figure 2.3.   CMOS inverter transistors conduct simultaneously a short-circuit current

Figure 2.4 illustrates the relations among the V_{in}, V_{out}, and $I_{short-circuit}$ of CMOS inverter with respect to time. More specifically, the left-hand side vertical axis denotes the input and output voltage (i.e. V_{in} and V_{out}), while the right-hand side vertical axis concerns the short-circuit current, $I_{short-circuit}$. These curves resulted from HSPICE simulation in $LEVEL = 49$ of a CMOS inverter of $0.5\mu m$ technology, with pMOS transistor width $W_p = 6\mu m$, nMOS transistor width $W_n = 4\mu m$, load capacitance $C_L = 100 fF$, input transition time $2ns$, and $V_{dd} = 5$ volts. Also, the gate oxide thickness was chosen to be $t_{ox} = 9.6 \times 10^{-9}m$, the threshold voltage of the pMOS transistor is equal to $V_{thp} = -0.9213$ volts, and the nMOS transistor to $V_{thn} = 0.7$ volts.

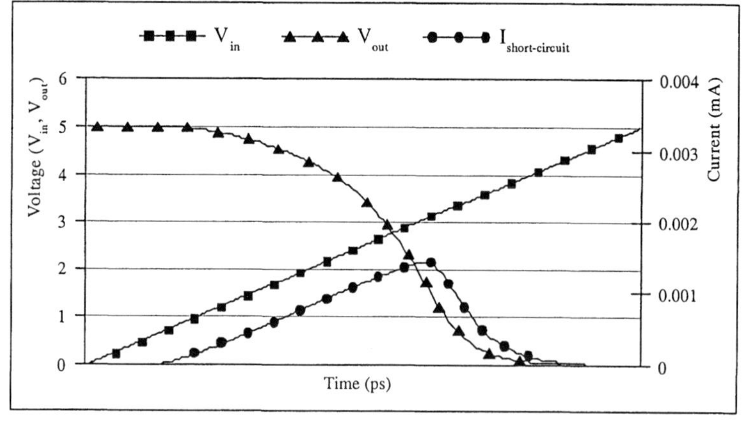

Figure 2.4.   CMOS inverter short-circuit current HSPICE simulation.

Here, it should be stressed again that the above measurements concern the case with small output capacitance. Similar results can be derived in the case with large output capacitance and smaller transition time. Since both transistors conduct for very small time interval, the resulting $I_{short-circuit}$ current is also smaller. However, the peak value of the supply current to charge up the output capacitance is larger. The reason for this the pMOS transistor remains in saturation during the whole input transition interval, as opposed to the previous case where the transistor leaves the saturation before the input transition is completed.

Reduction in the short-circuit power dissipation can be achieved by applying various techniques. From eq. 2.5 it is obvious that, reducing the transistor ratio W/L and scaling down the technology the switched capacitance and supply voltage are reduced. The $P_{short-current}$ is linearly proportional to the input signal rise and fall times and therefore, reducing the input transition times, the short-circuit current decreases. It should be stressed that short-circuit power dissipation exists in static CMOS logic families, but not in dynamic logic gates. Indeed, there is no direct current path between V_{dd} and ground in dynamic logic gates, because the precharged and evaluated transistors should never be simultaneously ON. Otherwise, they would function incorrectly [6].

2.2.3 Leakage Power Dissipation

The nMOS and pMOS transistors used in a CMOS logic circuit commonly have nonzero reverse leakage and subthreshold currents. Having a CMOS integrated circuit, which encompasses a very large number of transistors, these currents can contribute to the total power dissipation even when the transistors are not performing any switching action. The magnitude of the leakage currents depends mainly on the used technology parameters.

The leakage power dissipation, $P_{leakage}$, is caused by two types of leakage currents [3]:

- the reverse-bias diode leakage current at the transistor drains, and

- the subthreshold current through a turned-off transistor channel.

However, these current components are technologically-controlled and, thus, the designer can do a number of things their minimization. Diode leakage current occurs when a transistor is turned-off, and the other transistor ON charges up/down the drain of the former respect to its substrate potential. In Figure 2.5 is shown a pMOS transistor with a negative gate bias V_{dd} in respect to its substrate. Hence, the diode formed by the drain diffusion and the substrate is reverse-biased. We know from diode's theory that the reverse bias current is given by:

$$I_{leakage} = I_S \left(e^{\frac{V_{dd}}{V_T}} - 1 \right) \tag{2.6}$$

where I_S is the reverse saturation current, V_{dd} is the bias voltage, and $V_T = kT/q$ is the thermal voltage. For engineering purposes, we may assume that the leakage current is equal to the reverse saturation current.

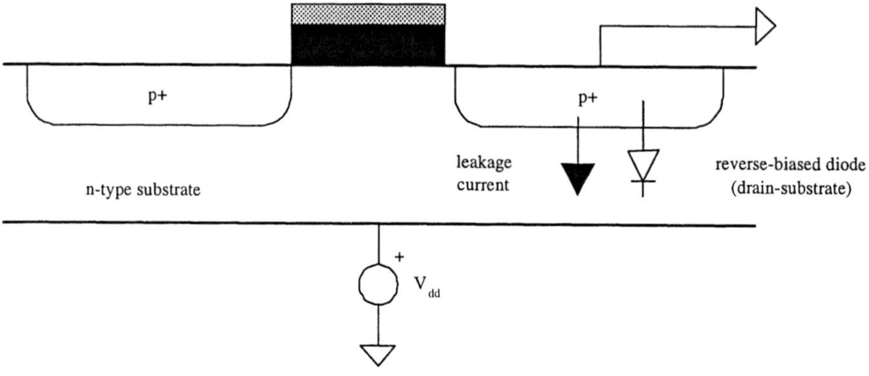

Figure 2.5.   The leakage current in a reverse-biased pMOS transistor.

The reverse saturation current is given by [7]:

$$I_S = qn_i^2 A \left(\frac{D_p}{N_d W_n} + \frac{D_n}{N_a W_p} \right) \tag{2.7}$$

where q is the electron charge, n_i is the intrinsic carrier concentration, A is the area of pn-junction diode (actually the drain area), D_p and D_n are the electron and hole diffusion coefficients, respectively, N_d and N_a are the donor and acceptor concentrations, respectively, W_p and W_n are the depletion layer widths of the p- and n-sides of pn-junction diode, respectively. For example, assuming an average drain of 10 μm^2, the total leakage current for one-million transistors is about 25 μA.

The second current component is the subthreshold leakage current, which occurs due to carrier diffusion between the source and the drain of MOS transistor, when the gate-source voltage, V_{gs}, exceed the weak inversion point, but it is still below the threshold voltage, V_{th}, above which the carrier drift mechanism is dominant. The behaviour of a MOSFET transistor in the subthreshold operation region is similar to a bipolar transistor, and the sub-threshold current is exponentially dependent on the gate-source voltage (Figure 2.6). The magnitude of the subthreshold current may increase significantly when the gate-to-source voltage is smaller than, but very close to, the threshold voltage of the transistor. This phenomenon results in a power dissipation due to subthreshold leakage, whose magnitude is comparable with the dynamic power dissipation. The concept of subthreshold leakage current is illustrated in Figure 2.7.

The current in the subthreshold region is given by [8]:

$$I_{ds} = K e^{(V_{gs}-V_{th})/nV_T} \left(1 - e^{-\frac{V_{ds}}{V_T}}\right) \qquad (2.8)$$

where K is a function of the technology, V_T is the thermal voltage, V_{th} is the threshold voltage and $n = 1 + \Omega t_{ox}/D$, where t_{ox} is the gate oxide thickness, D is the channel depletion layer width, the quantity $\Omega = \varepsilon_{Si}/\varepsilon_{ox}$. For $V_{ds} \gg V_T$, the quantity $\left(1 - e^{-V_{ds}/V_T}\right) \approx 1$; i.e., the drain-to-source leakage current does not depend on the drain-source voltage V_{ds}, for $V_{ds} \simeq 0.1$ volts.

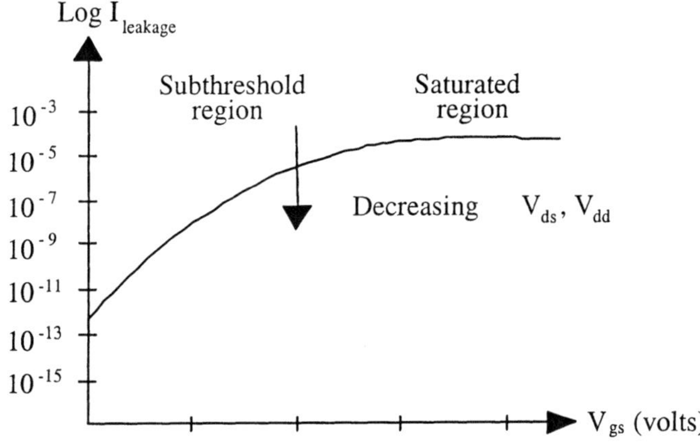

Figure 2.6. Subthreshold leakage with respect to gate-source voltage.

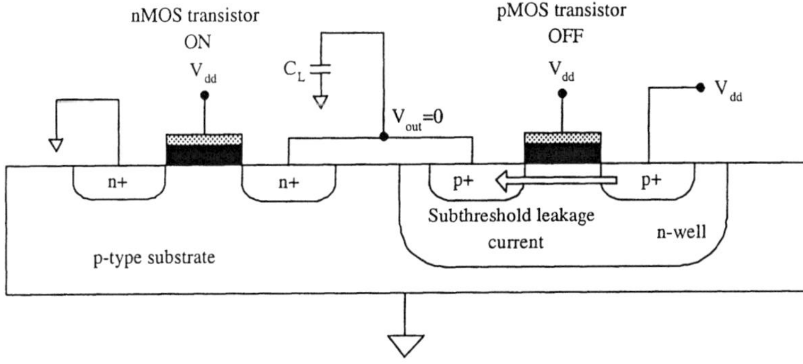

Figure 2.7. Subthreshold leakage current path in a CMOS inverter, assuming high V_{dd}.

2.2.4   Static Power dissipation

Ideally, in the steady state of CMOS circuits there is no static power dissipation, and this is the most attractive characteristic of CMOS technology. However, the actual operation of a CMOS circuit is slightly different. More specifically, we will deal with degraded voltage levels feeding into static complementary gates and pseudo-nMOS logic families. We will illustrate the phenomenon of static power dissipation due to degraded input signals through the example shown in Figure 2.8. A pass nMOS transistor drives an inverter. From basic CMOS circuit theory we know that, the voltage value at the node A is $V_{dd} - V_{thn}$, i.e. it is degraded. Since the inverter's input is high, i.e. $V_{dd} - V_{thn}$, its output should be low. However, the pMOS transistor will be weakly ON ($| V_{gs} - V_{thp} | \simeq 0$) and, thus, conducting static current from power supply to ground rails. The associated static power dissipation might be significant if the inverter operation frequency is low.

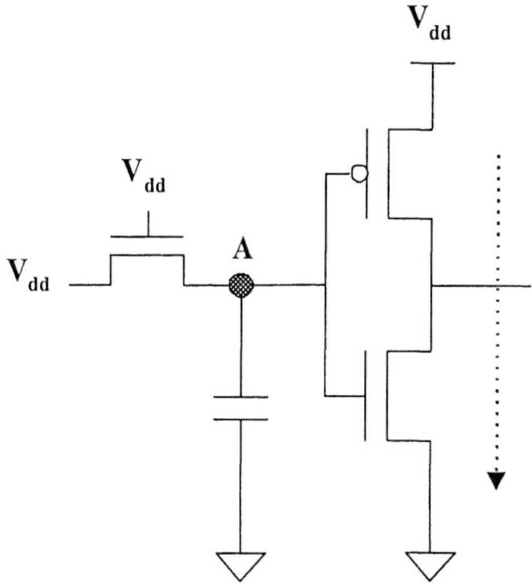

Figure 2.8.   Degraded voltage level as input signal to an inverter results in static power consumption.

A pseudo-nMOS logic gate consists of a single pMOS transistor, whose gate is always grounded, and a complex block of nMOS transistors, which actually implements the boolean function (Figure 2.9). It can easily be seen that there always exists a path from the power supply rail to the ground rail, because the pMOS transistor is always ON. Therefore, during the steady state there is static power dissipation. Whether a pseudo-nMOS logic family is appropriate for low power applications or not, depends on whether certain requirements, such as area savings and operation frequency, are met or not, and this should be the result of careful analysis of the respective trade-offs.

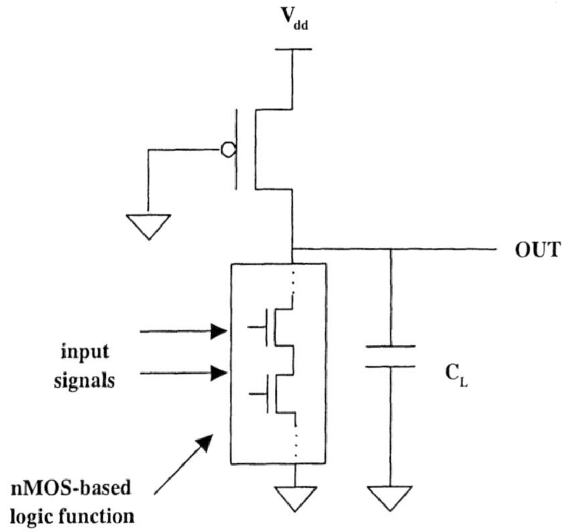

Figure 2.9. Pseudo-nMOS logic gate

The main characteristic of the pseudo-nMOS logic family is the reduced number of transistors required for realizing a logic function compared with the conventional static CMOS logic. Moreover, in a pseudo-nMOS logic, the current flows from V_{dd} to the ground only when the output is at the logic level '0'. In the opposite case, when the output is at the logic level '1', the nMOS transistor block are turned off and no static power is consumed, except leakage current. This characteristic may be useful in a low power design. More specifically, if the output signal exhibits very high transition probability being at the logic '1', it may be a good choice to implement a logic function with pseudo-nMOS logic. In contrary, if the output signal has very a very low transition probability, we may eliminate the nMOS block of a pseudo-nMOS circuit and substitute it with a load transistor of nMOS type. A typical example where a designer can use the above feature is the system reset circuitry. More specifically, the reset signal has extremely low transition probability, for instance during the power-on mode, which can benefit from such circuit technique.

2.3   Low Power Design with Multiple V_{dd} and V_{th}

Historically, VLSI designers derived circuits with a single supply voltage and single threshold voltage. Recently, multiple threshold voltages and multiple supply voltages are used as an attempt to reduce dynamic and leakage power dissipation. Extensive literature of low-voltage devices and circuits can be found in [10], [11]. The power savings arising from the selection of a multiple-V_{th} process and/or a multiple-V_{dd} should be weighted against the increased manufacturing costs, the available CAD tools, and the design complexity when optimizing both the process and circuit design.

The selection of multiple supply voltages on a CMOS circuit is one of the most effective approaches to reduce dynamic power especially for deep submicron design. In order to preserve performance, while also reducing power, dual-V_{dd} approach can be used assigning the low supply voltage, V_{ddL}, to gates on the critical paths, while high supply voltage, V_{ddH} to the gates of non-critical paths [12], [13]. Clustered Voltage Scaling (CVS) [13] is one structure proposed to implement dual-V_{dd} design. Specifically, in CVS approach, all paths follow the "route": primary inputs $\Rightarrow V_{ddH}$-supplied logic modules $\Rightarrow V_{ddL}$-supplied logic modules $\Rightarrow$ logic level converters. This approach leads to clusters of V_{ddH} and V_{ddL} logic modules. To assign the V_{ddL} modules, a depth-first-search algorithm is used from the primary outputs to the primary inputs. Also, the supply voltage of V_{ddL} should be around the 0.6 to 0.7 of V_{ddH} to achieve maximal power savings. Layout is another important issue and V_{ddL} and V_{ddH} logic modules should be separated because they have different n-well voltages. A row-by-row separation approach for standard cell and gate array is used [13]. Finally, the use of multiple supply voltages together with multiple threshold voltages results into significant improvements with respect to delay and V_{dd}.

As technology shrunk the supply voltage, threshold voltage should be scaled down to maintain performance. Equation 2.8 indicates that the subthreshold leakage current increases exponentially as threshold voltage reduces (for certain V_{dd}) and thus, the leakage power also increases. In [9] was reported that $P_{leakage}$ may be larger than dynamic power in near future. Usually, the cost of use of an additional threshold voltage is one mask.

The issue of multiple threshold voltages has been studied in [14], [15], [16]. More specifically, the assignment of low-threshold devices to critical paths was reported in [14]. In [15], threshold voltage is assigned on cell-by-cell basis aiming at the minimization of the used V_{thL} modules without any penalty in performance. The V_{thL} should be selected to meet the required frequency, while the V_{thH} to minimize the leakage. An integrated circuit of MPEG-4 video core processor with dual-V_{th} was designed [16]. Specifically, it consists of 3 million transistors of 0.3μm CMOS with $V_{thL} = 0.2$ volts for "active" operation and $V_{thH} = 0.2$ volts for "standby" operation. The external supply voltage is 3.3V, while the internal voltages are 2.5V and 1.75V.

Finally, the use of multiple supply voltages together with multiple threshold voltages results into significant improvements with respect to delay and V_{dd} [11], [17].

References

[1] J. Rabaey and M. Pedram, "Low Power Design Methodologies," Kluwer Academic Publishers, 1996.

[2] W. Nebel and J. Mermet, "Low Power Design in Deep Submicron Electronics," Kluwer Academic Publishers, 1997.

[3] A. Chadrakasan and R. W. Brodersen, "Minimizing Power Consumption in Digital CMOS Circuits", Proceedings of the IEEE, pp. 498-523, April 1995.

[4] D. Singh, R. Rabey, M. Pedram, F. Catthoor, S. Rajgopal, N. Sehgal, T. Mozden, "Power Conscious CAD Tools and Methodologies: A Perspactive", Proceedings of IEEE, Vol. 83, No. 4, pp. 570- 594, April 1995.

[5] A. Bellaouar and M.I. Elmarsry, "Low Power digital VLSI Design-Circuits and Systems," Kluwer Academic Publishers, Norwell, MA, 1995.

[6] N. Weste and K. Eshragian, "Principles of CMOS VLSI Design", Addison-Wesley, 1994.

[7] R. Howe and C.G. Sodini, "Microelectronics: An Integrated Approach," Prentice Hall Electronics and VLSI Series, 1997.

[8] S. Sze, "Physics of Semiconductor Devices," John Wiley & Sons, 1981.

[9] T. Kao and A. Chadrakasan, "Dual-Threshold Voltage Techniques fro Low-Power Circuits," in IEEE Journal of Solid-State Circuits, vol. 35, no. 7, pp. 1009-1018, July 2000.

[10] T. Kuroda and T. Sakurai, "Low-Voltage Technologies and Circuit," (invited paper) in "Low-Power CMOS design" Edited by A. Chadrakasan and R. W. Brodersen, pp. 61-65, 1998.

[11] Sematech SIA roadmap: http://public.itrs.net/

[12] K. Usami and M. Igarashi, "Low-Power Design Methodology and Applications Utilizing Dual Supply Voltages," in Proc. of ASP-Design Automation Conference 2000, pp. 123-128, Jan. 2000.

[13] K. Usami and M. Horowitz, "Clustered Voltage Scaling Technique For Low Power Design," in Proc. of Int. Symp. in Low Power Electronics and Design, (ISLPED), pp. 3-8, April 1995.

[14] T. Holloway et al, "0.18 μ m CMOS Technology for High Performance, Low Power, and RF Applications," in Proc. of Symp. on VLSI Technology Digest of Technical Papers, pp. 13-14, 1997.

[15] N. Kato, K. Yano, et al, "Multi-Threshold Voltage Design Methodology in Sub-2V Power Supply CMOS," in Trans. of IEICE Electronics, pp. 1747-1754, Nov. 2000.

[16] T. Kuroda, "Low-power CMOS digital design with dual embedded adaptive power supplies," in IEEE Journal of Solid State Circuits, no. 4, pp. 652-655, April 2000.

[17] D. Sylvester and H. Kaul, "Power-Driven Challenges in Nanometer Design," in IEEE Design & Test of Computers, pp. 12-22, Nov.-Dec. 2001.

Chapter 3

LOGIC LEVEL POWER OPTIMIZATION

George Theodoridis

University of Patras, Patras, Greece
theodor@ee.upatras.gr

Dimitrios Soudris

Democritus University of Thrace, Xanthi ,Greece
dsoudris@ee.duth.gr

Abstract   The reduction of logic circuit power consumption is a major priority of a designer attempting to meet implementation requirements. More specifically, the power optimization of a logic circuit implies the reduction of its switching activity of switched capacitance. The main features from a plethora of well-established techniques for logic optimization of combinatorial and sequential circuits are provided. Furthermore, the categorization of the existing techniques as well as the optimization design flow followed by a designer is provided.

Keywords: Combinatorial circuits, sequential circuits, technology-independent phase, technology depended phase, multi-level circuits, technology mapping.

3.1   Introduction

In recent years, an impressive growth of personal computing devices and wireless communication systems has been noted. This was the main reason that induced the VLSI designers to take into consideration the power consumption in addition to area and throughput. It is clear that in the absence of low-power design techniques, such devices will suffer from either a short operation battery life or a very heavy battery pack. Moreover, there are also some very important topics, such as reliability, hot spot detection, electromigration and packaging selection, which make the power consumption a major concern in VLSI design.

D. Soudris et al. (eds.), Designing CMOS Circuits for Low Power, 23–43.
© 2002 *Kluwer Academic Publishers. Printed in the Netherlands.*

Low-power design is taken place at all levels of the design flow starting from the system and reaching down to layout level [1].

In this chapter low-power design techniques applied at the gate level are presented. The chapter is organized as follows: In Section 3.2 the problem of low-power design at gate level is formulated, while in Section 3.3 the optimization algorithms for combinational circuits applied at the technology- independent step are presented. In Section 3.4 the technology-independent low-power optimization methods for sequential circuits are discussed. In Section 3.5 the technology-dependent optimization techniques are presented. Finally, the conclusions are given in Section 3.6.

3.2   Problem Formulation

Logic level optimization is the design task where an RT-Level circuit description is optimized in terms of some criteria such as area, time, and power. The output of this task is an optimized gate netlist.

A plethora of logic level optimization algorithms aiming at minimizing the circuit area and delay have been developed in the past [2]-[4]. Traditionally, logic level optimization consists of two basic steps; i) the technology-independent and ii) the technology-dependent optimization step. In the former step the circuit's Boolean description is optimized ignoring the technology in which the circuit will be implemented. Since the adopted technology is ignored, general metrics such as the number of literals of the Boolean functions and the number of levels of the Boolean network are used to estimate the area and delay of the circuit. Although this estimation is inaccurate, the technology-independent optimization step is important. First, the optimized Boolean network can be reused in a future technology change. Second, by decomposing the whole optimization problem in two sub-problems and solving each problem individually, the initial problem complexity is reduced, achieving better results. In the second step, the output of the technology independent optimization step (i.e. optimized Boolean network) is optimized considering the adopted technology.

Concerning the circuit type, the technology-independent optimization techniques are classified as: i) optimization techniques for combinational circuits and ii) optimization techniques for sequential circuits. In the case of combinational circuits, a further classification occurs, taking into account the number of the circuit levels. Two-level logic minimization targets at two-level circuit implementations, for instance Programmable Logic Arrays (PLAs) or Programmable Logic Devices (PLDs). Multi-level logic optimization targets to the cell-based designs such as standard cells and gate array of Field Programmable Gate Arrays (FPGAs) designs. Concerning the sequential part of a logic circuit, there exist two sub-steps. More specifically, the goal of the state assignment step is to encode the states of an Finite State Machine (FSM) aiming at the minimization of the circuit area, whereas the logic optimization step techniques, for instance retiming, aiming at the reduction of the circuit delay.

In the technology-depended step, the logic circuit is optimized considering the adopted technology library. Three sub-steps exist; the technology decomposition where the circuit is decomposed into basic gates (e.g. 2- or 3- input NAND/AND gates), the technology mapping where the circuit is mapped to the gates of the library and the post mapping optimization step where techniques, for instance rewiring, are taken place. The output of the technology-depended step is an optimized gate netlist, which is used for floorplaning, placement, routing, and layout optimization.

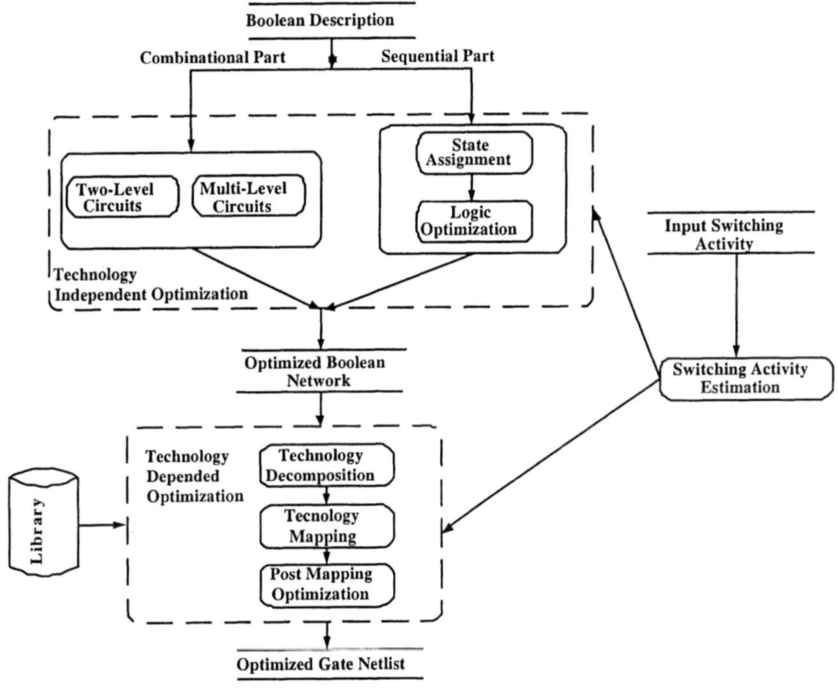

Figure 3.1. Logic Level Optimization Flow.

For all the above briefly mentioned steps a lot of techniques and algorithms have been developed in the past aiming at the optimization of a logic circuit in two axes, the occupied area and delay. Concerning the dynamic power consumption, the problem becomes a three-dimensional optimization problem (area-time-power). Thus, all the developed algorithms and techniques have to be modified to take into account the dynamic power dissipation of the circuit. Since the dynamic power consumption is proportional to the switching activity of the circuit, appropriate switching activity estimators are used to calculate quickly and with high accuracy the switching activity of the circuit, during optimization. The entire logic level optimization flow is shown in Fig. 3.1. In the next sections the low-power optimization techniques applied at each of the above-mentioned steps or sub-steps are presented and discussed.

3.3   Combinational Circuits Technology-Independent Optimization

3.3.1   Two-Level Logic Circuits

Two-level logic minimization for area is an important part of gate level synthesis for minimizing the area for programmable devices or multi-level circuits. There are many two-level Boolean logic minimization algorithms for area such as Espresso, Espresso Exact, Mini e.t.c. [4].

Recently, some methods for optimizing two-level combinational circuits for low-power have been presented. In [19] employing the statistical characteristics of the input signals, a method that modifies the Expand routine of Espresso has been proposed. A new cost-function was introduced for guiding the cube expansion so that a cover with reduced switching activity will be derived. However, there are some drawbacks: i) the static probability assumed to be equal to 0.5 for every input signal, ii) all input signals are spatiotemporallyuncorrelated, and iii) the power dissipation of the input signals is not taken into account.

In [20], considering signals with random probabilities and taking into account the power consumption of the input signals, the Power Prime Implicants (PPIs) was proposed to identify the set of all implicants that are sufficient and necessary for obtaining a minimum power solution. Using the set of PPIs, a minimum covering problem is solved to find the best solution in terms of power. However, since all signals are assumed spatiotemporallyuncorrelated the proposed solution is not accurate.

A more accurate method presented in [21], where the temporal correlation of the signals and the power consumption of the primary input signals are taken into consideration. The idea was to introduce input signals in the first logic level aiming to force the outputs of the AND gates to stay at low logic level for a number of combinations of the applied input vectors. Thus, a reduced switching activity and power consumption are achieved.

The following example describes the basic idea of the method. Let us assume the function $F = x_0x_1 + x_0x_2 + x_0x_3$ and its area-optimized implementation as shown in Fig. 3.2.

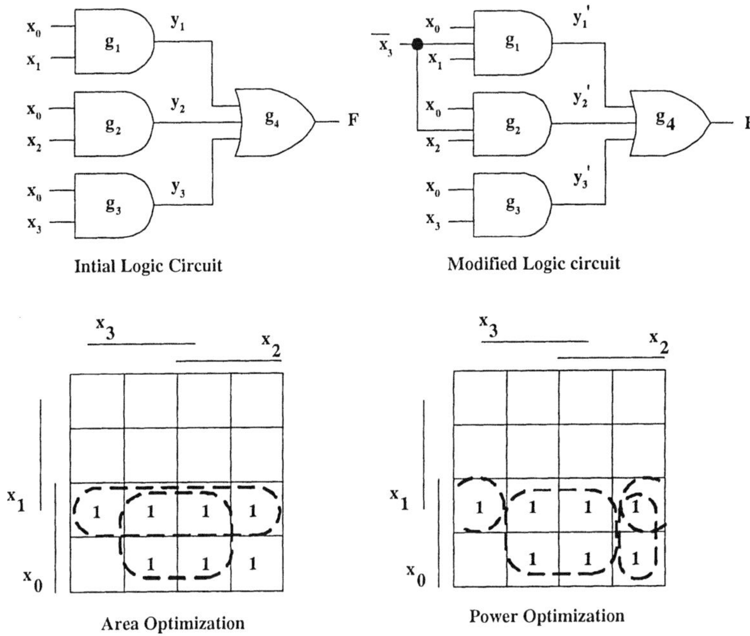

Figure 3.2. Low-Power Design for Two-Level Circuits

Let us also assume that the signal x_3 exhibits low transition probability and high static-1 probability, while the signals x_0, x_1, and x_2 are characterized by high transition probabilities. When $x_3 = 1$, the function can be expressed as $F_{(x_3=1)} = x_0x_1 + x_0x_2 + x_0 = x_0$. In this time interval the value of function, F, depends only on the value of the node y_3. However, energy is wasted due to the transitions of the nodes y_1 and y_2. Inserting the complement of signal x_3 into the first two terms of function, F, a new function,
$F' = x_0x_1\overline{x_3} + x_0x_2\overline{x_3} + x_0x_3$, is obtained. In that case, if $x_3 = 1$, the nodes y_1 and $y_2$remain at low logic level without performing transitions resulting in a power saving. In other words, as it is shown by the Karnaugh maps, the goal is to modify the cube expansion such that signal with low switching activity to appear as many time as it is possible in the first level of the circuit.

A formal method to determine the blocking variables (in above example is $\overline{x_3}$) and to classify them in groups according to their switching activity is proposed [21]. Also, an algorithm, which embodies the concept of the blocking variables and uses the main routines of ESPRESSO, has been introduced. After a series of iterations, a circuit with reduced power consumption results can be achieved. The number of iterations is user-specified, and depends on the upper and lower bounds of average power consumption per gate and the number of groups of logic variables with similar statistical parameters.

3.3.2   Multi-Level Circuits

A lot of techniques for optimizing multi-level combinational circuits in terms of area and delay have been presented [2]-[4]. These techniques are classified it two categories: i) the Boolean optimization techniques and ii) the algebraic optimization techniques. In particular, the Boolean techniques use the don't care set and complete set of De Morgan laws to optimize the circuit. On the other hand, the algebraic methods optimize the circuit without using the don't care set, while a sub-set of the De Morgan laws is used. Specifically, only distributive law applies, while the complement of a logic variable is not defined. Algebraic techniques are faster than the Boolean ones but their quality is smaller than that of Boolean techniques.

3.3.2.1   Boolean Techniques.   The use of don't care set (DC) maximizes the exploration space resulting in more efficient circuit implementations in the terms of area and time. However, when a local node area optimization is performed, the total power consumption of the circuit may be increased since the Boolean expressions of the transitive fanout nodes are changed. Therefore, efficient methodologies to overcome the above problem are required. Such a methodology was presented in [23].

The basic idea is to compute an appropriate subset of don't care terms for each circuit node, which can be used for local node optimization without increasing the switching activity of other circuit nodes beyond their current value. Thus, as nodes are optimized, a monotone decrease in the switching activity of the network is achieved. Let us consider the Boolean network shown in Fig. 3.3.

To realize the above goal, the impact of transformations on the function of node f must be exactly known when g is optimized. In this way the effect in the switching activity of node f, as function g is optimized, can be exactly measured.

The complete DC of node g consists of: i) the *External Don't Care set* (EDCs), ii) the *Satisfiability Don't Care set* (SDCs) and iii) the *Observability Don't Care set* (ODCs). The ODCs can be computed by the formula:

$$ODC_g = ODC_f + ODC_g^f \; \text{with} \, ODC_g^f = \overline{\frac{\partial f}{\partial g}} \qquad (3.1)$$

Given a node f and its fan-in node g, a subset of ODCs for node f called *Propagated Power relevant Observability Don't Care* ($PPODC_f$) is defined. Using $PPODC_f$ to compute ODC_g, it is ensured that any change in function of g does not increase the switching activity of f. $PPODC_f$ can be defined as follows:

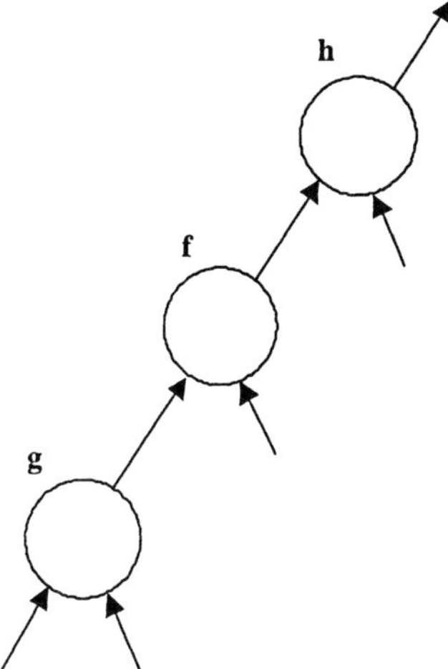

Figure 3.3. Power Relevant ODC computation for node g

$$PPODC_f = \begin{cases} \overline{f} \cdot ODCf & if\ p(f) > 0.5 \\ f \cdot ODCf & if\ p(f) \leqslant 0.5 \end{cases} \tag{3.2}$$

where $p(f)$ is the signal probability of f. Thus, the ODC of node g, is modified and the *Power relevant Observability Don't Care* (PODC) is defined as follows:

$$PODC_g = PPODC_f + ODC_g^f \tag{3.3}$$

Nevertheless, attention should be paid to ensure that the switching activity of any other node in the transitive fanout of g, for instance h, should not be increased when g is optimized. For this reason, the notion of *Monotone Power Relevant Observabilty Don't Care* ($MPODC_g$) for node g was introduced in a similar way. It is defined as the ODC conditions for g that guarantee changes in function of g do not increase the current switching activity of nodes in the transitive fanout cone of f (excluding f).

3.3.2.2 Algebraic Techniques.

Since logic circuits become more complicated, a very important concern of VLSI designers is the identification of common terms on their Boolean expression. It has been found that a kernel extraction method in such a circuit can result in successful area optimization [2]- [4].

A modified kernel extraction procedure, presented by [6] found providing lower power consumption. The idea is to calculate the power-saving factor for each candidate kernel according to the way its extraction affect the loading on its input lines and the amount of logic sharing.

Based on this concept, a more effective technique for both power and area reducing was proposed in [22]. An efficient power-saving factor, which assumes that nodes in a multilevel network can be represented by two level logic forms, is specified. This is similar to assumption made for calculating the literal saving cost of candidate kernels during algebraic operations. Additionally very efficient methods for *node elimination, factorization,* and *logic decomposition* are also proposed.

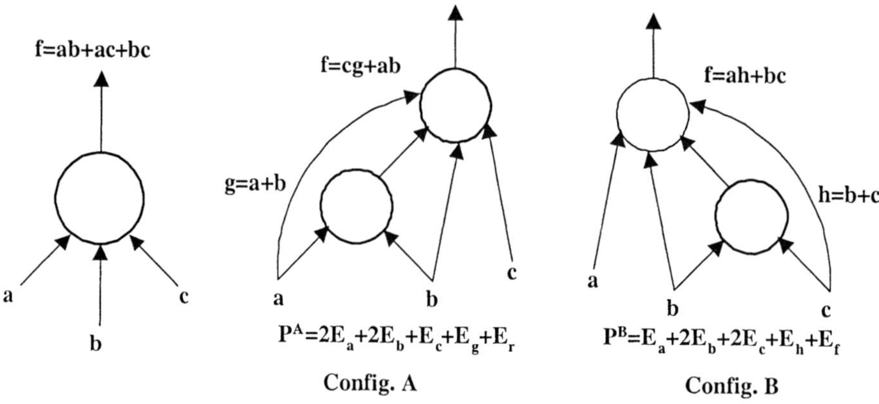

Figure 3.4.   Common Sub - Expression Extraction for Low - Power

Two alternative topologies computing the same logic function $f = ab + ac + bc$ are shown in Fig. 3.4. Note that these structures have the same number of literals in the factored form of their intermediate nodes. It can also be seen that if $E_a(sw)+E_g(sw)> E_c(sw)+E_h(sw)$, then $P^A > P^B$. If $prob(a)=0.5$ and $prob(b)=prob(c)=0.25$, then $P^A - P^B =13/128$. Generally, power savings of about 18% (compared to a minimum-literal network) are expected.

It is clear that an increase in the number of literals in a circuit results into increased switched capacitance and thus, more power consumption. Consequently, one should minimize a *lexicographic cost* (D_a, D_p) where D_a is the literal-saving factor and D_p is the power- saving factor. However, the above power-saving factor is very expensive to be computed, so it is better to calculate it only for a subset of candidate divisors (i.e. the top 10% of divisors in terms of their literal saving factors).

3.3.2.3   Guarded evaluation.   Guarded evaluation [24], [25] is a shut-down technique that does not require to synthesize additional logic to implement the shut-down mechanism; rather it exploits existing signals of the original circuit. The general structure of guarded evaluation architecture is illustrated in Fig. 3.5.

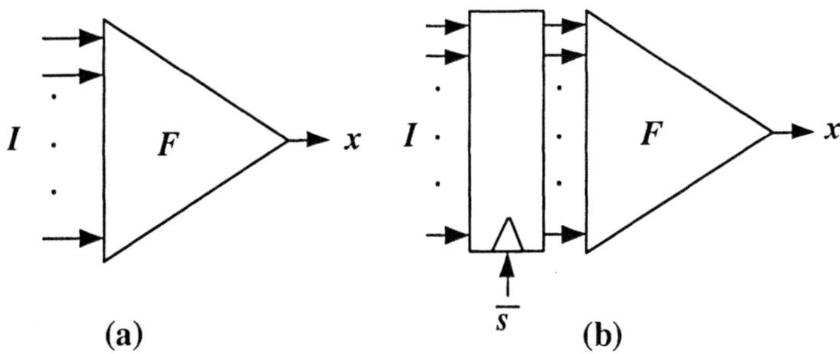

Figure 3.5. Guarded Evaluation

The approach is based on the placement of transparent latches with an enable signal in the input of each block of the circuit, which needs to be power managed. The main idea is to determine every clock cycle, which parts of the circuit compute results, which are used (i.e. there are observable in the output) and which are not. When a block should execute some useful computation, the enable signal makes the latches transparent. Otherwise, the latches retain their previous values, thus blocking any transition within the logic block. Formally this idea is described below:

$$s \Rightarrow ODC_x \ (i.e. \ \bar{s} + ODC_x \equiv 1)$$
$$t_l(s) < t_e(I)$$

(3.4)

where, ODC_x is the observability don't care for x, while $t_l(s)$ is the latest arrival time for s and $t_e(I)$ is the earliest arrival time for a signal I. In that case transparent latches can be added for reducing the switching activity of F. The latches are enabled when $s = 0$ and disabled when $s = 1$.

3.4 Sequential Circuits Technology-Independent Optimization

3.4.1 State Assignment

The state assignment problem can be stated in general as follows: *"Given the State Transition Graph (STG) of a Finite State Machine (FSM), assign binary codes in all states such that a given cost function G is minimized."*

In terms of power consumption, the goal is to encode the states of the FSM such that the number of transitions at the state lines in two successive clock periods to be minimized. Thus, combing the reduced switching activity of the state lines with a low-power implemented combinational logic, a significant power reduction can be achieved.

Since the goal is to minimize the number of transitions of the state lines, the Hamming distance between the codes assigned to pairs of states must be considered. In real designs, the transition probability from one state to another one is not the same for every pair of states. Therefore, the Hamming distance between pairs of states must be weighted with the corresponding transition probability. Thus, the cost function G is expressed as follows:

$$G = \sum_{\forall i,j} W_{i,j} H(S_i, S_j) \qquad (3.5)$$

where, $H(S_i, S_j)$ is the Hamming distance between the binary codes assigned to states S_i and S_j, while $W_{i,j}$ is a factor that represents the transition probability between the states S_i and S_j.

The state assignment algorithm reported in [6] computes the transition probabilities, p_{ij}, for a pair of states, (S_i, S_j), and labels with these values the corresponding edges of the STG. Afterwards, the problem is transformed to a graph optimization problem aiming to minimize a cost function similar to eq. (3.5). The objective during the encoding procedure is to minimize the number of state bits and therefore, the number of the flip-flops of the final circuit. Although, it is a well-worked strategy in terms of area it is not guaranteed that it results in low-power implementations. Fewer flip-flops imply a more complex combinational circuit to evaluate the next-state function. This may cause an increase of the switching activity in the combinational part resulting in increased total power dissipation.

Also, is known that the use of p_{ij} is an approximation of the exact state transition probability, P_{ij}. Specifically, the probability, p_{ij}, is computed considering only the input combinations, which cause a transition from S_i to S_j, ignoring the probability the machine being at state S_i. Interpreting the STG as a Markov chain [5], it is proved that the exact transition probability, P_{ij}, equals to $P_{ij} = P_i\,p_{ij}$, where P_i is the probability of the machine being at S_i. The exact state transition probabilities, P_{ij}, are derived by solving a system of linear equation called Chapman-Kolmogorov equations [5].

In [7] and [8] state assignment algorithms that use the, P_{ij}, probabilities have been presented. Thus, the estimation of the transition activity of the state lines is performed with high accuracy resulting in a more efficient state assignment. To consider the impact of state assignment in the consumed power of the combinational part, a number of heuristics were introduced. The key idea of those heuristics is that a combinational circuit optimized in terms of area is also characterized by low-power consumption. Therefore, the factor $W_{i,j}$, in eq. (3.5) is modified to take into account the exact state transition probabilities and the occupied area of the combinational circuit. The number of literals of the Boolean functions of the combinational logic is used as an area metric. Linear Integer programming was used for solving the state assignment problem in [7], while a genetic algorithm was employed in [8].

In [9] a more precise algorithm has been proposed. In this approach accurate power models that consider both the switching activity of state lines and the loading factor of the combinational part are introduced. Also, appropriate cost functions are proposed when the combinational part is implemented either as a two- or a multi-level gate network.

Besides the presented approaches additional interesting techniques have been published. The interested reader can be referred to [10], [11], and [12].

3.4.2   Logic Optimizations

3.4.2.1   Precomputation .   This optimization technique is based on selectively pre-computing of the output logic values of a circuit one clock cycle before they are required, and then use the pre-computed values to reduce the internal switching activity of the combinational logic in the successive clock cycle [13].

A simple, on this idea based structure is shown in Fig. 3.6, where the inputs of sequential block A have been partitioned into two sets, corresponding to registers R_1 and R_2, and the output of block A is the input of register R_3. The Boolean functions g_1 and g_2 serve as the predictor functions according to the following equations:

$$g_1 = 1 \Rightarrow f = 1$$
$$g_2 = 1 \Rightarrow f = 0$$

(3.6)

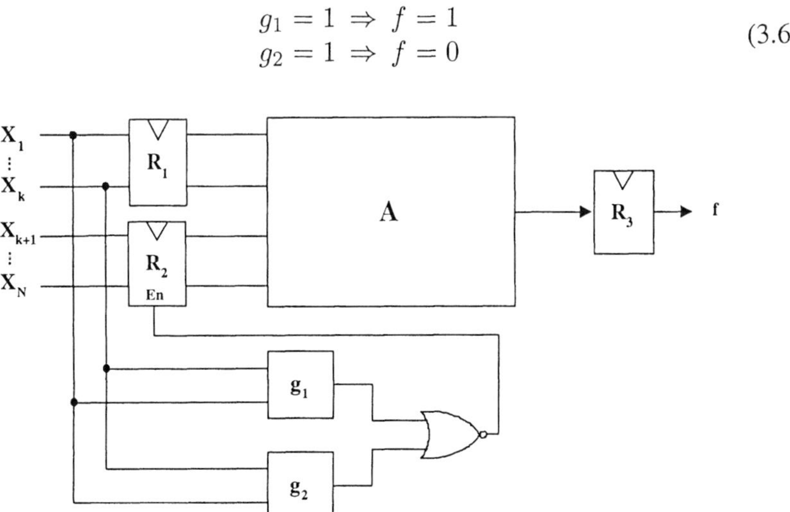

Figure 3.6.   Precomputation Structure for Sequential Circuits

The physical meaning of (6) is that if function g_1 or g_2 equals 1, the value of the output function, f, is fully determined. However, the Boolean variables of functions g_1 and g_2 are a subset of the input signals of block A. Thus, the remaining signals, $(X_{k+1}, ..., X_N)$, can be frozen. It must be stressed that is not allowed both g_1 and g_2 to be evaluated to 1.

Consequently, if the logic level of g_1 or g_2 is high, during clock cycle T, then the enable signal of register R_2 is low. Thus, the outputs of R_2, during clock cycle $T+1$ are not changed. Since the output of R_1 is updated the function f is evaluated correctly. As a subset of the inputs of block A changes, the switching activity of this block is reduced, implying a remarkable power saving. However, since functions g_1 and g_2 occupy extra area and they consumes additional power. Therefore, attention should be paid in the construction of g_1 and g_2 and appropriate trade off analysis should be taken place.

3.4.2.2   Retiming.   A novel method for reducing the power consumed in pipelined sequential circuits has been proposed in [14]. The method is based on *retiming*, which is a technique that repositions the flip-flops of the circuit resulting into the minimization either the area or the delay of the circuit.

Since a flip-flop output makes at most one transition when the clock is asserted, the idea is to place a flip-flop in a circuit node with high glitching activity and high load capacitance. Thus, glitches are not propagated to the transitive fan-out of the node resulting in a reduction of the total switching activity, as shown in Fig. 3.7.

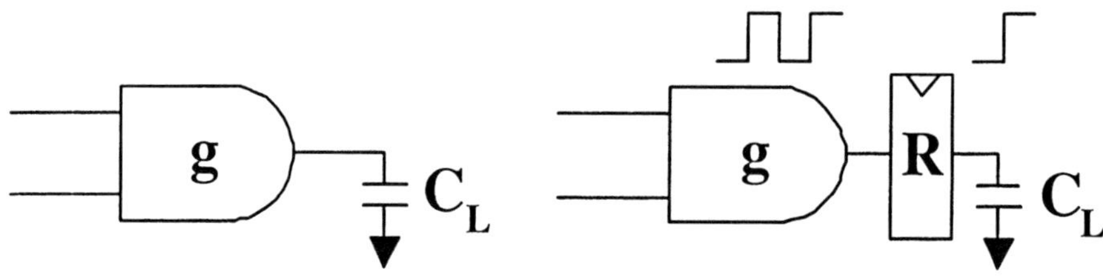

Figure 3.7.   Retiming for Low-power

However, attention should be paid because the switching activity of some nodes of the circuit may be changed due to retiming, which may result in an increase of the power consumption. Also, the number of the used registers should be minimized since the power dissipation of the registers and clock line are not negligible. Finally, attention should also be paid to preserve the timing behavior of the circuit when retiming for low-power is performed.

3.4.2.3   Synthesis of FSMs with Gated Clocks .   This technique, presented in [15]-[18], refers to the power reduction that can be achieved in FSM circuits by using gated clocks. The key idea is that during the operation of an FSM there are conditions where the state and the output of the FSM do not change. Hence, clocking the circuit in the corresponding time intervals wastes power both in the combinational logic and registers. Thus if we can detect the idle conditions of the FSM, we can also stop the clock during the corresponding time intervals. The benefit by using gated-clock is twofold: First, when the clock is stopped no power is consumed by the combinational logic since its inputs remains unchanged. Second, no power is consumed by the flip-flops and gated-clock line.

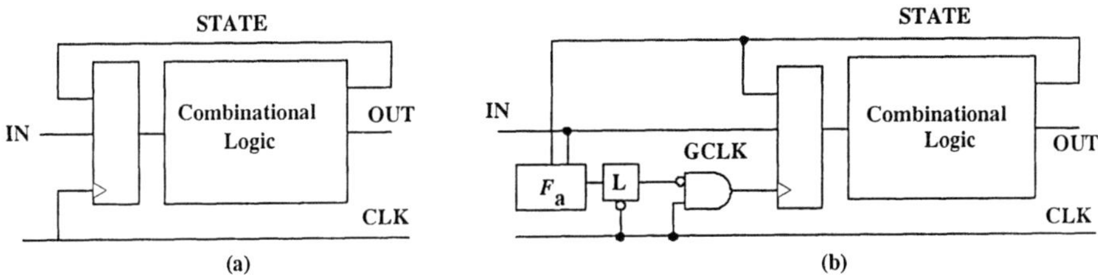

Figure 3.8.   (a) Single clock, flip-flop based FSM. (b) Gated- clock version

The flip-flop-based architecture is modified by setting a new *activation signal, GCLK,* as shown in Fig. 3.8, whose purpose is to selectively stop the local clock for the FSM when the machine is idle. The combinational circuit, F_a, which provides the activation signal, uses as its inputs the primary inputs and the state lines of the machine. It has been found that the application of this technique to such circuits provides power savings ranging between 10-30%. Since the function F_a consumes power, it is recommended to select a subset of all idle conditions such that this subset takes place with high probability during the circuit operation.

3.5   Technology-Dependent Optimization

3.5.1   Path Balancing

The way the gates of a logic circuit are interconnected can strongly affect the overall switching activity, and hence the power dissipation. For example, timing skew between signals in a circuit can cause spurious transitions (glitches) resulting in extra power. To reduce the possible spurious activity in a circuit, delay of all true paths that converge at each gate must be balanced, as shown in Fig. 3.9, where the logic function *f=abcd* is implemented on two alternative ways (chain structure, and tree structure). It can also be seen that the tree implementation of function f provides glitches elimination, thus reducing effectively the total power dissipation.

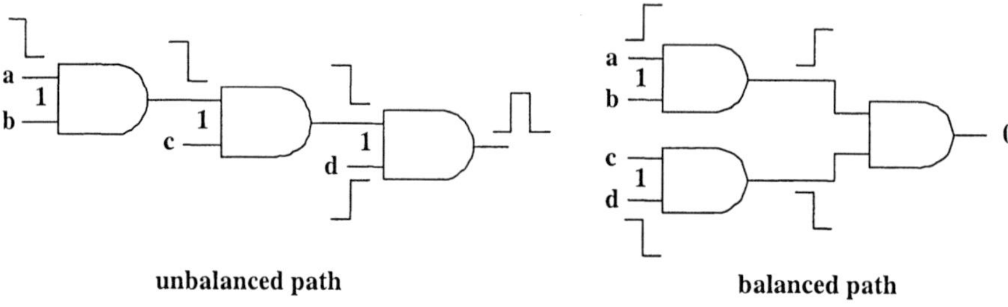

Figure 3.9.   Path Balancing for Glitching Reduction

Path balancing can be achieved before technology mapping by selective collapsing and logic decomposition or after technology mapping by delay insertion and pin reordering. The advantage of this technique is that by selectively collapsing the fan-ins of a node, the arrival time at the output of the node can be changed. Logic decomposition and extraction can be performed so as to minimize the level difference between the inputs of the nodes that are driving high capacitive nodes. Additionally, by inserting variable-delay buffers in a circuit, the delays of all paths in the circuit can be made equal. The issue in delay insertion is to use the minimum number of delay elements to achieve the maximum reduction in glitching activity. Path delays may sometimes be balanced by an appropriate signal to the pin assignment. This is possible, because the delay characteristics of CMOS gates vary as a function of the input pin that is causing a transition at the output.

3.5.2   Technology Decomposition

The next step during logic synthesis of a network is to convert the network to another, which only contains two-input AND/NAND and inverter gates. This step, named technology decomposition, is very useful for network synthesis and is carried out before the mapping of the network, according to the current cell library, takes place. Therefore, a decomposition scheme that minimizes the total switching activities of the network is a good starting point for power-efficient technology mapping.

Given the switching activity at each input of a node, Tsui et al. [26] suggested a technique for AND decomposition of this node, which reduces the total switching activity in the resulting two-input AND structure under a zero-delay model. The idea is to inject the high switching activity inputs into the decomposition model as late as possible, as shown in Fig. 3.10 where two different decomposition structures for the four-input AND gate are depicted.

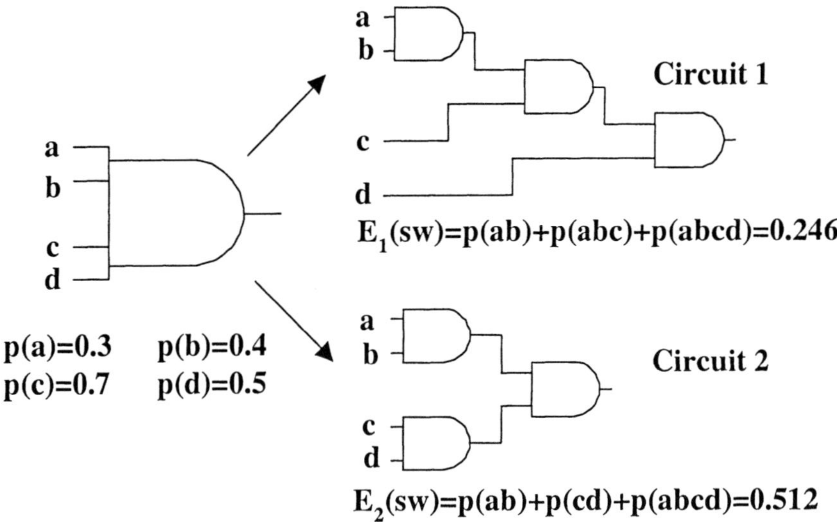

Figure 3.10. Technology Decomposition for Minimizing Switching

Note that signal *d*, which has the highest switching activity, is injected last in configuration A, thus implying better power performance for this configuration. This technique has been found being optimal for dynamic CMOS circuits, but also produces very good results for static CMOS circuits. In general, the low-power technology decomposition procedure reduces the total switching activity in the circuits by 5% over the conventional balanced tree decomposition method

3.5.3 Technology Mapping

Technology mapping refers to the process of binding a given Boolean network to the gates included in a target cell library. In [26]-[28] some design techniques for low-power consumption during technology mapping have been proposed. The main concept is to hide nodes with high switching activity inside the gates, thus they can drive smaller load capacitance, as shown in Fig. 3.11.

According to [26], the whole process consists of two steps. The first step requires the computation of power-delay curves (power consumption versus arrival time) of all nodes in the network. The second step produces the mapping solution according to the previous curves and the required times at the primary inputs. This method has been proved implying an 18% power saving at the expense of a 16% increase in area without any penalty in network performance. In other words, we can say that the power-delay mapper reduces the number of high switching activity sub-networks at the expense of increasing the number of them having low switching activity. In addition, it reduces the network average load.

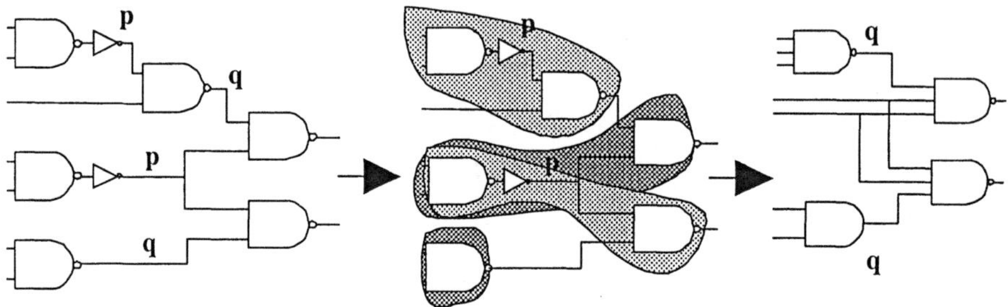

Figure 3.11.   Technology Mapping for Minimizing Switching Activity

Although the approach mentioned above refers to mapping for zero-delay circuits, an extension to a real-delay model is considered in [26], resulting in optimum power solutions. According to [26], every point on the power-delay curve of a specific node uniquely defines a mapped subnet from the circuit inputs up to the node. The principle is to compute each such point with the probability waveform for the node in the corresponding mapped subnet. Hence, the total power cost, owing to steady-state transitions and hazards, of a candidate match can be calculated from the computed power-delay curves at the inputs of the gate and the power-delay characteristics of the gate itself.

3.5.4   Post mapping Optimization

The major drawback of low-power technology-independent optimization is that the estimation of the power consumption, which is used to guide the optimization, of the Boolean network is inaccurate. This arises from the fact that the final structure of the circuit and the load capacitance of each node are unknown, since the technology mapping is not applied. On the other hand, the post mapping optimization allows performing power and timing analysis on the mapped circuit, which provides more realistic estimates. However, the exploration freedom for optimization is reduced after the circuit mapping. It is assumed that the mapped circuit meets the timing constraint and thus, transformations that do not change drastically the circuit structure and timing behavior are allowed.

The main idea of the post mapping low-power optimization (often called re-synthesis for low-power) is based on the redundancy addition and removal. Specifically, adding new gates and interconnections to the mapped circuit some existing connections become redundant resulting into their elimination. If the eliminated interconnections include high power consumption nodes, a remarkable power reduction can be achieved.

Consider the circuit of Fig. 3.12, where for simplification reasons zero-delay gate model is assumed. Also, it is assumed that the circuit signals are uncorellated, which means that the switching activity, E_i, of a node, i, is equal to $E_i = 2p_i(1 - p_i)$. It is also assumed that the input capacitance for every gate is C_g, while the static probabilities of the input lines are $p_a = p_b = p_d = p_e = 1/2$ and $p_b = 1/4$.

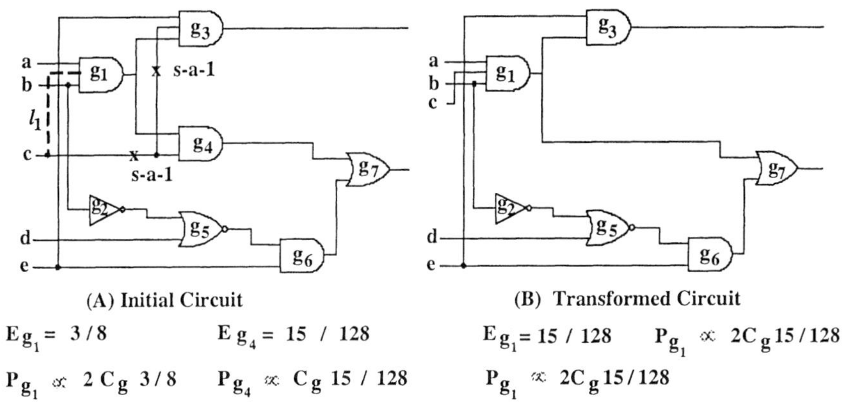

$$E_{g_1} = 3/8 \qquad E_{g_4} = 15/128 \qquad E_{g_1} = 15/128 \qquad P_{g_1} \propto 2C_g 15/128$$

$$P_{g_1} \propto 2C_g 3/8 \qquad P_{g_4} \propto C_g 15/128 \qquad P_{g_1} \propto 2C_g 15/128$$

Figure 3.12. Post Mapping Power Optimization

Considering a new connection, l_1, is added to the circuit, this connection is redundant and thus, the functionality is preserved. However, adding l_1 connections from c to g_3 and g_4 become redundant (i.e. s-a-1 faults on these two lines are undetectable). After removing the new redundancies the transformed circuit is obtained. Since the gate g_4 has been removed in the transformed circuit and the switching activity of g_1 has been reduced, the power consumption of the transformed circuit is smaller than that of the initial circuit. Based on this idea a lot of post-mapping low-power transformations were proposed.

In [30] an Automatic Test Pattern Generator (ATPG) based approach was proposed. This technique is able to identify permissible transformations on the network that may reduce power consumption. The method of finding the permissible transformations is simulation-based. The implications are classified in three clauses and bit-parallel fault simulation is performed to eliminate most of the invalid clauses. The remaining potentially valid clauses are combined to create different combinations, each of which is checked for validity using ATPG. If additional clauses are assumed, the number of combinations that should be considered, increase dramatically.

In [31] a sophisticated learning mechanism to find the satisfiability and observability implications of the target nodes was introduced. Such implications are used to identify circuit transformations that add and remove connections in the circuit aiming at removing high power consumption nodes. The major benefit of this method is that it uses learning procedures based on symbolic calculations to find the permissible implications. It offers great advantages in terms of complexity compared with ATPG-based methods.

In [32] a similar approach was proposed. The method is based on the addition and removal of redundant connections. The main difference of this method and other implication-based rewiring methods for power optimization is that the procedure for identifying the permissible implications is switching activity oriented. Instead of searching for all permissible implications and then re-evaluating their impacts on power, the method finds transformations that are useful for power reduction.

It should be stressed that the timing behavior of the circuit is a strict constraint in all the above-mentioned methods. The reported experimental results show that power saving up to 70% can be achieved preserving the timing behavior of the circuit.

Another interesting post mapping technique based on gate resizing has been presented in [29]. This technique consists of replacing some gates of the circuit with other gates having smaller area and therefore smaller capacitive load. Given that the power dissipated by a gate is proportional to its load, reducing that load leads to a power reduction. However, smaller gates are also slower; thus to preserve the timing behavior of the circuit only gates not belonging in the critical path are resized. However, an accurate timing analysis is strongly required to identify non-critical gates in the circuit. Specifically, the arrival, the required and the slack times should be computed with high accuracy determining how many gates can be resized without changing the original speed of the circuit. To achieve this *Algebraic Decision Diagrams* (ADDs) [33] are used to compute symbolically the timing behavior of the circuit.

3.6   Conclusions

The reduction of switching activity topic of combinatorial and sequential logic circuits was addressed. The existing techniques were categorized whether are technology independent or not. Also, specific steps were determined, which should be followed by a designer to reach an efficient solution. However, the existing power optimizing techniques were concerned only with the logic behaviour of a circuit and but not with the issue of interconnect. Current technology advances lead to deep submicron dimensions where the impact of the interconnect on the power consumption is significant. Consequently, future logic synthesis approaches should take the effects interconnect into account.

References

[1] A. Chadrakasan, S. Sheng, and R. W. Brodersen, "Low- Power CMOS Design", in *IEEE Journal* of *Solid- State Circuits*, vol. 27, no. 4, pp. 472-484, April 1992.

[2] G. De Micheli, " Synthesis and Optimization of Digital Circuits", *McGraw-Hill*, 1994.

[3] S. Devadas, A. Gosh, and K. Kreutzer, "Logic Synthesis", *McGraw-Hill*, 1994.

[4] R. K. Brayton, G. Hatchel, and A. Sangiovanni-Vincentelli, "Multi-Level Synthesis", in *IEEE Proceedings*, vol. 78, pp. 264-300, February 1990.

[5] K. Triverdi, "Probability and Statistics with Reliability, Queuing, and Computer Science Applications", *Prentice-Hall*, 1982.

[6] K. Roy and S. Prasad, "Circuit Activity Based Synthesis for Low Power Reliable Operations", in *IEEE Trans.* on *VLSI*, Vol. 1, No 4., pp.503-513, Dec. 1993.

[7] L. Benini, G. De Micheli, "State Assignment for Lowe Power Dissipation", in *IEEE Journal* of *Solid-State Circuits*, Vol. 30, No. 3, pp. 258-267, March 1995.

[8] E. Olson and S. M. Kang, "Low-Power State Assignment for Finite State Machines", in *Proc.* of *Int. Workshop on Low Power Design*, pp. 63-68, 1994.

[9] C. Tsui, M. Pedram, C. Chen and A. Despain, "Low Power State Assignment Targeting Two- and Multilevel Logic Implementations", in *IEEE Trans.* on *CAD*, Vol. 17, No. 12, pp. 1281-1291, Dec 1998.

[10] S-H. Chow, Y-C. Ho, T. Hwang and C.L. Liu, "Low Power Realization of Finite State Machines- A Decomposition Approach", in *ACM Trans.* on *Design Automation of Electronic Systems*, Vol. 1, No. 3, pp. 315-340, July 1996.

[11] B. Kumthekar, I-H. Moon, and F. Somenzi, "A Symbolic Algorithm for Low Power Sequential Synthesis", in *Proc.* of *ISLPED*, pp. 56-61, 1997.

[12] J. Monteiro and A. Oliveira, "Finite State Machine Decomposition for Low-Power", in *Proc.* of *DAC*, pp. 758-763, 1998.

[13] M. Alidima, J. Monteiro, S. Devadas, A. Ghosh and M. Papaefthumiou, "Precomputation-Based Sequential Logic Optimization for Low Power", in *IEEE Trans.* on *VLSI*, Vol. 2, No. 4, pp. 426-435, Dec 1994.

[14] J. Monteiro, S. Devadas, and A. Ghosh, "Retiming Sequential Circuits for Low Power:, in *Proc* of *ICCAD*, pp. 398-402, 1993.

[15] L. Benini, P. Siegel, and G. De Micheli, "Saving Power by Synthesizing Gated Clocks for Sequential Circuits", in *IEEE Design & Test of Computers*, pp. 32-41, Winter 1994.

[16] L. Benini, and G. De Micheli, "Automatic Synthesis of Low-Power Gated-Clock Finite-State Machines", in *IEEE Trans.* on *CAD*, Vol. 15, No. 6, pp. 630-643, June 1996.

[17] L. Benini, G. De Micheli, E Macii, M. Poncino and R. Scrasi, "Symbolic Synthesis of Clock-Gating Logic for Power Optimization of Synchronous Controllers", in *ACM Trans.* on *Design Automation of Electronic Systems*, Vol.4, No. 4, pp. 351-375, Oct 1999.

[18] L. Benini, G. De Micheli, A. Lioy, E. Macii, G. Odasso, and M. Poncino, "Synthesis of Power-Managed Sequential Components Based on Computation Kernel Extraction", in *IEEE Trans.* on *CAD*, Vol. 20, No. 9, pp. 1118-1131, Sept. 2001.

[19] S. Vrudhula and H.Y. Xie, "Techniques for CMOS Estimation and Logic Synthesis for Low Power", in proc. of ISLPED, pp. 21-26, 1994.

[20] S. Iman and M. Pedram, "Two-Level logic Minimization for low-Power", in Proc of *ICCAD*, pp. 433-438, 1995.

[21] G. Thedoridis, S. Theoharis, D. Soudris, and C. E. Goutis, " A Method for Minimizing the Switching Activity in Two-Level Logic Circuits", in IEE Proc. on Computers & Digital Techniques, Vol. 145, No. 5., pp. 357-363, Sept. 1998.

[22] S. Iman and M. Pedram , "Logic Extraction and factorization for Low Power", in Proc. of *DAC*, pp. 248-253, 1995.

[23] S. Iman and M. Pedram," An approach for Multilevel Logic Optimization Targeting Low Power", in *IEEE Trans.* on *CAD*, Vol. 15, No. 8, pp. 889-901, Aug. 1996.

[24] V.Tiwari, S. Malik, and P. Ashar, "Guarded evaluation: pushing power management to logic synthesis/design" in *IEEE Trans.* on *CAD*, Vol. 17, No. 10, pp. 1051-1060, Oct. 1998.

[25] T. Kitahara and R. K. Brayton, "Low Power Synthesis via transparent Latches and Observability Don't Cares", in *Proc.* of *ISCAS*, pp. 1516-1519, 1997.

[26] C.-Y. Tsui, M. Pedram and A. Despain, "Power Efficient Technology Decomposition and Mapping under Extended Power Consumption Model", in *IEEE Trans.* on *CAD*, Vol. 13, No. 9, Sept. 1994.

[27] V.Tiwari, P. Ashar, and S. Malik, "Technology Mapping for Low Power in Logic Synthesis", in *Integration, The VLSI Journal*, July 1996.

[28] B. Lin and H. De Man, "Low-Power Driven Technology Mapping under Timing Constraints", in *Proc.* of *ICCAD*, pp. 421-427, 1993.

[29] R. Iris Bahar, H. Cho, G. Hatchel, E. Macii and F. Somenzi, "Symbolic Timing Analysis and Resynthesis for Low-Power of Combinational Circuits Containing False Paths", in *IEEE Trans.* on *CAD*, Vol. 16, No. 10, pp. 1101-1115, Oct. 1997.

[30] B. Rohfleisch, A. Kolbl, and B. Wurth, "Reducing Power Dissipation after Technology Mapping by Structural Transformation" in Poc. of DAC, pp.789-794, June 1996.

[31] R. Iris Bahar, E. Lampe, and E. Macii, "Power Optimization of technology-Dependent Circuits Based on Symbolic Computation of Logic Implications", in *ACM Trans.* on *Design Automation of Electronic Systems*, Vol.5, No. 3, pp. 267-293, July 2000.

[32] Q. Wang, S. Vrudhula, G. Yeap, and S. Ganguly, "Power Reduction and Power-Delay Trade Offs Using Logic Transformations", in *ACM Trans.* on *Design Automation of Electronic Systems*, Vol.4, No. 1, pp. 97-121, Jan. 1999.

[33] R. Iris Bahar, E. Frohm, C. Ganoa, G. Hatchel, E. Macii, A. Pardo, and F. Somenzi, "Algebraic Decision Diagrams and their Applications", in Int. Rep. VLSI/CAD Group, Dept. Elec. Comp. Eng., Univ. of Colorado, Boulder, Apr. 1993.

Chapter 4

CIRCUIT-LEVEL LOW-POWER DESIGN

Spiridon Nikolaidis

Aristotle University of Thessaloniki, Thessaloniki, Greece

snikolaid@physics.auth.gr

Alexander Chatzigeorgiou

University of Macedonia, Thessaloniki, Greece

achat@uom.gr

Abstract   It is in the nature of digital design that any logic function can be implemented by different transistor structures. These alternatives present different characteristics in terms of speed and power dissipation, depending on the capacitance at the internal nodes, the fan-in requirements, the operating conditions and the structure itself. Speed and low power dissipation are in the most cases contradictory factors but there are also exceptions, where design for high speed results in low power circuits. Circuit designers have to be aware of this diversity of structures and the available possibilities that can be exploited to optimally correlate high speed and low power dissipation.

Keywords: Circuit design, Logic styles, Latches, Flip-flops, Transistor sizing

4.1   Introduction

CMOS technology has been established as the basic technology for the fabrication of integrated circuits mainly because of its inherent low power characteristics. CMOS is the only circuit design technique, which does not consume any static power except leakage. Power is only consumed during output switching. However, the continuously increasing requirements for speed and functionality tend to lead classic static CMOS logic to the limits of acceptable power consumption. Different CMOS logic styles and special circuit design techniques have been proposed for improving the power characteristics as well as the speed.

D. Soudris et al. (eds.), Designing CMOS Circuits for Low Power, 45–69.
© 2002 *Kluwer Academic Publishers. Printed in the Netherlands.*

In this chapter the power characteristics of different logic styles and alternative implementations of basic digital circuits are reviewed and compared. However, it is not always possible to define an absolute winner in this comparison. This becomes more difficult when the other design targets, speed and area, are taken into account. Many factors can influence the efficiency of each of the proposed techniques. For example, signal probabilities are the determinative factor for the use or not of dynamic logic in a design. Pseudo-nMOS could reduce power only if it is used in complex logic function designs switching at high frequencies. The need or not of both outputs of a flip-flop can be the determinative factor for the selection of the appropriate circuit.

The rest of the chapter is organized as follows: Section 4.2 discusses several logic styles in terms of performance, area and power consumption. Section 4.3 provides an overview of latches and flip-flops focusing on power characteristics. Techniques for reducing power dissipation based on transistor sizing and reordering are presented in Section 4.4, while in Section 4.5 energy issues in the design of drivers for large loads are discussed. Section 4.6 concludes this chapter.

4.2   Logic Style

The scope of this section is to introduce the basic logic styles for designing digital integrated circuits and to discuss the influence of each logic style on speed, size and power dissipation. Each paragraph includes a *"Concept"* section describing the structure and operation of each logic style and a *"Power Considerations"* section discussing the power consumption relative to other logic families. Although not all logic styles are discussed, the advantages and drawbacks of each style regarding power consumption, provide an insight to the forces that increase or reduce the energy dissipation in every circuit design.

4.2.1   Static Logic

Concept.   Static logic is the most widely used logic style for digital CMOS designs due to its excellent properties in many areas. The term *static* means that the output of a logic gate at every time point is connected through a low-resistance path to the power supply rails [1]. In this subsection only *complementary* or *conventional* static CMOS gates will be studied: A static CMOS gate consists of a *pull-up network* and a *pull-down network* which provide conditional connection to V_{dd} or V_{ss}, respectively. A typical static logic gate is the NAND2 gate shown in Figure 4.1. In case the applied inputs A, B cause the function $F = \overline{A \cdot B}$ to evaluate to false, (i.e. A and B are both set to logic "1"), the output node is pulled down through the path formed by the series of nMOS transistors. In case one of the applied inputs is set to logic "0", the output node

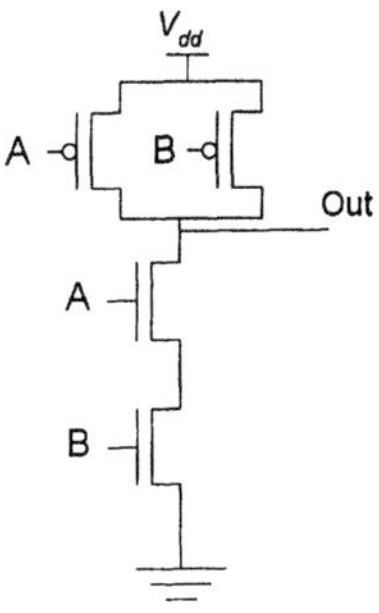

Figure 4.1.   Two-input NAND gate in static CMOS logic style

is pulled high through the path formed by the corresponding *p*MOS transistor which is turned on.

The widespread acceptance of static CMOS logic compared to other logic styles, is mainly due to the following characteristics [1], [2]:

- Ratioless logic.

- High noise margins which offer low sensitivity to noise.

- Sufficient speed, especially for small gates.

- Comparable rise and fall times under appropriate scaling.

- Ease of design.

Power Considerations.   Complementary CMOS logic is suitable for building low-power circuits, mainly because no power is consumed when a gate is in steady-state operation (i.e. non transient operation). That is because no direct path between V_{dd} and *GND* exists, when the output is not switching. However, a CMOS gate suffers from the other three power components, namely dynamic, short-circuit and leakage power [3].

Assuming that a CMOS gate drives a load with capacitance C_L, its output undergoes on average α transitions from logic "0" to logic "1" per clock cycle, has a supply voltage V_{dd} and operates at a frequency f_{clk}, the dynamic power is given by:

$$P_{dyn} = \alpha C_L V_{dd}^2 f_{clk} \tag{4.1}$$

Short-circuit power exists when there is a direct conducting path from supply to ground during the period in which the output changes state. If transistors are properly sized, short-circuit power is less than 10% of the total power. Recent techniques attempt to operate the circuits at supply voltages which are less than the sum of the *p*MOS and *n*MOS threshold voltages to essentially eliminate any short-circuit currents [3].

The third power component is leakage power and has two primary sources: a) leakage currents due to reverse bias diodes which are present at a transistor's drain and b) subthreshold leakage currents due to carrier diffusion between source and drain, when the gate-to-source voltage of a transistor is below the threshold voltage. Although leakage power is small compared to other sources of power dissipation, it should be carefully taken into account when a circuit operates in stand-by mode for large periods of time (e.g. in sleep mode for portable devices).

A variation of static logic, called *branch-based logic* (BBL) [4] provides a layout optimization for low-power by reducing parasitic node capacitances. Logic cells are designed exclusively with branches which are implemented by transistor in series between supply and output. Apart from being robust for deep submicron technologies, BBL has been used for building a cell library with a limited number of standard cells. This new approach, achieves improved speed and lower power consumption compared to conventional libraries with a large number of complex gates, due to better optimizations enabled by the limited choice of cells [5]. A BBL 16-bit adder has been shown in [6] to have lower static and dynamic power dissipation compared to an equivalent complementary CMOS adder.

4.2.2   Dynamic Logic

Concept.   Dynamic (precharged) logic is often used in CMOS circuits to reduce the transistor count, to increase speed and to avoid static power consumption, which is present in pseudo-*n*MOS logic [1]. In Figure 4.2 an example of a dynamic gate is shown. The *n*MOS *pull-down network*, which implements the logic function, is similar to the *n*MOS network in complementary CMOS logic. Dynamic gates are clocked and are based on the sequence of two phases: *precharge* and conditional *evaluation* phase.

Precharge.   When clock is low, the output node is precharged to V_{dd} by the *p*MOS transistor. The *n*MOS transistor whose gate is connected to *Clk* is cut-off and therefore no DC current flows regardless of the values of the input signals.

Evaluation.   When clock is high, transistor M_p is off, while transistor M_n is turned on. Depending upon the value of inputs I_1, I_2, ..., I_n, a conditional path between *Out* and GND is created, discharging the output node causing a low output signal. If such a path does not exist, the output node remains precharged causing a high output value. Consequently, once *Out* is discharged, it cannot be charged again and as a result the input to the gate can make at most

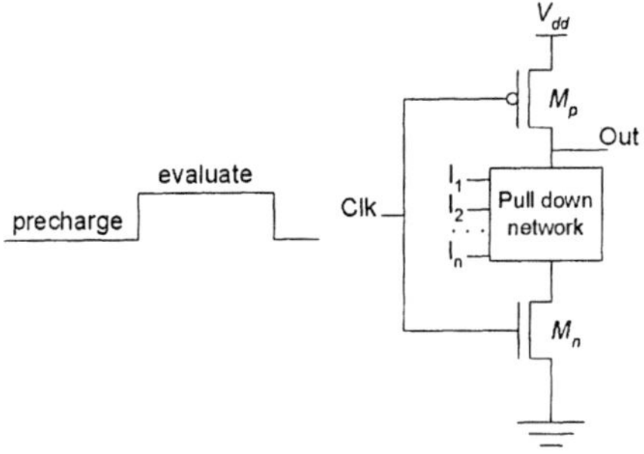

Figure 4.2. Dynamic logic gate

one transition during the evaluation phase. If this condition is not met, charge redistribution occurs, which might corrupt the output node voltage [2]. This inherent drawback does not allow single-phase dynamic gates to be cascaded.

Power Considerations. In a dynamic implementation power is consumed during the precharge phase each time the output capacitor is discharged in the preceding cycle. This is equivalent to stating that power is consumed every time the output equals 0, independent of the preceding or following values! [1]. In contrast to static CMOS gates, a dynamic gate can consume power even if the inputs remain constant. Consequently, for dynamic logic, the output transition probability which determines the dynamic power consumption depends only on the signal probabilities (The probability that a given signal has a particular value). For uniformly distributed inputs, this means that the transition probability is [3]:

$$p_{0\to1} = \frac{N_0}{2^N} \tag{4.2}$$

where N_0 is the number of 0's for the output signal in the truth table and N the number of inputs. As an example for a NOR2 gate, the dynamic power consumption of the dynamic implementation is $P_{NOR} = 0.75C_L V_{dd}^2 f_c$, while for the static implementation power consumption is significantly smaller $P_{NOR} = (3/16)C_L V_{dd}^2 f_c$. However, the capacitance being switched in dynamic logic is smaller than in a static implementation and total power should include the power dissipated in driving the capacitance of the clock lines. Table 4.1 summarizes the advantages and disadvantages of static vs. dynamic logic.

Two and four-phase clocking strategies that have been developed [2] to overcome the problem of cascading dynamic gates need up to 8 clock signals and they are clearly not suited for low power operation. To correct the problem of cascading gates, modifications to the basic dynamic logic style have been proposed, such as DOMINO [7], *np*-CMOS [8] and NORA [9] logic.

Table 4.1.   Static versus Dynamic Logic [3]

	Static logic	**Dynamic logic**
Glitches	30% energy increase	intrinsically does not have this problem
Short-circuit currents	less than 10% of the total power	
Parasitic capacitance		fewer transistors => reduced switched capacitance
Switching activity	depends on previous state	does not depend on previous state. Generally, higher activity factor.
Power down models	effectively used	not well suited
Clock Power	no clock	due to gate capacitance of precharge MOS transistor

4.2.3   Pass-Transistor Logic

Concept.   One approach to reduce the physical capacitance in a digital circuit and in this way lower the power consumption, is to use transfer gates over conventional CMOS gates to implement logic functions. Boolean functions are implemented as a network of switches, realized by pass-transistors. Logic operation is based on the fact that a series connection of pass transistors implements an AND function, while a parallel connection an OR function. A XOR logic gate in pass transistor logic is shown in Figure 4.3

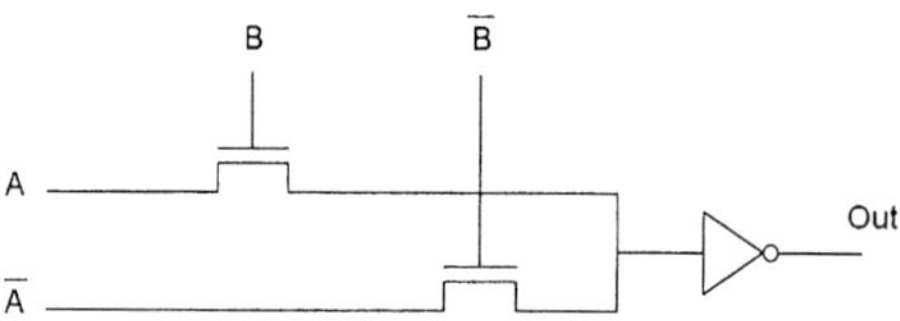

Figure 4.3.   XOR gate in pass-transistor logic

Although pass-transistor logic styles are relatively expensive for simple monotonic gates, they are very efficient in terms of transistor count for designs that employ the XOR and MUX operation [10]. The implementation of a XOR logic gate in CMOS counts 12 transistors while the XOR implementation shown above only 4. The most efficient realization of the full adder counts

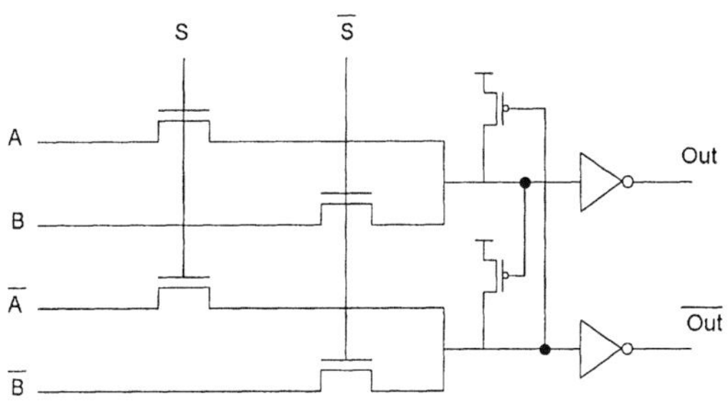

Figure 4.4. Two-input multiplexer in CPL

28 transistors [2], while the most compact pass transistor implementation (in LEAP [11]) only 24 (including 6 transistors for the three input inverters, 4 transistors for the two output buffers and 2 level-restoring transistors). However, pass transistor logic presents the inherent problem of the threshold drop across a transistor, which causes static power dissipation in the following stages and imposes the addition of level restoring transistors. It should be stressed that *n*-channel MOS pass-transistors are not well suited for low supply voltages (below 1V).

The most important pass-transistor logic style is *Complementary Pass Transistor Logic* (CPL) [3], [12] consisting of two *n*MOS logic networks (one for each signal rail), two small pull-up *p*MOS transistors for level restoration and two output inverters for the complementary output signals. The basic and minimal CPL gate structure is the two-input multiplexer shown in Figure 4.4. CPL logic is ratioless and high noise margins enable reliable operation even at low voltages [1], [10]. Also, CPL gates present a good output driving capability due to the output inverters and a fast differential stage due to the cross-coupled *p*MOS pull-up transistors. Moreover, small input loads reduce the overall capacitance switched and consequently the power consumption is lower and rise/fall times are faster. As an example, this aspect of CPL is attractive in high-performance applications, for instance, multipliers [12].

Power Considerations. Throughout the literature, CPL or related pass-transistor logic styles have been propagated as low-power logic styles mainly due to the fact that CPL gates count fewer transistors, have smaller transistor sizes and thus, smaller node capacitances. Generally, for pass-transistor logic, significant power reduction can be achieved if the problem of threshold drop and static power dissipation of the output inverter is properly addressed, for example, by reducing the threshold voltage [3]. In [12] it has been reported that a pass-gate family adder with zero-threshold pass transistors at a supply voltage of 4V consumes 30% less energy than a conventional static design. In [13] a CPL full adder implementation based on transmission gates is presented,

Table 4.2. Full Adder Simulation Results [10]

Logic Family	Delay (ns)		Power (mW)		PT (norm)	
	3.3V	1.5V	3.3V	1.5V	3.3V	1.5V
CMOS	1.89	7.88	32.9	6.4	1.00	1.00
CPL	1.39	8.33	34.1	6.0	0.76	0.99

providing 50% energy savings compared to the conventional CMOS adder. For a multiplier based on this full-adder and using a modified Booth algorithm power savings of 18% and speed improvement of 30% have been reported.

However, a more rigorous exploration in [10] revealed that static CMOS proves to be superior to all pass-transistor logic styles in both delay and power for all logic gates except for the full adder at higher supply voltages. In the same paper it was concluded that pass transistor logic styles are not the best choice for low power design; however only small circuits have been evaluated and only CPL logic style is analyzed in detail. Results for a full adder in terms of delay, power and power-delay product are presented in Table 4.2. It should be mentioned that the full adder is based on XOR gates and 2:1 multiplexers, which are gates perfectly suited for pass-transistor implementation. Moreover, the adoption of pass-transistor logic should consider the fact that the number of full adders in most digital circuits, is limited compared to other logic gates and flip-flops.

One of the conclusions in [10] is that Single-Rail Pass-Transistor Logic (SPL) (which will be discussed in the next section) is a viable alternative if low power and compatibility with cell-based design is of concern.

The popularity of pass-transistor logic has increased during the last few years and is proved by the large number of designs with increased performance and low power consumption [12], [14], [15]. Moreover, many research efforts resulted in automated synthesis methodologies that target pass transistor logic starting from high-level, technology independent design specifications, such as Hardware Description Languages (HDL) [11], [16].

4.2.4   Single-Rail Pass-Transistor Logic

Concept.   Single-Rail Pass-Transistor Logic (SPL) is also known as Single-ended Pass-Transistor Logic and more often refer to as LEAP (LEAn integration with Pass-Transistors) [11] and offers a promising approach in low power circuit design [17]. SPL is the simplest member of the pass-transistor logic family and like CPL uses only *n*MOS transistors in the pass network. Unlike CPL it does not implement the second pass-transistor network for the complementary

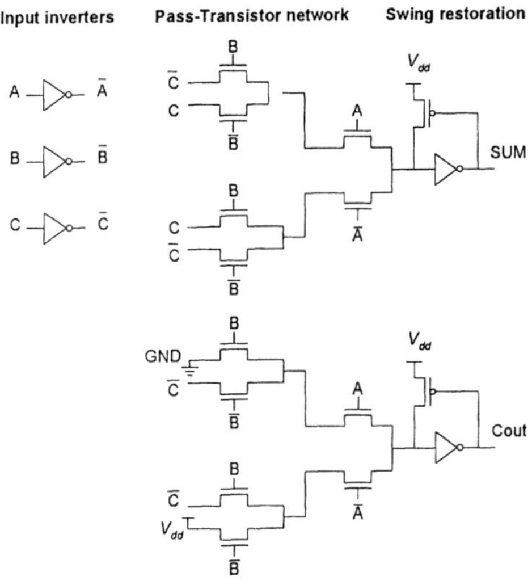

Figure 4.5. SPL full adder implementation

signals, which are generated locally if required. The SPL implementation of a full adder is shown in Figure 4.5. Three main components of an SPL circuit can be distinguished:

- Input inverters that buffer inputs and generate all signals for the pass-transistor network.

- The pass-transistor network that implements the logic function. Because SPL uses only n-type transistors, the voltage swing at the end of the network will be 0V to V_{dd}-V_{TN} (where V_{TN} is the nMOS transistor threshold voltage).

- The output buffers for speed improvement including a weak pMOS transistor for voltage level restoration and elimination of short-circuit currents in the output inverter.

According to simulations, for the optimum power-delay product, an output buffer must be inserted every three [18] or four [17] stages, with possibly more buffers on the critical path. However, the swing restoration structure only works for $V_{dd} > V_{TN} + |V_{TP}|$, because a degraded logic "1" due to threshold voltage drop would otherwise prevent the nMOS transistor of the inverter and with that the restoring pMOS transistor from turning on.

As it can be observed, the basic element in the pass-transistor network is the two-input multiplexer, while each multiplexer corresponds to a node in the BDD (Binary Decision Diagram) representation of the logic function [19].

SPL logic has the following advantages:

- The SPL cell library has no more than 10 components. Basically, it has only the three main components mentioned above.

- Most functions and especially those with many inputs or functions based on XOR and multiplexers can be implemented very efficiently in terms of transistor count.

- The pass-transistor network contains only nMOS transistors resulting in a compact layout with fast operation.

- Circuits synthesis can be automated starting from BDD representations of the Boolean functions.

Power Considerations. Because of the extensive use of nMOS transistors that suffer from threshold drop, power consumption in SPL circuits is very sensitive to the minimum voltage of operation. In [11] it is claimed that low-voltage performance of nMOS-based pass-transistor logic is better than the conventional CMOS circuits down to 1V, when the threshold voltage is 0.4 V. For lower threshold voltages subthreshold leakage may become a severe constraint against low-power design for both SPL and CMOS. [11] reports for a seven-input, four-output random logic a reduction in power consumption around 15.5 % with a 31% increase in speed, while for a 4-b Adder/Subtractor circuit there is no significant power reduction through the use of SPL. More extensive simulation results [17] indicate that at 3.3V supply voltage and 0.35μm technology, the power consumption of the SPL implementation of a full adder and a 4-b adder is about half of the standard cell CMOS design. For the same conditions, the delay of the SPL circuit is worse and becomes significantly greater for lower supply voltages. [10] reports that SPL does not work for low supply voltages (1.5 V) while its superiority compared to CMOS could not be confirmed for higher voltages, for a full adder circuit.

As a final conclusion, it can be stated that the power-delay performance of SPL logic has to be investigated for future deep submicron technologies with lower supply voltage. In addition, it should be kept in mind that full adder comparisons favor pass-transistor based logic styles due to the XOR gates and multiplexers involved, which are perfectly suited for pass-transistor implementation.

4.2.5   Other Logic Styles

4.2.5.1   Pseudo-nMOS.

The pseudo-nMOS logic style is attractive when designing complex gates with large fan-ins, because the pull-up network is replaced by a single load transistor. The pull-down nMOS network that implements the logic function, is similar to that of a conventional CMOS gate. A NOR gate implemented in pseudo-nMOS logic is shown in Fig. 4.6(a). For a gate with N inputs, only $N+1$ transistors are required, resulting in smaller area

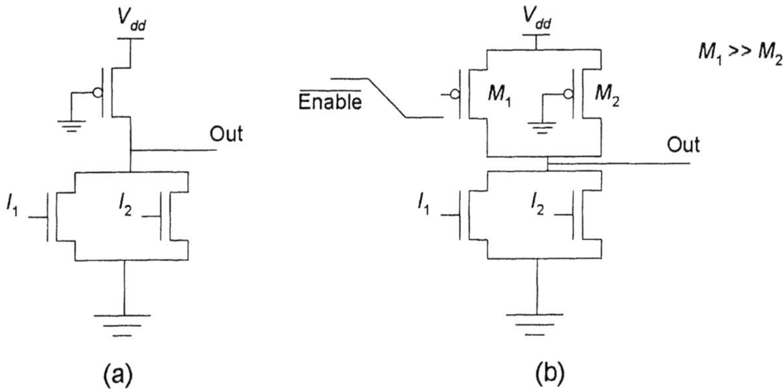

Figure 4.6.   (a) Pseudo-*n*MOS NOR gate, (b) Pseudo-*n*MOS NOR gate with adaptive load

and smaller parasitic capacitance. Moreover, the preceding gate has a smaller load capacitance, since each input is connected only to a single transistor. The inherent problem is that to generate consistent high and low logic levels, the logic has to be *ratioed*; that is, transistor sizes have to be selected carefully to ensure correct operation [2].

Pseudo-*n*MOS is not suitable for low power operation because of static power dissipation that occurs whenever the pull-down network is turned on. According to [1] a minimal-size gate consumes approximately 1mW of static power. Therefore, a 100,000 gate circuit consumes 50W, assuming that half of the gates are in low-output mode. Pseudo-*n*MOS can reduce power consumption only for complex logic functions switching at high frequencies where savings to the dynamic power component due to reduced capacitances are dominant.

An improved load can be constructed if a second large *p*MOS pull-up device is connected in parallel to the grounded one, Figure 4.6(b). When the circuit should be in standby mode consuming limited power, transistor M_1 is turned off. The small *p*MOS device (transistor M_2) represents a large resistance and as a consequence the static current is small and the output value sufficiently low. When the output should make a low-to-high transition, the large *p*MOS transistor is turned on resulting in a large current drive and a fast transition. This circuit is well suited for gates that should switch only during certain time periods, as for example in address decoders.

4.2.5.2   Differential Voltage Logic Styles.

Differential Cascade Voltage Switch Logic (DCVSL) is another approach that eliminates the static current in pseudo-*n*MOS logic style, based on dual rail coding [20]. The principle of this logic style is shown in Figure 4.7. Each gate generates both polarities of the output signal, while the pull-down networks are complementary and implement the required logic function and its inverse. The basic advantage of DCVSL

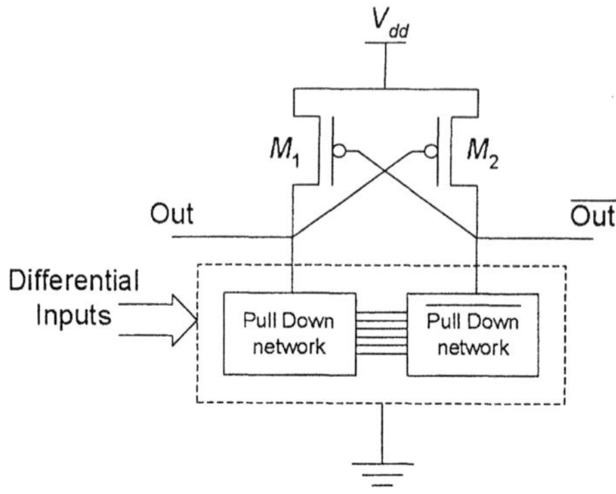

Figure 4.7.   Basic Principle of DCVSL logic

over conventional CMOS is faster switching due to reduced output capacitance. Compared to pseudo-nMOS the DCVSL exceeds in that there is no static power consumption; however, the current during switching increases due to the large pull-up transistor. The fact that two pull-down networks are being used can be seen as both an area penalty and as an advantage since complementary signals are available eliminating the need for additional inverting stages.

A dynamic version of a DCVSL logic can also be implemented [2]. However, this is a power hungry logic style and not suitable for low power devices. A novel DCVSL family, namely Differential Current Switch Logic (DCSL) was proposed in [21]. DCSL is suitable for implementing high fan-in gates and aims at reducing power consumption by restricting internal node voltage swings (down to 1V for 5V supply voltage) in the evaluation nMOS tree, without any performance degradation. Moreover, once evaluation is complete, a DCSL gate does not respond to its inputs, resembling a latch followed by a combinatorial circuit. For moderate tree heights it is claimed that DCSL reduces power dissipation by a factor of two compared to DCVSL gates.

DCSL however, is not free from its own problems:

- In common with precharged differential logic it has a high activity factor

- The gate is sensitive to noise because of the strong positive feedback.

- Simple AND/NAND cannot be implemented efficiently.

- Balanced layout techniques are required to ensure correct operation.

Finally, Charge Recycling Differential Logic (CRDL), proposed in [22] reduces power consumption by using some of already-used charge for precharge, while speed is comparable to that of conventional dynamic logic circuits. In a CRDL gate the charge which is used for logic evaluation in a cycle is recycled to establish a precharge value in the next cycle. For ideal conditions, assum-

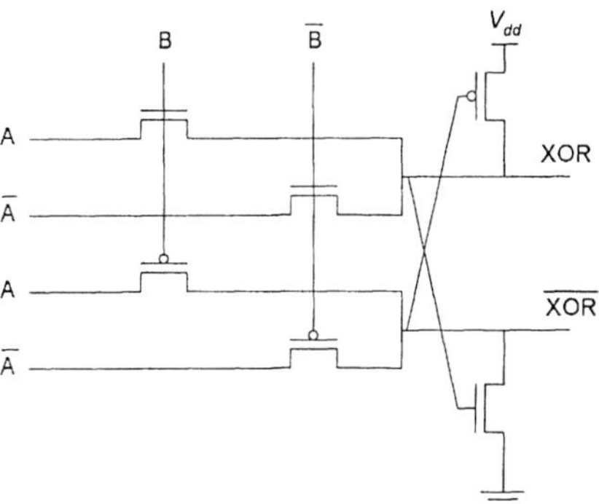

Figure 4.8. XOR gate in PPL

ing that precise half-supply precharge level is achieved, the amount of power consumed is 50% that with a full-swing technique. Experimental results for a 0.8μm, 5V CMOS Manchester carry chain indicate a 27% improvement in power-delay product compared to DCVSL.

4.2.5.3 Push-Pull Pass-Transistor Logic. Push-Pull Pass-Transistor logic (PPL) [23] is similar to CPL. However, its pass transistor network is not composed solely of nMOS transistors, but employs two complementary pass-transistor networks (Figure 4.8). Although the pMOS and nMOS pass-transistor networks suffer from threshold voltage drop, if one network is responsible for transferring an input value to the output, the complementary network turns the corresponding pull-up or pull-down transistor on, restoring the output level to its nominal value. This push-pull action eliminates the need for output buffers with restoring transistors. PPL is claimed to be a good low power choice for logic style: In [23] simulation results for a 40 stage full adder indicate a power-delay product of only 60% compared to SPL. Measurement results for multipliers realized in 0.8μm 3.3V CMOS technology show that the power-delay product for the PPL implementation is only 42% of the CMOS multiplier, 63% of the CPL implementation and 78% of the SPL one.

4.2.6 Logic Styles: Discussion

It is definitely not possible to conclude that a specific logic style is the optimum in terms of both performance and power consumption. Even if some experimental results indicate that one logic style offers more advantages than the others, the range of experiments performed, the technology which has been used, and the variation of design/architectural parameters does not permit a safe conclusion. Ideally, a number of benchmark circuits should be simulated and implemented for the same technology and exercised under identical conditions to possibly define the most appropriate solution, expressed by the minimum power-delay product.

Bearing in mind the previous observations, it could be claimed that SPL has been a promising logic style in the era of low power designs. It offers the advantage of reduced transistor count for complex functions, compact layout and low power consumption. However, the latter is put into question for low supply voltages, especially for future deep submicron technologies with supply voltages below 1V.

Concerning power consumption, analysis should not focus only on the primary sources of power consumption, i.e. dynamic power due to charging or discharging node capacitances. Short-circuit power dissipation, energy consumption due to glitching activity, sub-threshold currents and the ability to benefit from future developments (such as ultra-low supply voltages) should be taken into account.

From a comparison of several logic styles in [1] it appears that static logic remains the most reliable logic style: It is a simple, robust and relatively low power technique, offers ease of design, and has the advantage of being supported by the majority of Electronic Design Automation (EDA) tools currently available, a fact which is possibly the most significant barrier for the widespread usage of other logic styles.

4.3   Latches and Flip-Flops

Power consumption of the clock distribution network in synchronous systems has usually been a substantial part of the system total power consumption. This power is dissipated in the clock distribution network and clocked registers which consist of flip-flops or latches. In order to save power the clocked capacitance in such circuits has to be minimized.

4.3.1   Latches

Dynamic latches are the simplest and most efficient timing circuits. In Figure 4.9(a) the classic latch is shown. This is the fastest latch [24]. It can also be seen as a transmission gate followed by an inverter. It is very fast because it behaves as a true single stage. The C2MOS latch, shown in Figure 4.9(b) is very similar to the classic one. This is slower but most robust than the classic and additionally it is slightly more power efficient because there is no contact at the intermediate nodes. Both of these have four clocked transistors (including the inverter) three of which are loading the clock input. The circuit shown in Figure 4.9(c) is a True Single Phase Clocked (TSPC) half latch [25]. This latch will only isolate high inputs at clock low. The non-precharged TSPC latch, shown in Figure 4.9(d), consists of two of the TSPC half-latch circuits. The

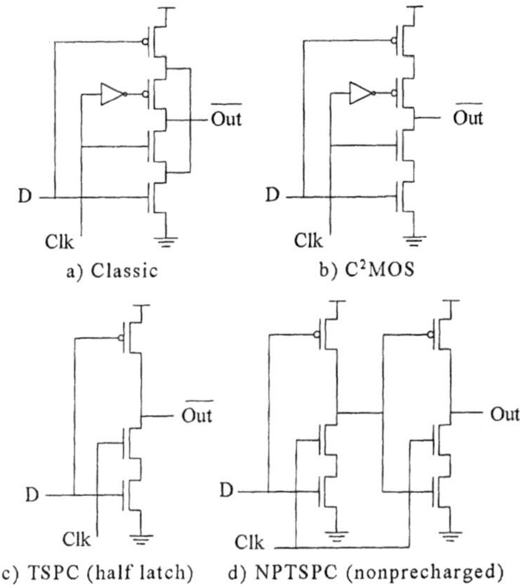

Figure 4.9. Dynamic latches

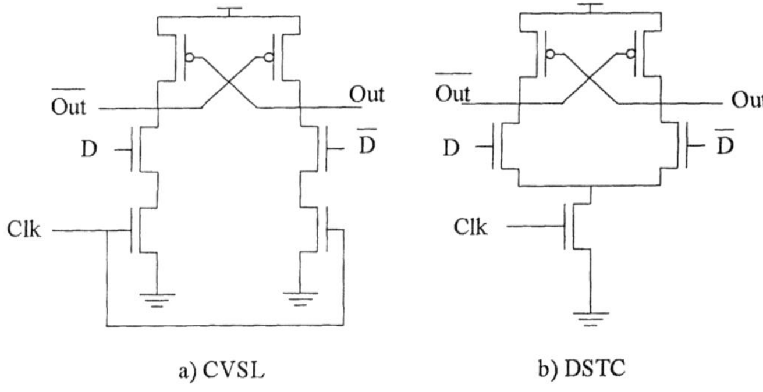

Figure 4.10. Double rail dynamic latches

last latch is slower than the classic one, more robust and has only two clocked transistors.

Another class of latches using dual rail data are shown in Figure 4.10. Figure 4.10(a) shows a latch derived from CVSL logic [2]. This latch depends on transistor ratios, i.e. the n-transistor must be stronger than the p-transistors in order to flip the latch. In Figure 4.10(b) the Dynamic Single Transistor Clocked (DSTC) latch is shown, which uses only a common clocked transistor to save power [26]. This latch is both fast and power-efficient, but sensitive to input glitches when in hold state.

According to simulation results reported in [27], the Non Precharged True Single Phase Clocked latch (NPTSPC) has the lowest power consumption of

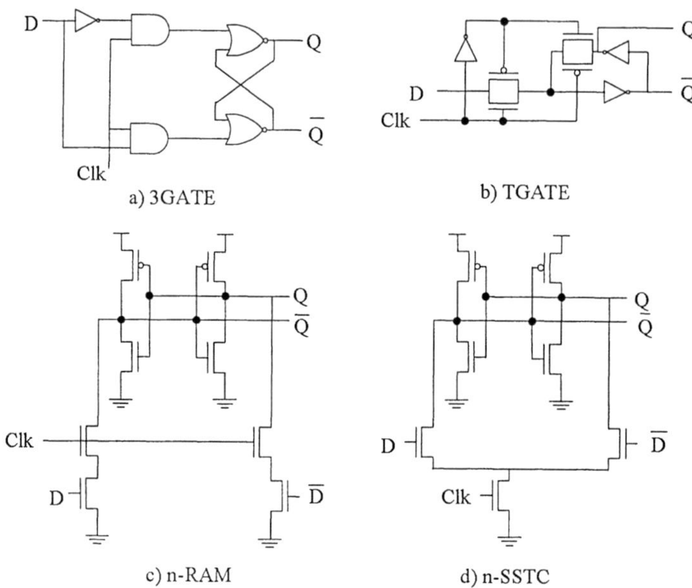

Figure 4.11.   Static latches

all the above mentioned latches. This is explained by its very few clocked transistors (only two) and the lack of precharging. The precharged version of TSPC and PTSPC [27] is next. All others have comparable power consumption.

Static latches always need some form of positive feedback (like a cross-coupled inverter pair)", automatically making them have complementary outputs [2]. In Figure 4.11(a) a single-input latch based on the classical set-reset flip-flop, 3GATE latch, is shown. This latch looks complex but has relatively low power consumption due to only four clocked transistors. In Figure 4.11(b) the classical transmission gate based static latch, TGATE latch, using six clocked transistors, is presented. Both of these are used as basis for flip-flops in standard CMOS cell libraries, the latter being more common than the former. Therefore Both are very often used in industry. Another static latch is the RAM type, based on the 6-transistor SRAM memory cell (Figure 4.11(c)) [2]. Like the CVSL latch, this latch depends on transistor ratios. Finally, in Figure 4.11(d) a static version of the DSTC latch is given.

The n-SSTC and the *n*-RAM latches present the lowest power consumption and the highest speed. The *n*-SSTC demonstrates a small advantage at speed compared to the *n*-RAM latch. However, the *p* versions of these latches present considerably worse performance. The reason for this is that they depend on transistor ratios, which means that when the *p*-transistors should be stronger than the *n*-transistors, their sizes (and so the parasitic capacitance) have to

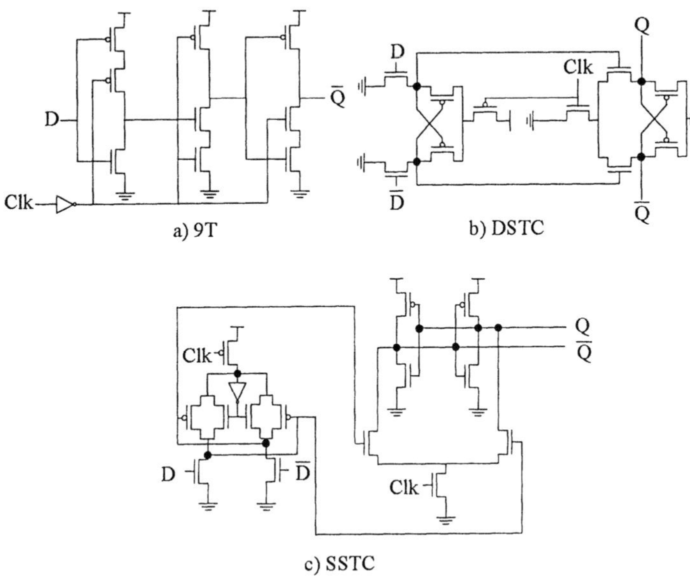

Figure 4.12.   Flip - Flops

increase substantially. Of the other two, the 3GATE latch has somewhat less power consumption than the TGATE latch, but worse speed [27].

4.3.2   Flip - Flops

Dynamic flip-flops can be constructed by cascading two latches with different polarities. This scheme is directly used for classic, C2MOS and NPTSPC latches forming corresponding flip-flops. A very efficient flip-flop can be constructed by combining a *p*-half TSPC latch and a PTSPC latch (9T latch) as shown in Figure 4.12(a). An inverter can be added at the output of these circuits to achieve a complementary output. Concerning dual rail latches a very efficient flip-flop with only two clocked transistors can be constructed based on the DSTC latch (Figure 4.12(b)). According to simulation results presented in [27] the DSTC, NPTSPC and 9T flip-flops have the lowest power consumption. Also, the NPTSPC and 9T flip-flops can improve considerably their speed if they are designed without complementary outputs. The other two flip-flops have comparable power consumption with the classic to be faster.

Static flip-flops can be formed by static latch pairs, e.g. for the latch types 3GATE and TGATE. There is also a static version of the DSTC flip-flop called SSTC flip-flop and shown in Figure 4.12(c) [26]. Like the dynamic version, it uses only two clocked transistors, making it a good candidate for very-low power consumption. From a power point of view, the SSTC flip-flop is obvi-

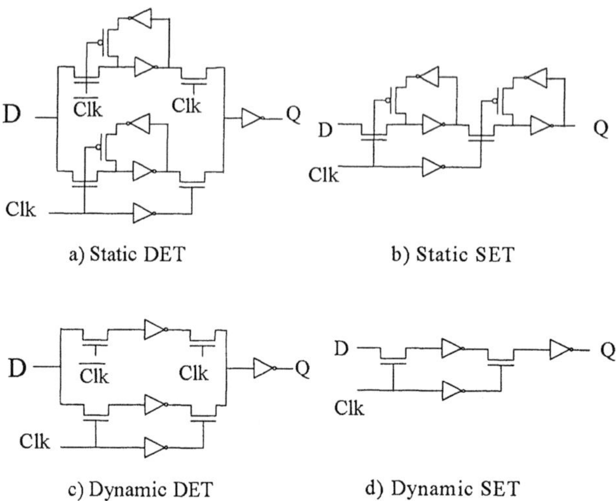

Figure 4.13.   Double edge triggered flip-flop

ously superior to the others. It also has a nearly minimum delay. The low-power consumption of this flip-flop is due to simplicity and very few clocked transistors. Among the others, the 3GATE flip-flop presents lower power consumption than TGATE but a larger delay. A very valuable comparison of the performance and power characteristics of flip-flops can be found in [28].

4.3.2.1   Double edge triggered flip-flop.   The Double Edge Triggered (DET) flip-flop is triggered on both edges of the clock pulses instead of only one edge. Using DET flip-flops the clock frequency can be halved for the same data rate, thus reducing the power dissipation on the clock distribution network.

In [29] the DET and Single Edge Triggered (SET) flip-flops, static and dynamic, shown in Figure 4.13 are compared. Both the SET and the DET flip-flops have two D-type latches. In the DET flip-flop the latches are arranged in parallel while in the SET flip-flop they are placed serially.

DET flip-flops have been shown to have lower energy requirements (~20%) than SET flip-flops, with only a limited overhead in complexity. This has been confirmed by analytical and simulation means [29]. DET and SET flip-flops have comparable maximum data rates – the dynamic SET flip-flops being slightly faster, but requiring a clock operating at twice the frequency as a dynamic DET flip-flop. Simulation results also indicated that dynamic flip-flops consume less energy than static flip-flops. In addition to the energy savings possible in the devices themselves, system level energy savings are also possible with the use of DET flip-flops. A conservative example provides encouraging results in terms of the energy savings (~17%) possible via DET flip-flops.

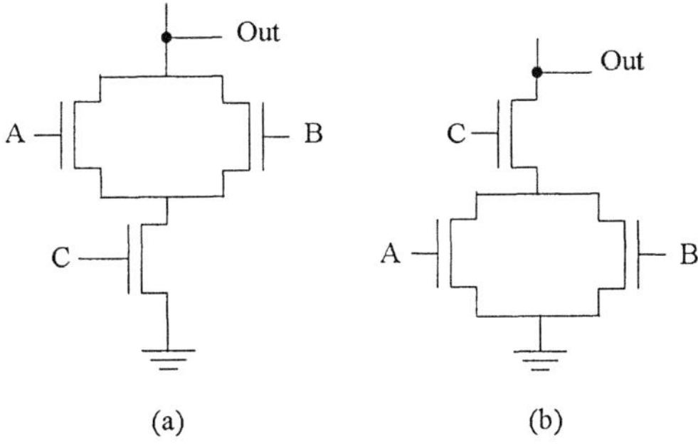

Figure 4.14. Transistor Reordering

4.4 Transistor Sizing and Ordering

4.4.1 Transistor ordering

The relative placement of the transistors in a serially connected MOSFET chain does not alter the functionality of the chain in the circuit. In complex gates where groups of transistors may be connected in series, flexibility also often exists in the relative placement of the transistors. This freedom in transistor placement can be exploited to achieve CMOS circuits with reduced power dissipation. Transistor order in a CMOS gate strongly determines the switching activity, and hence, the power dissipation in the internal nodes of the gate. By reordering transistors in a CMOS gate, power dissipation can be reduced in two ways: minimizing the drain-source capacitance and by signal probability algorithms to reduce transitions [30], [31].

To minimize the drain-source capacitance, nodes with the highest capacitance have to be placed closest to the supply and ground rails as it is shown in Figure 4.14(b). However, this contradicts signal probability based algorithms [30]. From the point of view of signal probability, after various simulations, a scheme with a pull down network as in Figure 4.14(a) was reported to be better in [30].

In [31] the effect of transistor reordering in simple and complex CMOS gates is studied. The reduction in power dissipation compared with the worst case configuration and the average power dissipation was reported to be 15.1% and 7.2%, respectively. In [30] more complex structures (MUX, ADDER and ALU) are analyzed and a 12% reduction of the average power dissipation was reported with a 4% increase in delay.

A set of simple transistor reordering rules is presented in [32]. Experimental results show that the presented technique typically reduces power by about 10% on average, but in some cases the improvement can be as great as 35%.

4.4.2   Transistor sizing

Appropriate sizing of the transistors in CMOS circuits can be applied for minimizing the power consumption under a given delay constraint. Two kinds of algorithms have been proposed for transistor size optimization :

i) Algorithms that start with a circuit that satisfies the timing constraint and reduce the size of the gates to reduce power dissipation.

ii) Algorithms that start by performing an initial power-optimal sizing on each gate. If the power minimal layout satisfies the delay constraint, the process is terminated. Otherwise, the power delay optimal sizing is applied to transistor sizes on the critical paths until the timing target is met. An algorithm of this type is presented in [33]. This algorithm is more complex than previous ones because it takes into account not only the power dissipation which is due to the charging of the circuit capacitance but also the short circuit power dissipation.

Another interesting conclusion of [33] is that the active area is not a reliable indicator of the power consumption of a circuit. It is shown that the power consumption of a CMOS circuit is a convex function of the active area and the objective of minimizing the power dissipation for a transistor sizing algorithm is different than that of minimizing the active area.

4.5   Drivers for large loads

Quite often it is necessary to drive high capacitance loads with reasonable speed and/or with short rise and fall times. Long rise/fall times can lead to large short-circuit power consumption in the following stages. Examples of high capacitance nodes are: clock networks, clock drivers, long buses, long interconnects and chip outputs. The standard method to drive big loads is to use a tapered inverter chain as shown in Figure 4.15. The optimal buffer design

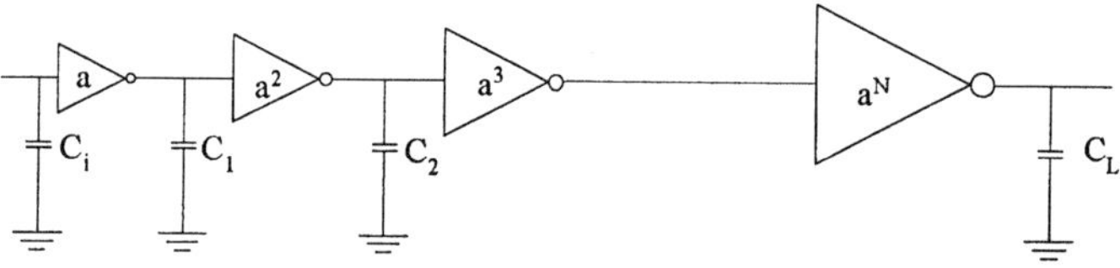

Figure 4.15.   Tapered buffers

scales consecutive stages in an exponential fashion. An analytical solution for the optimum value of the scaling factor in a uniformly tapered buffer, which minimizes the power-delay product, is given in [34]. Computer simulation shows that about 15-35% savings in power-delay product can be achieved by designing with this factor compared with that designing for minimizing the propagation delay. It is also found that there exists a non-uniform tapering factor, which gives a global optimum condition for a minimum power-delay

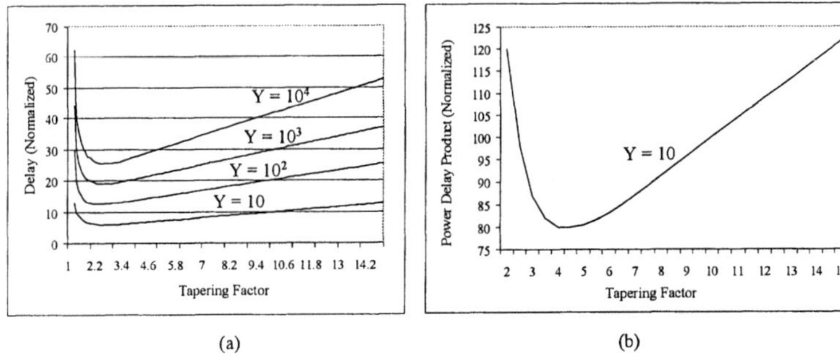

Figure 4.16. (a) Propagation delay, (b) power-delay product vs. tapering factor

product. Results obtained indicate that the tapering factor of the last stage must be larger than those of previous stages. Although, compared with uniform tapering, non-uniform tapering shows about 8% improvement in dynamic switching energy, the improvement becomes smaller (3-5%) for total switching energy that takes both dynamic and short-circuit power into account. Therefore, it is recommended to use a uniform buffer, since it is much simpler and provides better insight into the optimization of power-delay product.

Let us study the chain of Figure 4.15, which is loaded by C_L and uses N stages with a tapering factor of a. For $Y = C_L/C_i$ where C_i is the input capacitance of the first inverter of the chain, it is:

$$a^N = Y \tag{4.3}$$

$$N = \frac{\ln Y}{\ln a} \tag{4.4}$$

The total delay time is given by [34]:

$$\tau_{tot} = \frac{\ln Y}{\ln a} a\tau_i \tag{4.5}$$

where τ_i is the propagation delay of the first inverter in the chain. This relation is illustrated in Figure 4.16(a) for various values of Y. Differentiating (4.5) with respect to α the optimum value of a is obtained, $a_{opt}=e=2.72$. This minimum can be observed on the curves presented in Figure 4.16(a).

The total capacitance of the chain is given as [34]:

$$C_{tot} = \frac{a\,(Y-1)}{a-1} C_i \tag{4.6}$$

Since total dynamic power is proportional to total capacitance, the power-delay product of an inverter chain is given by:

$$P_{tot}\tau_{tot} \propto \frac{a \ln Y}{\ln a} \frac{a(Y-1)}{a-1} \qquad (4.7)$$

Differentiating this equation with respect to a the optimum value of power-delay product is derived, leading to $a = 4.25$. This can be observed in Figure 4.16(b), where the normalized power-delay product is represented as a function of the tapering factor.

From equation 4.6 it is observed that as the tapering factor increases, the total capacitance and thus the power consumption decreases. This means that the fewer buffers there are, the smaller the power consumption is (the smallest power consumption results when only one inverter is used). Instead of minimizing the power-delay product of the driver chain, by choosing a larger value for a, the value of excess power will be lower. For example, if a is increased from 3.5 to 9, the power consumption overhead will be reduced from 80% to 25% of the load power, at the cost of a delay increase of 20% [35]. However, for very large value of α, the delay will be too large (Figure 4.16(a)).

4.6   Conclusions

An attempt has been made to describe some of the basic alternative techniques which are contained in the toolbox of the engineer who designs circuits targeting low power dissipation. Different logic styles, circuit structures and circuit design techniques have been presented and their power characteristics discussed and compared. The comparison of these techniques was based on the results from real implementations presented in the literature.

The results are often contradictory, especially in the case of implementations with different logic styles, which makes the selection of the optimum solution in terms of power consumption very difficult. Things are worse when performance is also considered as a design parameter. Only in case of designs which employ specific operations (e.g. MUXes, XORs) some logic styles (e.g. pass transistor logic, SPL) present a clear advantage. Although static CMOS still remains a very efficient solution, it could be claimed that SPL has also been a promising logic style in the era of low power operation. However, the efficiency of SPL for deep submicron devices with very low supply voltages should be further investigated.

The power consumption of latches and flip-flops is strongly dependent on the clock load and on the data activity. Low power circuits should therefore have as few clocked transistors as possible. When comparing different circuit topologies for latches and flip-flops no correlation between power and speed has been found. It is therefore quite possible to combine low power with high speed. The NPTSPC topology for the dynamic latches and flip-flops and the SSTC for the static versions seem to incorporate very good characteristics in terms of both low power consumption and high performance.

Finally, power savings can be achieved by appropriate transistor ordering and sizing. Also, since driving of large loads means increased power consumption, appropriate tapered buffer design can lead to significant power reduction.

References

[1] J. M. Rabaey, Digital Integrated Circuits: A Design Perspective, Prentice Hall, Upper Saddle River, 1996.

[2] N. H. E. Weste, and K. Eshraghian, Principles of CMOS VLSI Design, Addison Wesley, Reading, 1994.

[3] A. P. Chandrakasan, and R. W. Brodersen, Low Power Digital CMOS Design, Kluwer Academic Publishers, Boston, 1995.

[4] C. Piguet, J.-M. Masgonty, S. Cserveny, E. Dijkstra, Low-Power Low-Voltage Digital CMOS Cell Design, in Proc. of 4th Int. Workshop on Power and Timing Modeling, Optimization and Simulation, Barcelona, Spain, October 1994.

[5] J.-M. Masgonty, S. Cserveny, C. Arm, P.-D. Pfister, C. Piguet, Low-Power Low-Voltage Standard Cell Libraries with a Limited Number of Cells, in Proc. of 11th Int. Workshop on Power and Timing Modeling, Optimization and Simulation, Yverdon, Switzerland, September 2001.

[6] A. Néve, D. Flandre, Branch-Based Logic for High-Performance Carry-Select Adders in 0.25 μm Bulk and Silicon-On-Insulator CMOS Technologies, in Proc. of 11th Int. Workshop on Power and Timing Modeling, Optimization and Simulation, Yverdon, Switzerland, September 2001.

[7] R. Krambeck et al., High-Speed Compact Circuits with CMOS, IEEE Journal of Solid State Circuits, vol. SC-17, no. 3, pp. 614-619, June 1982.

[8] C. M. Lee and E. W. Szeto, Zipper CMOS, IEEE Circuits and Systems Magazine, pp. 10-16, May 1986.

[9] N. Gonclaves and H. J. DeMan, NORA: a racefree dynamic CMOS technique for pipelined logic structures, IEEE Journal of Solid State Circuits, vol. SC-18, no. 3, pp. 261-266, June 1983.

[10] R. Zimmermann, and W. Fichtner, Low-Power Logic Styles: CMOS Versus Pass-Transistor Logic, IEEE Journal of Solid-State Circuits, vol. 37, no. 7, pp. 1079-1090, July 1997.

[11] K. Yano, Y. Sasaki, K. Rikino, K. and Seki, Top-Down Pass-Transistor Logic Design, IEEE Journal of Solid-State Circuits, vol. 31, no. 10, pp. 792-803, October 1996.

[12] K. Yano, T. Yamanaka, T. Nishida, M. Saito, K. Shimohigashi, and A. Shimizu, A 3.8-ns CMOS 16x16-b Multiplier Using Complementary Pass-Transistor Logic, IEEE Journal of Solid-State Circuits, vol. 25, no. 2, pp. 388-393, April 1990.

[13] I. S. Abu-Khater, A. Bellaouar, and M.I. Elmasry, Circuit Techniques for CMOS Low-Power High-Performance Multipliers, IEEE Journal of Solid-State Circuits, vol. 31, no. 10, pp. 1535-1546, October 1996.

[14] U. Ko, P. T. Balsara, and W. Lee, Low-Power Design Techniques for High-Performance CMOS Adders, IEEE Transactions on VLSI Systems, vol. 3, no. 2, pp. 327-333, June 1995.

[15] M. Suzuki, N. Okhubo, T. Shinbo, T. Yamanaka, A. Shimizu, K. Sasaki, and Y. Nakagome, A 1.5-ns 32-b CMOS ALU in Double Pass-Transistor Logic, IEEE Journal of Solid-State Circuits, vol. 28, no. 11, pp. 1145-1150, November 1993.

[16] A. Jaekel, S. Bandyopadhyay and G.A. Jullien, Design of Dynamic Pass-Transistor Logic Circuits Using 123 Decision Diagrams, IEEE Trans. on Circuits and Systems-I: Fundamental Theory and Applications, vol. 45, no. 11, pp. 1172-1181, November 1998.

[17] M. Munteanu, P. A. Ivey, N. L. Seed, M. Psilogeorgopoulos, N. J. Powell, I. A. Bogdan, Single Ended Pass-Transistor Logic - A Comparison with CMOS and CPL, in Proc of X IFIP International Conference on VLSI (VLSI'99), Lisbon, Portugal, December 1999.

[18] V. Bertaco et al., Decision Diagrams and Pass Transistor Logic Synthesis, Technical report No: CSL-TR-97-748, Computer Systems Laboratory, Stanford University, December 1997.

[19] R. E. Bryant, Graph-Based Algorithms for Boolean Function Manipulation, IEEE Transactions on Computers, vol. C-35, no. 8, pp. 677-691, August 1986.

[20] L. G. Heller et al., Cascode Voltage Switch Logic: A Differential CMOS Logic Family, ISSCC Digest of Technical Papers, pp. 16-17, Februrary 1984.

[21] D. Somasekhar, K. Roy, Differential Current Switch Logic: A Low Power DCVS Logic Family, IEEE Journal of Solid-State Circuits, vol. 31, no. 7, 981-991, July 1996.

[22] B.-S. Kong, J.-S. Choi, S.-J. Lee, and K. Lee, Charge Recycling Differential Logic for Low Power Application, IEEE Journal of Solid-State Circuits, vol. 31, no. 9, pp. 1267-1276, September 1996.

[23] W.-H. Paik, H.-J. Ki, and S.-W. Kim, Low Power logic design using push-pull pass-transistor logics, International Journal of Electronics, vol. 84, no. 5, pp. 467-478, 1998.

[24] W. Bowhill et al., A 300MHz 64b quad-issue CMOS microprocessor, in ISSCC Digest of Technical Papers, pp. 182-183, 1995.

[25] J. Yuan and C. Svensson, High speed CMOS circuit technique, IEEE Journal of Solid-State Circuits, vol. 24, no. 1, pp. 62-70, February 1989.

[26] J. Yuan and C. Svensson, New single-clock CMOS latches and flip-flops with improved speed and power savings, IEEE Journal of Solid-State Circuits, vol. 37, no. 1, pp. 62-69, January 1997.

[27] C. Svensson and J. Yuan, Latches and Flip-Flops for Low Power Systems, Low Power CMOS Design, ed. A. Chandrakasan, IEEE Press, 1998.

[28] V. Stojanovic, V.G. Oklobdzija, Comparative Analysis of Master-Slave Latches and Flip-Flops for High-Performance and Low-Power Systems, IEEE Journal of Solid-State Circuits, vol. 34, no. 4, pp. 536–548, April 1999.

[29] R. Hossain, L.D. Wronski and A. Albicki, Low power design using double edge triggered flip-flops, IEEE Transactions on VLSI Systems, vol. 2, no. 2, pp. 261-265, June 1994.

[30] E. Mussol and J. Cortadela, Optimising CMOS Circuits for Low-Power using Transistor Reordering, Department of Computer Architecture, Universidad Politecnica de Catalunya, Barcelona, Spain, 1995.

[31] R. Hossain, M. Zheng and A. Albicki, Reducing power dissipation in CMOS circuits by signal probability based transistor reordering, IEEE Trans. Computer Aided Design and Integrated Circuits and Systems, vol. 15, no. 3, pp. 361-368, March 1996.

[32] W. Shen, J. Lin, F. Wang, Transistor reordering rules for power reduction in CMOS gates, Asia South Pacific Design Automation Conference (ASPDAC), July 1995.

[33] M. Borah, R.M. Owens and M. J. Irwin, Transistor sizing for low power CMOS circuits, IEEE Transactions on Computer Aided Design of Integrated Circuits and Systems, vol. 15, no. 6, pp. 665-671, June 1996.

[34] J. Choi and K. Lee, Design of CMOS tapered buffer for minimum power-delay product, IEEE Journal of Solid-State Circuits, vol. 29, no. 9, pp. 1142-1145, September 1994.

[35] J. M. Rabaey, M. Pedram, Low Power design methodologies, Kluwer Academic Publishers, Boston, 1996.

Chapter 5

CIRCUIT TECHNIQUES FOR REDUCING POWER CONSUMPTION IN ADDERS AND MULTIPLIERS

Labros Bisdounis

INTRACOM S.A., Athens, Greece
lmpi@intracom.gr

Dimitrios Gouvetas

Odysseas Koufopavlou

University of Patras, Rio, Greece
odysseas@ee.upatras.gr

Abstract   An important issue in the design of VLSI Circuits is the choice of the basic circuit approach and topology for implementing various logic and arithmetic functions such as adders and multipliers. In this chapter, several static and dynamic CMOS circuit design styles are evaluated in terms of area, propagation delay and power dissipation. The different design styles are compared by performing detailed transistor-level simulations on a benchmark circuit (ripple carry adder) using HSPICE, and analyzing the results in a statistical way. After the comparison between the different design styles, a number of well known types of adders (ripple carry, carry skip, carry lookahead, carry select etc.) are compared in terms of propagation delay, number of gates and logic transition's average number. Furthermore, power measurements and comparisons for a number of well-known multipliers are provided. Based on the results of the provided analysis some of the tradeoffs that are possible during the design phase in order to improve the circuit power-delay product are identified.

Keywords: circuit design techniques, circuit macroblocks, adders, circuit styles

D. Soudris et al. (eds.), Designing CMOS Circuits for Low Power, 71–96.
© 2002 *Kluwer Academic Publishers. Printed in the Netherlands.*

5.1   Introduction

Much of the research efforts of the past years in the area of digital electronics has been directed towards increasing the speed of digital systems. Recently, the requirement of portability and the moderate improvement in battery performance indicate that the power dissipation is one of the most critical design parameters [1]. The three most widely accepted metrics to measure the quality of a circuit or to compare various circuit styles are area, delay and power dissipation. Portability imposes a strict limitation on power dissipation while still demands high computational speeds. Hence, in recent VLSI systems the power-delay product becomes the most essential metric of performance.

The reduction of the power dissipation and the improvement of the speed require optimizations at all levels of the design procedure. In this chapter, the proper circuit style and methodology is considered. Since, most digital circuitry is composed of simple and/or complex gates, we study the best way to implement adders in order to achieve low power dissipation and high speed. Several circuit design techniques are compared in order to find their efficiency in terms of speed and power dissipation. A review of the existing CMOS circuit design styles is given, describing their advantages and their limitations. Furthermore, a four-bit ripple carry adder for use as a benchmark circuit was designed in a full-custom manner by using the different design styles, and detailed transistor-level simulations using HSPICE [2] were performed. Also, various designs and implementations of four multipliers are analysed in the terms of delay and power consumption. Two ways of power measurements are used.

Conventional static CMOS has been a technique of choice in most processor design. Alternatively, static pass transistor circuits have also been suggested for low-power applications [3]. Dynamic circuits, when clocked carefully, can also be used in low-power, high speed systems [4]. However, several other design techniques need to be applied and evaluated along with these circuit styles in order to improve the speed and reduce the power dissipation of VLSI systems. In this chapter we study eight different CMOS logic styles:

- Conventional Static CMOS - *CSL*,

- Complementary Pass-transistor - *CPL* [5],

- Double Pass-transistor - *DPL* [6],

- Static and Dynamic Differential Cascode Voltage Switch - *DCVSL* [7,8],

- Static Differential Split-level - *SDSL* [9],

- Dual-Rail Domino - *DRDL* [10,11], and

- Enable/disabled CMOS Differential - *ECDL* [12].

The rest of the chapter is structured as follows. In the next section a brief introduction of the power dissipation and the delay in CMOS circuits is given. In section 3, the CMOS adder logic styles and their characteristics are described in details. The different adder logic styles are compared in terms of speed, power dissipation and silicon area, in section 4. Also, the power-delay product of the designs is considered, due to the importance of this metric in modern VLSI applications. Comparison results among different realizations of a 16-bit adder in terms of area, delay, and power are presented in Section 6. The next section provides results for four implementations of multipliers. Finally, the main points are summarized in Section 7 of conclusions.

5.2   Power and Delay in CMOS Circuits

Since the objective is to investigate the tradeoffs that are possible at the circuit level in order to reduce power dissipation while maintaining the overall system throughput, we must first study the parameters that affect the power dissipation and the speed of a circuit. It is well known that one of the major advantage of CMOS circuits over single polarity MOS circuits, is that the static power dissipation is very small and limited to leakage. However, in some cases such as bias circuitry and pseudo-nMOS logic, static power is dissipated. Considering that in CMOS circuits the leakage current between the diffusion regions and the substrate is negligible, the two major sources of power dissipation are the switching and the short-circuit power dissipation [1],

$$P = p_f \, C_L \, V_{dd}^2 \, f \; + \; I_{sc} \, V_{dd}, \qquad (5.1)$$

where p_f is the node transition activity factor, C_L is the load capacitance, V_{dd} is the supply voltage, f is the switching frequency. I_{sc} is the current which arises when a direct path from power supply to ground is caused, for a short period of time during low to high or high to low node transitions [13]. The switching component of power arises when energy is drawn from the power supply to charge parasitic capacitors. It is the dominant power component in a well designed circuit and it can be lowered by reducing one or more of p_f, C_L, V_{dd} and f, while retaining the required speed and functionality.

Even though the exact analysis of circuit delay is quite complex, a simple first-order derivation can be used [14,15] in order to show its dependency of the circuit parameters

$$T_d \; \propto \; \frac{C_L \, V_{dd}}{K \, (V_{dd} - V_{th})^{\alpha}}, \qquad (5.2)$$

where K depends on the transistors aspect ratio (W/L) and other device parameters, V_{TH} is the transistor threshold voltage, and α is the velocity saturation index which varies between 1 and 2 (α is equal to 1.4 for the 1.5μm process technology which is used in the experiments of the next section). Since a quadratic improvement in power dissipation may be obtained by lowering the supply voltage (equation (5.1)), many researchers have investigated the effects of lowering the supply voltage in VLSI circuits. Unfortunately, reducing the supply voltage reduces power, but the delay increases (equation (5.2)) with the effect being more drastic at voltages close to the threshold voltage [16]. Equations (5.1) and (5.2) indicate that by reducing the node parasitic capacitance in a CMOS circuit, the power dissipation is reduced and the circuit speed is increased.

5.3 CMOS Circuit Design Styles

In the following, the circuit design styles are described using the full adder circuit, which is the most commonly used cell in arithmetic units. Also, their characteristics in terms of power dissipation and delay are investigated.

5.3.1 Conventional Static CMOS Logic - CSL

Conventional Static CMOS logic is used in most chip designs in the recent VLSI applications. The schematic diagram of a conventional static CMOS full adder cell is illustrated in Figure 5.1. The signals noted with "-" are the complementary signals. The pMOSFET network of each stage is the dual network of the nMOSFET one. In order to obtain a reasonable conducting current to drive capacitive loads the width of the transistors must be increased. This results in increased input capacitance and therefore high power dissipation and propagation delay.

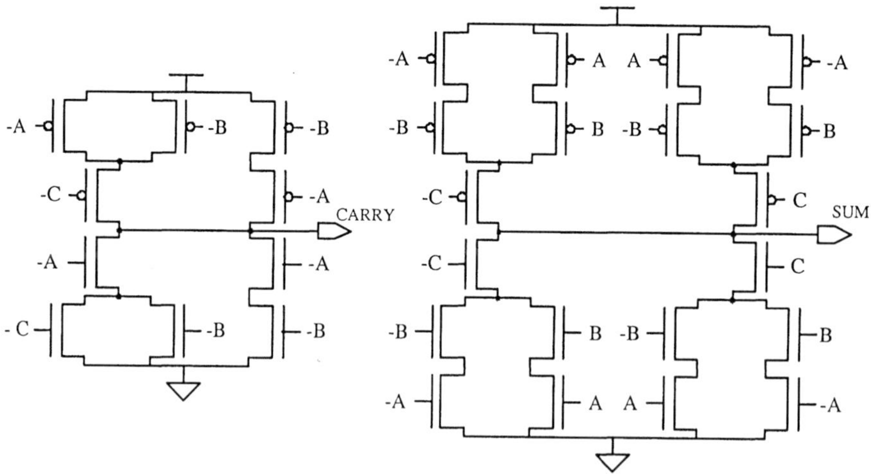

Figure 5.1. Conventional static CMOS full adder

5.3.2   Complementary Pass-Transistor Logic - CPL

The main concept behind CPL [5] is the use of only an nMOSFET network for the implementation of logic functions. This results in low input capacitance and high speed operation. The schematic diagram of the CPL full adder circuit is shown in Figure 5.2. Because the high voltage level of the pass-transistor outputs is lower than the supply voltage level by the threshold voltage of the pass transistors, the signals have to be amplified by using CMOS inverters at the outputs. CPL circuits consume less power than conventional static circuits because the logic swing of the pass transistor outputs is smaller than the supply voltage level. The switching power dissipated from charging or discharging the pass transistor outputs is given by

$$P_D = V_{dd} V_{swing} C_{node} f,$$ (5.3)

where $V_{swing} = V_{dd} - V_{thn}$. In the case of conventional static CMOS circuits the voltage swing at the output nodes is equal to the supply voltage, resulting in higher power dissipation. To minimize the static current due to the incomplete turn-off of the pMOSFET in the output inverters, a weak pMOSFET feedback device can also be added in the CPL circuits of Figure 5.2, in order to pull the pass-transistor outputs to full supply voltage level. However, this will increase the output node capacitance, leading to higher switching power dissipation and higher propagation delay.

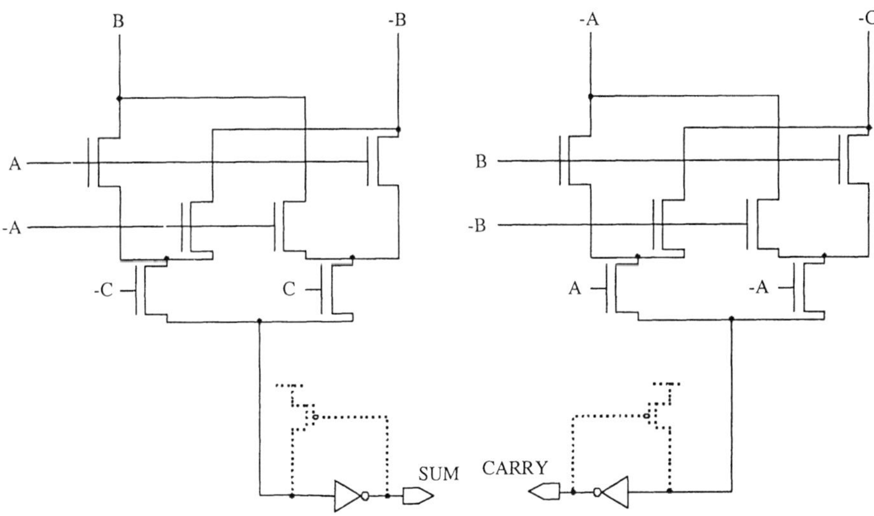

Figure 5.2.   Complementary pass-transistor full adder

5.3.3   Double Pass-Transistor Logic - DPL

DPL [6] is a modified version of CPL. The circuit diagram of the DPL full adder is given in Figure 5.3. In DPL circuits full-swing operation is achieved by simply adding pMOSFET transistors in parallel with the nMOSFET transistors. Hence, the problems of noise margin and speed degradation at reduced supply voltages which are caused in CPL circuits due to the reduced high voltage level, are avoided. However, the addition of pMOSFETs results in increased input capacitances.

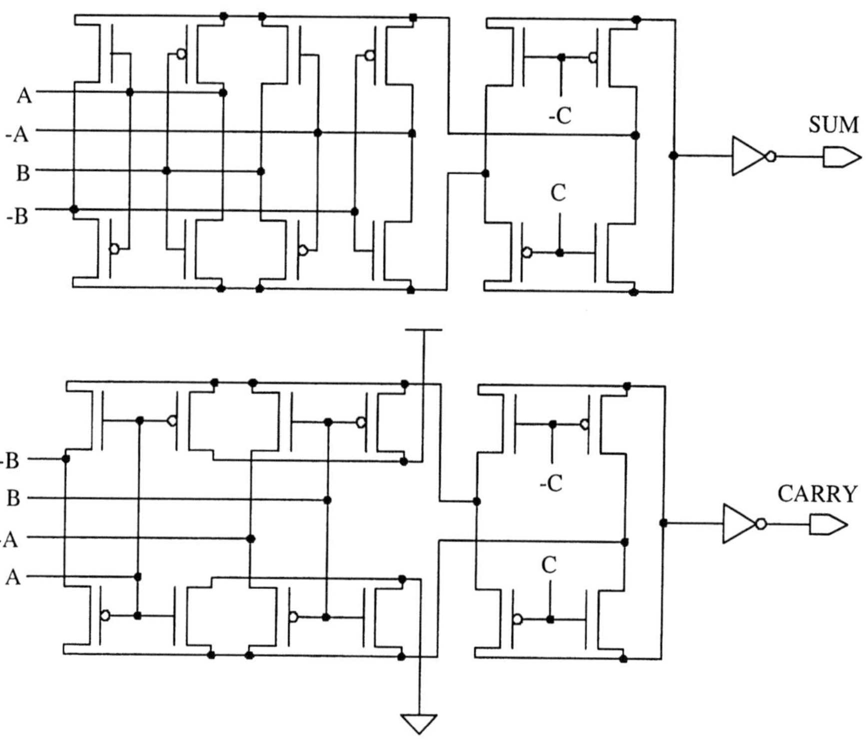

Figure 5.3.   Double pass-transistor full adder

5.3.4   Static Differential Cascode Voltage Switch Logic - SDCVSL

Static DCVSL [7], is a differential style of logic requiring both true and complementary signals to be routed to gates. Figure 5.4 shows the circuit diagram of the static DCVSL full adder. Two complementary nMOSFET switching trees are constructed to a pair of cross-coupled pMOSFET transistors. Depending on the differential inputs one of the outputs is pulled down by the corresponding nMOSFET network. The differential output is then latched by the cross-coupled pMOSFET transistors. Since the inputs drive only the nMOSFET transistors of the switching trees, the input capacitance is typically two or three times smaller than that of the conventional static CMOS logic.

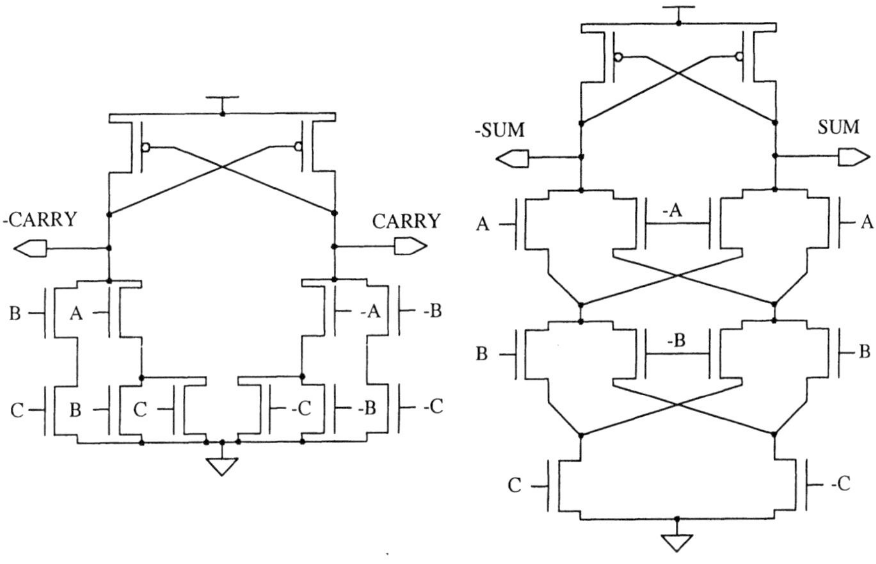

Figure 5.4.   Static differential cascode voltage switch full adder

5.3.5   Static Differential Split-level Logic - SDSL

A variation of the differential logic described above is the Static DSL [9]. The SDSL full adder circuit diagram is illustrated in Figure 5.5. Two nMOS-FET transistors with their gates connected to a reference voltage ($V_{ref} = (V_{dd}/2) + V_{thn}$, V_{thn}: nMOSFET threshold voltage) are added to reduce the logic swing at the output nodes. The output nodes are clamped at the half of the supply voltage level. Thus, the circuit operation becomes faster than standard DCVSL circuits. However, due to the incomplete turn-off of the cross-coupled pMOSFET transistors, SDSL circuits dissipate high static power dissipation. Also, the addition of two extra nMOSFET transistors per gate results in area overhead.

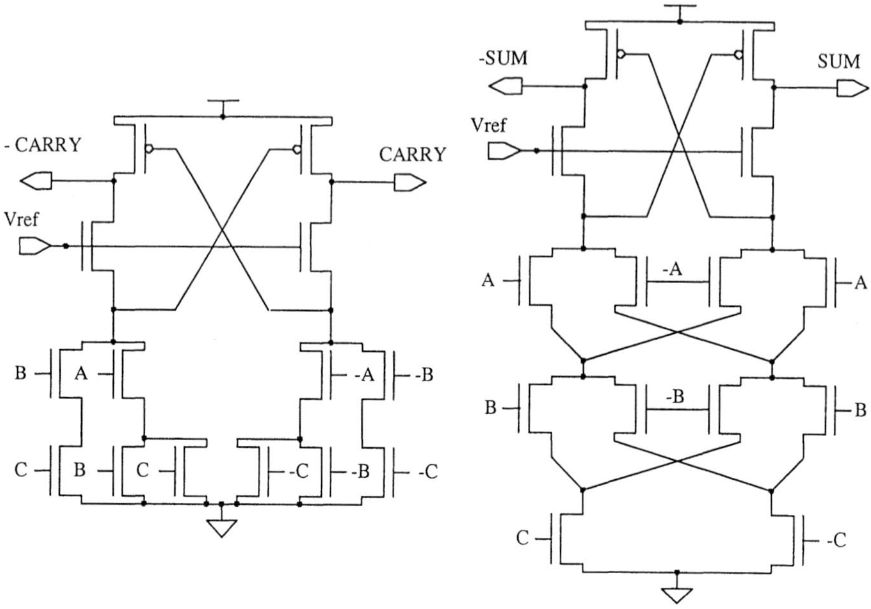

Figure 5.5.   Static differential split-level full adder

5.3.6   **Dual-Rail Domino Logic - DRDL**

Dual-Rail Domino Logic [10,11] is a precharged circuit technique which is used to improve the speed of CMOS circuits. Figure 5.6 shows a Dual-Rail Domino full adder cell. A domino gate consists of a dynamic CMOS circuit followed by a static CMOS buffer. The dynamic circuit consists of a pMOSFET precharge transistor and an nMOSFET evaluation transistor with the clock signal (CLK) applied to their gate nodes, and an nMOSFET logic block which implements the required logic function. During the precharge phase (CLK = 0) the output node of the dynamic circuit is charged through the precharged pMOSFET transistor to the supply voltage level. The output of the static buffer is discharged to ground. During the evaluation phase (CLK = 1) the evaluation nMOSFET transistor is *ON*, and depending on the logic performed by the nMOSFET logic block, the output of the dynamic circuit is either discharged or it will stay precharged. Since in dynamic logic every output node must be precharged every clock cycle, some nodes are precharged only to be immediately discharged again as the node is evaluated, leading to higher switching power dissipation [1]. One major advantage of the dynamic, precharged design styles over the static styles is that they eliminate the spurious transitions and the corresponding power dissipation. Also, dynamic logic does not suffers from short-circuit currents which flow in static circuits when a direct path from power supply to ground is caused. However, in dynamic circuits, additional power is dissipated by the distribution network and the drivers of the clock signal.

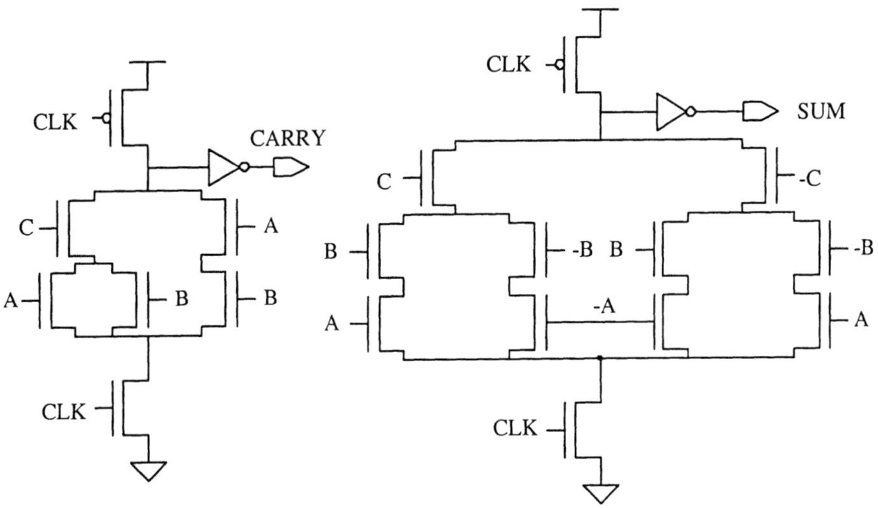

Figure 5.6.   Dual-rail domino full adder

5.3.7   Dynamic Differential Cascode Voltage Switch Logic - DDCVSL

Dynamic DCVSL [8], is a combination between the domino logic and the static DCVSL. The circuit diagram of the dynamic DCVSL full adder is given in Figure 5.7. The advantage of this style over domino logic is the ability to generate any logic function. Domino logic can only generate noninverted forms of logic. For example, in the design of a ripple carry adder, two cells must be designed for the carry propagation, one for the true carry signal and another for the complementary one (in Figure 5.6, the cell for the true carry signal is only shown, but the one for the complementary signal is also required). Using DCVSL to design dynamic circuits will eliminate p-logic gates because of the inherent availability of complementary signals. The p-logic gates usually cause long delay times and consumes large areas.

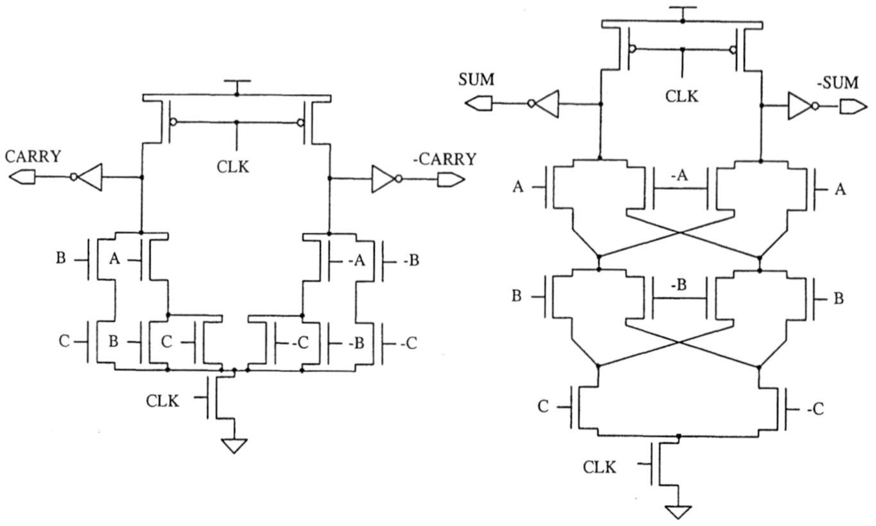

Figure 5.7. Dynamic differential cascode voltage switch full adder

5.3.8   Enable/disabled CMOS Differential Logic- ECDL

ECDL [12] is a self-timed differential logic which is used in the case of implementing logic functions using iterative networks. It uses extra signals to indicate the beginning and ending of a function evaluation, in order to improve the circuit speed. The structure of the ECDL full adder is illustrated in Figure 5.8. The signals $Done_{i-1}$ and $Done_i$ are the input and output self-timing control signals. During the disabled state, $Done_{i-1}$ has a value of logic one, which discharges both the true and the complementary outputs to logic zero. During the enabled state, $Done_{i-1}$ changes to logic zero and the topmost pMOSFET transistor (Figure 5.8) is *ON* to provide power to the inverters below. Then, depending on the logic of the differential nMOSFET network, a path exists from one of the output nodes to ground, holding that node to ground while leaving the other output node to be driven to logic one. One major advantage of the ECDL circuits is that there is no minimum clocking frequency requirement. However, ECDL circuits suffer from extra power dissipation due to the inverters which are needed to change the polarity of the output nodes. Also, their complex pull-up circuitry leads in extra silicon area.

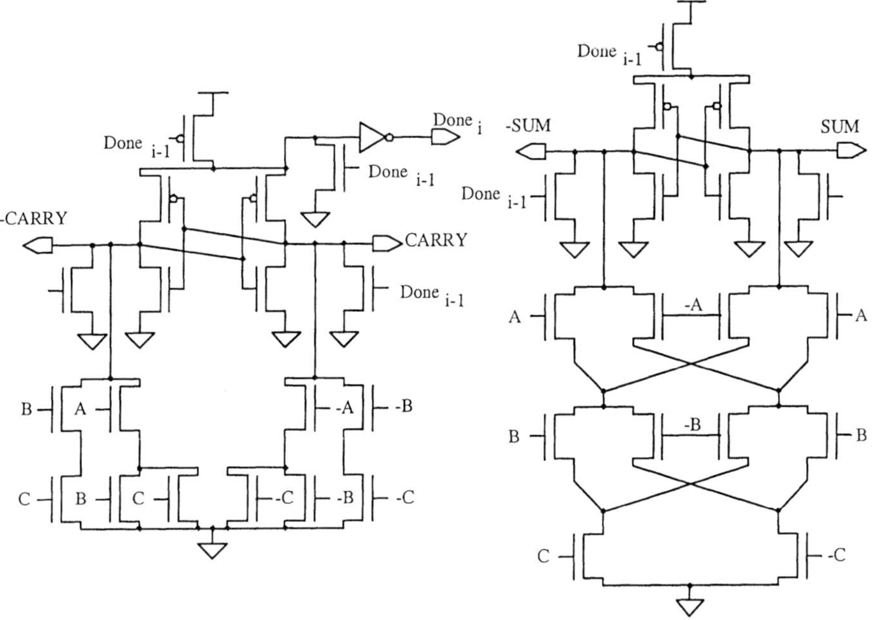

Figure 5.8.   Enable/Disable CMOS differential full adder

5.4   Power, Delay and Area Comparisons of a 4-Bit Ripple Carry Adder

The experimental results described in this section were obtained using a four-bit ripple carry adder. A general block diagram of the adder is illustrated in Figure 5.9.

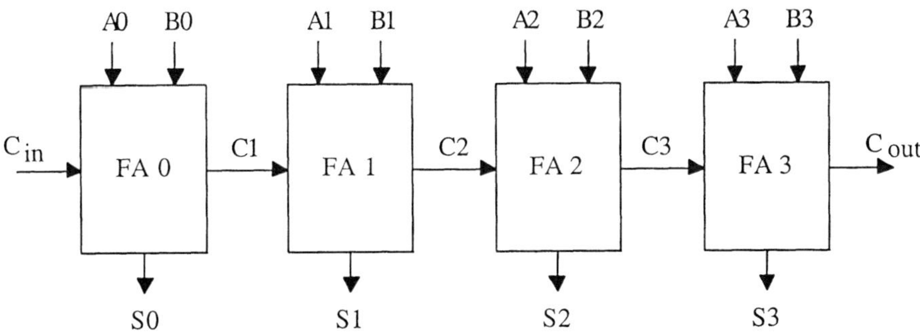

Figure 5.9.   Block diagram of the four-bit ripple carry adder

The circuit was designed in a full custom manner for all the design styles described in the previous section, using a 1.5μm CMOS process technology. The channel width of the transistors was 4.8μm for the nMOSFETs, and 9.6μm for the pMOSFETs. The design was based on the full adder cells presented in Figures 5.1 to 5.8.

Figure 5.10 shows the layout of the conventional static four-bit ripple carry adder, as an example of the designed circuits.

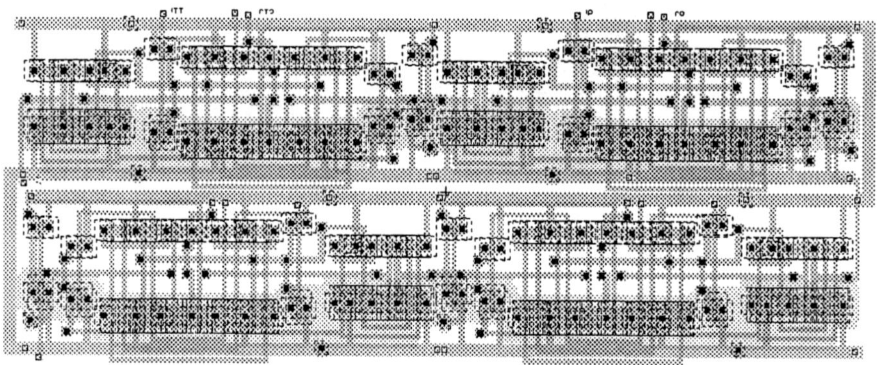

Figure 5.10.   Layout of the conventional static four-bit ripple carry adder

In Table 5.1 the adder silicon area and the number of the transistors for each design style are given. Although no extensive attempts were made to minimize area, the numbers presented are a good indication of the relative areas of the eight adder implementations, which account not only for the transistors, but for the interconnections as well. For example, even though DPL adder has fewer transistors than the CSL one, it has longer interconnections, which is reflected by its large area. Dynamic design styles and styles which uses control signals (such as ECDL) occupy extra area for the routing of the clock and the control signals. The smallest area is occupied by the CPL circuit, which has fewer transistors and shorter interconnections than the other adder implementations.

After the design of the layouts, circuit equivalents were extracted for a detailed circuit simulation using HSPICE [2] to obtain the power and delay measurements. In our experiments, a supply voltage of 5Volts is used. All measurements were obtained with each input supplied through a driver consisting of two minimum-sized inverters in series, and each output node driving a minimum-sized inverter load.

The estimation of power dissipation is a difficult problem because of its data dependency, and has received a lot of attention [17]. Some direct simulative power estimation methods have been proposed [18,19], which are expensive in

Table 5.1.   Area and number of transistors of the four-bit ripple carry adder implementations

Design Style	Adder Area ($\times 10^4$ μm^2)	No. of Transistors
CSL	5.42	144
CPL	4.46	88
DPL	6.52	136
SDCVSL	5.19	114
SDSL	6.39	130
DRDL	6.48	146
DDCVSL	7.22	154
ECDL	7.65	166

terms of time. Also, several power estimation methods have been proposed, where possibilities are used to solve the pattern-dependence problem. However, in order to achieve good accuracy, the spatial and temporal correlations between internal nodes should be modeled [20,21]. An alternative way is the use of statistical methods [22,23,17], that combines the accuracy of simulation-based techniques with the speed of probabilistic approaches.

In this chapter, the statistical approach proposed by Burch *et al.* [22] is used in order to estimate the power dissipation of our designs. Using the powermeter sub-circuit proposed by Kang [18], HSPICE can measure the average power consumed by a circuit given a set of input transitions and a time interval. In the method, the inputs are randomly generated and statistical mean estimation techniques are used to determine the final result. In our case for each adder design we use 200 independent, pseudorandom input transition samples, and the power consumed for each sample is monitored by HSPICE. All simulations were carried out at 27°C, with an input frequency of 50MHz in order to accommodate the slowest adder. The power dissipation measures do not include the power consumed by the drivers and the loads. In Figure 5.11, the probability distributions of the power dissipation per addition derived from the measurements, for the eight adder implementations, are shown. Since the data inputs are independent, power can be approximated to be normally distributed [22]. This conclusion can also be extracted from the curves of Figure 5.11.

Hence, the mean power dissipation is given by

$$\bar{P} \ \pm \ t_{\alpha/2}\frac{s}{\sqrt{N}}, \tag{5.4}$$

where $\bar{P}$ is the sample average, s is the standard deviation, N is the number of samples, and $t_{\alpha/2}$ is obtained from the $t-$distribution for a $(1-\alpha)\%$ confidence interval [24]. The mean power dissipation of the eight adder implementations using the simulation results and the equation (4) is given in Table 5.2.

The number of the required samples is extracted using the stopping criterion [22] of the above method

$$\frac{t_{\alpha/2}\, s}{\bar{P}\,\sqrt{N}} \;<\; e, \tag{5.5}$$

where e is the desired percentage error in the power estimate. The error in our statistical power analysis for $N = 200$ and 95% confidence interval ($t_{\alpha/2} = 1.96$) is less than 7%. In Table 5.2, the percentage error for each adder design is also given. For the four last designs the error is quite small because of the high normality of their distributions which leads to small standard deviation.

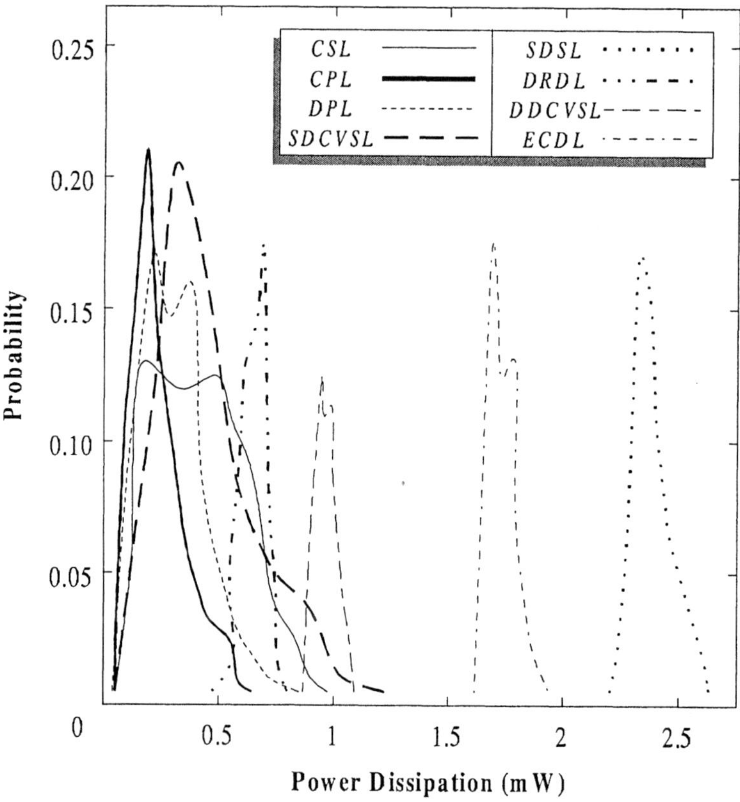

Figure 5.11.   Power dissipation histograms

The delay of each design was measured directly from the output waveforms generated by simulating the adder using HSPICE for the worst case inputs, that is, inputs which cause the carry to ripple from the least significant bit position to most significant bit position. The worst case delays of the eight adder designs are listed in the fourth column of Table 5.2. As mentioned in Section 5.1, the most essential metric of performance in modern VLSI applications is the power-delay product. By multiplying each power measurement with the worst case delay, we can found the mean power-delay product of the designs using a method similar to that used for the mean power dissipation. Hence, the mean power-delay product is given by

$$\overline{P \times D} \ \pm \ t_{\alpha/2}\frac{s}{\sqrt{N}}, \tag{5.6}$$

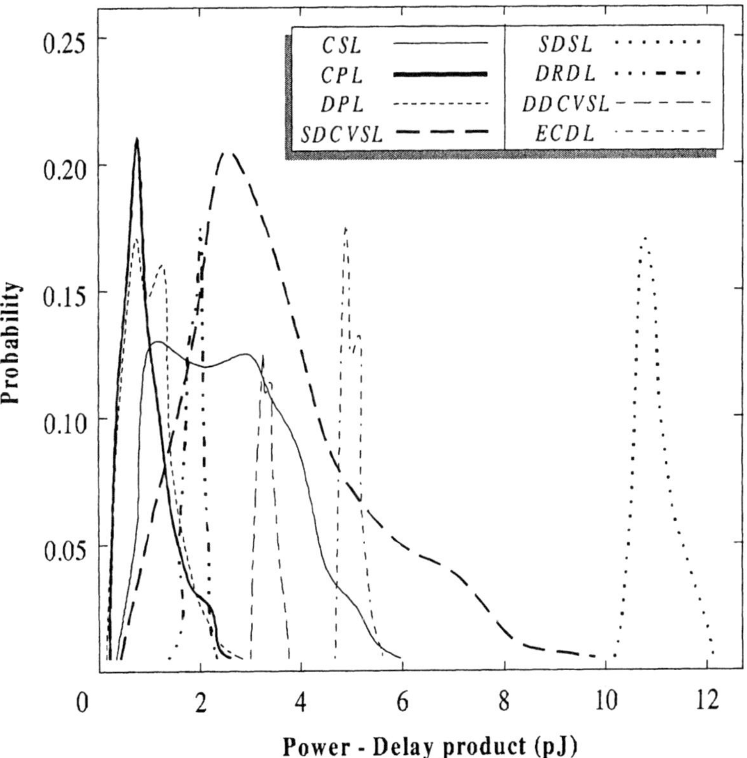

Figure 5.12.   Power-delay product histograms

where $\overline{P \times D}$ is the sample average power-delay product. The mean power-delay product values of the eight adder designs are listed in Table 5.2, and the probability distributions of the power-delay product are shown in Figure 5.12.

Table 5.2.   Power dissipation, delay and power-delay product of the four- bit ripple carry adder implementations

Adder Design Style	Mean Power Dissipation per addition (mW)	Statist. Error (%)	Worst Case Delay (nsec)	Mean Power-Delay Product per addition (pJ)
CSL	0.422 ± 0.0302	6.1	6.125	2.585 ± 0.1850
CPL	0.238 ± 0.0208	4.8	4.042	0.962 ± 0.0841
DPL	0.305 ± 0.0263	6.9	3.345	1.020 ± 0.0879
SDCVSL	0.432 ± 0.0362	6.5	7.986	3.450 ± 0.2891
SDSL	2.383 ± 0.0129	0.6	4.606	10.976 ± 0.0594
DRDL	0.641 ± 0.0091	1.4	2.909	1.865 ± 0.0265
DDCVSL	0.957 ± 0.0074	0.8	3.453	3.304 ± 0.0255
ECDL	1.721 ± 0.0096	0.6	2.892	4.977 ± 0.0278

As we can see in the probability distributions of Figure 5.12, the curves of the dynamic designs (DRDL and DDCVSL) are shifted to the right, because of the power dissipated due to the precharge cycles. The same phenomenon occurs in the ECDL adder due to the power dissipation of its disabled state. The shifting to the right of the SDSL adder curve is caused because of the high static power which is dissipated due to the incomplete turn-off of the cross-coupled pMOSFET transistors. The other static design styles are more power efficient compared to the dynamic circuits.

The static DCVSL circuit consumes more power than the conventional static circuit due to the difference of the charging and discharging times of its output nodes. The asymmetry in the rise and fall times of the potential at these output nodes will prolong the period of current flow through the latch during the transient state, thus increasing the power dissipation.

It can be obtained from the results of Table 5.2, that the dynamic circuits exhibit an increase in speed compared to the conventional static circuit. Comparing the dynamic logic styles, Domino logic has better power-delay product characteristics (Figure 5.12). The circuit operation in the SDSL circuit becomes faster than the standard SDCVSL circuit, due to the reduced logic swing at the output nodes, but in the cost of high static power dissipation. ECDL circuit is the faster one, but consumes high switching power due to the inverters which are needed to change the polarity of the outputs.

The design styles which use pass-transistor logic (CPL and DPL) are the best in terms of power dissipation. CPL circuit consumes lower power than the DPL one, because of its lower parasitic capacitance. On the contrary, DPL circuit is faster than the CPL, because the addition of pMOSFET transistors in parallel with the nMOSFET transistors results in higher circuit drivability. Also, DPL avoids the problems of noise margin and speed degradation at reduced supply voltages which are caused in CPL circuits. As shown in Figure 5.12 and in Table 5.2, the two styles exhibit similar power-delay product characteristics, and they are the most efficient for low-power and high-speed applications.

The mean power dissipation and the propagation delay values of the eight adder implementations are summarized in Figure 5.13. The fast adder circuits lie to the left of the figure , and those with low power consumption lie toward the bottom of the figure .

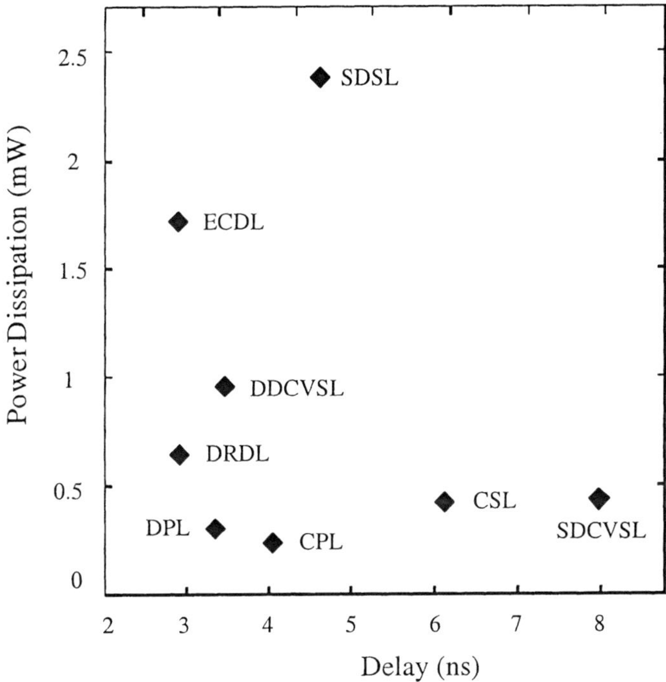

Figure 5.13.   Power dissipation versus delay of the adder implementations

5.5   Adders

In static CMOS the dynamic power dissipation of a circuit depends primary on the number of transitions per unit area. As a result, the average number of logic transitions per addition can serve as the basis of comparing the efficiency of a variety of adder designs. If two adders require roughly the same amount of time and roughly the same number of gates, the circuit which requires fewer logic transitions is more desirable as it will require less dynamic power. This is only a first order approximation as the power also depends on switching speed, gate size, fan-out, output loading e.t.c.

The following types of adders were simulated: Ripple Carry, Constant Block Width Single-level Carry Skip, Variable Block Width Multi-level Carry skip, Carry Lookahead, Carry Select, and Conditional Sum. Table 5.3 presents the worst case number of gate delays, the number of gates, and the average number of logic transitions for the six 16-bit adder types. All the gates are assumed to have the same delay, regardless of the fan-in or fan-out.

Table 5.3. Worst Case Delay, Number of Gates, and Average Number of Logic Transitions for a 16-bit Adder

Adder Type	Worst Case Delay (in gates units)	Number of Gates	Average Number of logic Transitions
Ripple Carry	36	144	90
Constant Block Width Single-level Carry Skip	23	156	102
Variable Block Width Multi-level Carry skip	17	170	108
Carry Lookahead	10	200	100
Carry Select	14	284	161
Conditional Sum	12	368	218

5.6   Multipliers

The majority of the real life applications, such as microprocessors and digital processing implementations, require the computation of the multiplication operation. Specifically, speed, area and power efficient implementation of a multiplier is a very challenging problem. Here, four well-known multipliers: i) the Array Multiplier [25], ii) the Split Array Multiplier [26], iii) Wallace Tree Multiplier [27] and iv) the Radix-4 Modified Booth Recoded Wallace Tree Multipliers [28], are studied in terms of power consumption.

Two kinds of measurements and comparisons in the terms of different design parameters are performed providing to the designer a plethora of alternative implementations. Particularly, we provide SPICE-like measurements with respect to the average logic transitions as well as the power consumption. The second kind of measurements are performed following a typical high-level analysis flow and is based on a state-of-the-art CAD framework.

5.6.1 SPICE-like Power Measurements of Multipliers

The multipliers were described using only AND, OR, and INVERT gates. The simulation was made using a program called CazM [29], which is similar to SPICE. Each multiplier was fed with 1.000 pseudorandom inputs. For the sake of completeness, the carry save array multiplier and the Wallace tree is presented in the following section, in order to briefly describe the architectures of the most common multipliers. The former is a representative paradigm of array multipliers, while the Wallace tree is an efficient way to add multiple partial products together.

A gate level simulation with 10.000 pseudorandom inputs, enabled the gathering of average number of gate-output transitions for each multiplier. During each input, the number of gates that switch output states is recorded, and an average number of gate-output transitions per multiplication are computed at the end of the simulation. Table 5.4 presents the results.

Table 5.4. Average Number of Gate-Output Transistions

Average Number of Gate-Output Transitions			
Multiplier Type	**8-bit**	**16-bit**	**32-bit**
Array	570	7224	99906
Split array	569	4874	52221
Wallace	549	3793	20055
Modified booth	964	3993	19542

The average power dissipation per multiplication is shown in Table 5.5.

Table 5.5. Avarage Power Dissipation from CAzM

Multiplier Type	**Power (mW)**	**Logic Transitions**
Array	43.5	7224
Split array	38.0	4874
Wallace	32.0	3793
Modified booth	41.3	3993

These results were obtained by simulating the multiplication of 1000 pseudorandom inputs with a clock period of 100 ns. The results vary significantly. The Wallace multiplier, which presents the lower power dissipation, is neither the smallest nor the slowest one.

5.6.2   High-Level Power Characterization of Multipliers

The second power estimation procedure is illustrated in Figure 5.14. The first

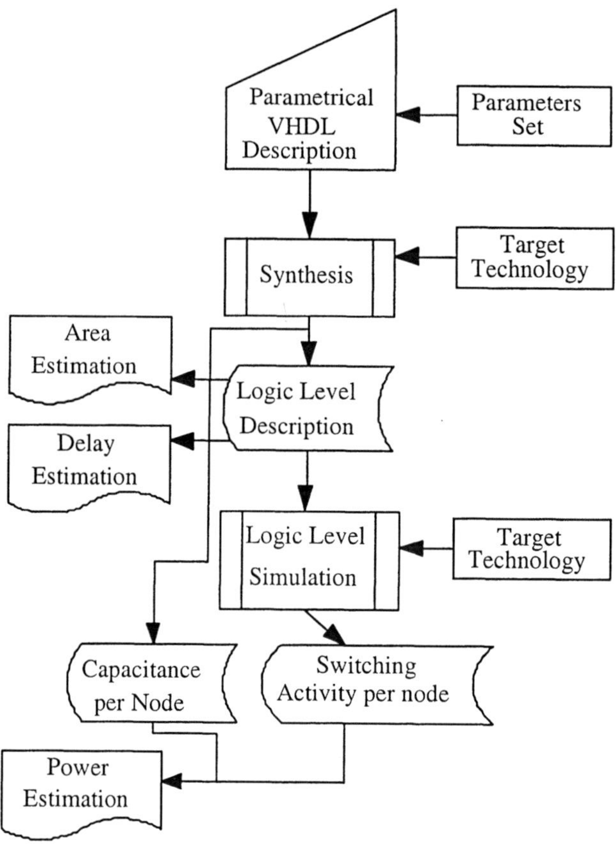

Figure 5.14.   High level Characterization Flow

step is the logic synthesis of the parameterized and structural VHDL description of the arithmetic modules. Here, a 0.6-micron process, AMS standard-cell library has been used. For power characterization, only the dynamic power dissipation, which forms the dominant component of the total power, is taken into account [30]. Specifically, the activity per node, resulting via logic-level simulation that takes place in a second step, is combined with the capacitance per node, to compute the power for a certain input vector, according to

$$Power = \sum_{i=1}^{N} C_{load_i} \, V_{dd}^2 \, f \, E_i \qquad (5.7)$$

where C_{load_i} is the capacitance at node i, V_{dd} is the power supply voltage , f is the frequency and E_i is the activity factor at node i. The term $f \cdot E_i$ of Eq. 5.7 is actually the number of transitions from logic '1' to logic '0' per time unit for the node i, which is equal to the ratio of number of node transitions from logic '1' to logic '0', divided by the total number of input vectors:

$$f \cdot E_i = f_{1 \to 0} = \frac{\#trans_{1 \to 0_i}}{\#vectors} \qquad (5.8)$$

From Eq. 5.7 and 5.8, the power is:

$$Power = \frac{V_{DD}^2}{\#vectors} \sum_{i=1}^{N} C_{load_i} \#trans_{1 \to 0_i} \qquad (5.9)$$

Following this procedure, the power estimation errors are in the range of 10-25% [31], compared with SPICE transistor-level simulator. However, the accuracy of the estimates suffices for the purpose of comparing alternative module architectures, since its relative evaluation is of importance and not the absolute accuracy. The 8-bit wide input modules were simulated with 50.000 random vectors, the 16-bit modules with 100.000 vectors, the 32-bit modules with 150.000 random vectors, and the 64-bit modules with 200.000 vectors. It should be stressed here, that the energy figures, given later in the characterization sections of the arithmetic components, correspond to the average energy per operation. The difference of the two characterization procedures in the number of test vectors is significant.In this section, power measures for the synthesized multipliers, namely the carry save array, the Booth encoded Wallace tree and the non-Booth encoded Wallace tree, will be presented, while analysis of results and comparisons with previous work are made.

Table 5.6 presents the power estimation of the synthesized multipliers. The measurements of power are normalized by frequency, reflecting the fact of simulation, with different operating frequencies. In this way, a representative power measure is given, for every kind of multiplier and for every bit width. As it is shown in Table 5.6, the most power-efficient multiplier for small bit-widths (less than 32 bits) is the carry save array, but with a small difference, compared to the Wallace tree with non-Booth encoding. In the 64-bit implementations, the Wallace tree with non-Booth encoding multiplier is the most power efficient choice. This fact is explained by the glitches arising from the ripple of carries of the array multiplier for large bit-widths. Finally, for all bit-widths, the Wallace tree with Booth encoding multiplier has the worst power dissipation.

Table 5.6.   Power dissipation estimates

	Power (mW)			
Multiplier Type/ Multiplier Width	**8-bit**	**16-bit**	**32-bit**	**64-bit**
Carry Save Array	0.3084	2.2484	3.057023	20.96759
Wallace Tree (Booth Encoded)	0.7868	3.1204	5.384	23.7164
Wallace Tree (Non Booth Encoded)	0.5488	2.5992	4.2212	18.8588

For comparison purposes, Table 5.7 shows the results presented in [29], considering 16-bit implementations of array and Wallace tree multipliers. These designs were described only by AND, OR, and INVERT gates. The implementation technology was a 2-level metal 2-μm process. It can be seen that the array multiplier consumes more energy than the Wallace tree multiplier, which is contradictory to corresponding values shown in Table 5.6. Only in the 64-bit case, the array multiplier consumes more energy than the Wallace tree multiplier. The reason for this is the spurious transitions that occur by the rippling of carries for the array multiplier of the 64-bit implementation, which cannot compensate the interconnect area/capacitance switched by the array multiplier. For the remaining cases, the factors of the greater interconnect and cell area for the Wallace multiplier dominate the power performance of these two kinds of multipliers.

Table 5.7.   16-bit Multiplier Average Power Dissipation [29]

Multiplier	**Power (mW)**	**Logic transitions**
Carry Save Array	43.5	7224
Wallace Tree	32	3793

The Wallace multiplier with Booth encoding dissipates the most power, while it is not the largest. It is the fastest multiplier for bit-widths larger than 16 bits and can be assumed that Booth encoding is a rather power-hungry operation.

Finally, Table 5.8 depicts the *Power × Delay product* of the multipliers at 1MHz frequency. More specifically, the carry save array multiplier exhibits the worst product for all bit-widths, except the 8-bit, due to the large delay and the substantial power consumption Although the non-Booth Wallace tree multiplier is the largest, it shows the best *Power×Delay product* for every bit-width. Where speed is of great interest, especially if large bit widths are required, and chip area is not a problem, the non-Booth encoded Wallace tree multiplier is the best candidate for selection.

Table 5.8. Power-Delay product of Multipliers at 1MHz

Multiplier Type/ Multiplier Width	Power*Delay (mW*ns)			
	8-bit	16-bit	32-bit	64-bit
Carry Save Array	4,13256	53,51192	141,9987	2090,049
Wallace Tree (Booth Encoded)	4,980444	25,96173	58,09336	312,345
Wallace Tree (Non Booth Encoded)	3,172064	19,8059	44,95578	251,9536

5.7 Conclusions

In this chapter, the most common kinds of adders and multipliers have been characterized in terms of power, using either a traditional low-level design flow paradigm, which is rather tedious and incompatible with modern design flows, but provides the most accurate results, or a high-level design flow paradigm, which is commonly used.

A four-bit ripple carry adder was used, as the benchmark circuit. All the circuits have been designed in a full-custom manner, and simulated using HSPICE. A statistical approach was used in order to analyze the simulation results. It has been shown that the circuits which use pass-transistor logic (CPL and DPL) exhibit better power and the power-delay product characteristics compared to other design styles.

The array multiplier is power-efficient for small bit widths. Its power consumption grows in proportion to the cube of the word size. The Wallace multiplier is less regular, but is more power efficient, while its power dissipation grows with the square of the word size.

The speed of the synthesized optimized carry look-ahead is traded-off for the worst energy power consumption among all the investigated adders, which have been synthesized. As opposed to the speed optimized architecture of the carry look-ahead adder, a non-optimized architecture is power-efficient, though much slower. Power-efficient is the ripple carry adder, too.

References

[1] A. Chandrakasan, R. Brodersen, Low Power Digital Design, Kluwer Academic Publishers, 1995.

[2] Meta-Software, HSPICE User's Manual - Version 96.1,1996.

[3] K. Yano, Y. Sasaki, K. Rikino, K. Seki, "Top-Down Pass-Transistor Logic Design". IEEE Journal of Solid-State Circuits, vol.31, pp. 792-803. 1996

[4] MIPS Technologies, "R4200 Microprocessor Product Information", MIPS Technologies Inc., 1994

[5] K. Yano, T. Yamanaka, T. Nishida, M Saito, K. Shimohigashi, A. Shimizu, "A 3.8-ns CMOS 16×16-b Multiplier Using Complementary Pass-Transistor Logic", IEEE Journal of Solid-State Circuits, vol.25, pp. 388-395, 1990.

[6] M. Suzuki, N. Ohkubo, T. Shinbo, T. Yamanaka, A. Shimizu, K. Sasaki, Y. Nakagome, , 1993, "A 1.5-ns 32-b CMOS ALU in Double Pass-Transistor Logic", IEEE Journal of Solid-State Circuits, vol.28, pp. 1145-1151, 1993.

[7] L. Heller, W. Griffin, J. Davis, N. Thoma, "Cascode Voltage Switch Logic: A Differential CMOS Logic Family", Proceedings of IEEE International Solid-State Circuit Conference, pp. 16-17, 1984.

[8] K. Chu, D. Pulfrey, "Design Procedures for Differential Cascode Voltage Switch Circuits", IEEE Journal of Solid-State Circuits, vol.21, pp. 1082-1087, 1986.

[9] L. Pfennings, W. Mol, J. Bastiaens, J. Van Dirk, "Differential Split-level CMOS Logic for Subnanosecond Speeds", IEEE Journal of Solid-State Circuits, vol.20, pp. 1050-1055, 1985.

[10] R. Krambeck, C. Lee, H Law, "High-Speed Compact Circuits with CMOS", IEEE Journal of Solid-State Circuits, vol.17, pp. 614-619, 1982.

[11] V. Oklobdzija, R. Montoye, "Design-Performance Trade-offs in CMOS-Domino Logic", IEEE Journal of Solid-State Circuits, vol.21, pp. 304-309, 1986.

[12] S. Lu, S., "Implementation of Iterative Networks with CMOS Differential Logic", IEEE Journal of Solid-State Circuits, vol.23, pp. 1013-1017, 1988.

[13] L. Bisdounis, O. Koufopavlou, S. Nikolaidis, "Accurate Evaluation of CMOS Short-Circuit Power Dissipation for Short-Channel Devices", Proceedings of IEEE International Symposium on Low Power Electronics and Design, pp. 189-192, 1996.

[14] A. Bellaouar, M. Elmasry, Low-Power Digital VLSI Design: Circuits and Systems, Kluwer Academic Publishers, 1995.

[15] T. Sakurai, A. Newton, "Alpha-Power Law MOSFET Model and its Applications to CMOS Inverter Delay and Other Formulas", IEEE Journal of Solid-State Circuits, vol.25, pp. 584-594, 1990.

[16] S. Sun, P. Tsui, "Limitation of CMOS Supply-Voltage Scaling by MOSFET Threshold-Voltage", IEEE Journal of Solid-State Circuits, vol.30, pp. 947-949, 1995

[17] F. Najm, "A Survey of Power Estimation Techniques in VLSI Circuits", IEEE Transactions on VLSI Systems, vol.2, pp. 446-455, 1994

[18] S. Kang, "Accurate Simulation of Power Dissipation in VLSI Circuits", IEEE Journal of Solid-State Circuits, vol.21, pp. 889-891, 1986

[19] G. Yacoub, W. Ku, "An Enhanced Technique for Simulating Short-Circuit Power Dissipation" IEEE Journal of Solid-State Circuits, vol.24, pp. 844-847, 1989.

[20] S. Devadas, K. Keutzer, J. White, J., "Estimation of Power Dissipation in CMOS Combinational Circuits using Boolean Function Manipulation", IEEE Transactions on Computer-Aided Design of Integrated Circuits and Systems, vol.11, pp. 373-383, 1992.

[21] P. Schneider, U. Schlichtmann, B. Wurth, "Fast Power Estimation of Large Circuits", IEEE Design and Test of Computers Magazine, vol.13, 70-78, 1996.

[22] R. Burch, F Najm, P. Yang, T. Trick, "A Monte Carlo Approach for Power Estimation", IEEE Transactions on VLSI Systems, vol.1, 63-71, 1993.

[23] M. Xakellis, F. Najm, "Statistical Estimation of the Switching Activity in Digital Circuits", Proceedings of ACM/IEEE Design Automation Conference, 728-733, 1994.

[24] I. Miller, J. Freund, R. Johnson, Probability and Statistics for Engineers, Prentice Hall, 1990.

[25] A.D. Pezaris, "A 40ns 17-bit by 17-bit array multiplier" in IEEE Trans. On Computers, vol. C-20, 1971, pp. 442-447.

[26] J. Iwamura, K. Siganuma, M. Kimura, and S. Taguchi, "A CMOS/SOS multiplier," in Proceedings of ISSCC, 1984, pp. 92-93.

[27] C.S. Wallace, "A suggestion for a fast multiplier," in IEEE Trans. On Electronic Computers, vol. EC-13, 1964, pp. 14-17.

[28] O.L. MacSorley, "High-speed arithmetic in binary adders," IRE Proceedings vol.EC-11, June 1962, pp. 67-91.

[29] T. K. Callaway and E. E. Swartzlander, "The Power Consumption of CMOS Adders and Multipliers," Low-Power CMOS Design, IEEE Press, pp. 218-224, (*invited paper*).

[30] J. Rabaey, Digital Integrated Circuits: A Design Perspective, Prentice Hall, 1996.

[31] Power Compiler User Guide, Synopsys Online Documentation (SOLD), v1999.10

Chapter 6

COMPUTER ARITHMETIC TECHNIQUES FOR LOW-POWER SYSTEMS

Vassilis Paliouras

Thanos Stouraitis
University of Patras, Rio, Greece
{ paliuras,thanos } @ee.upatras.gr

Abstract   This chapter investigates techniques that reduce the power dissipated in digital systems based on alternative arithmetic schemes. By exploiting transformations that alter both the data representation and the corresponding operators, it is found that power dissipation can be significantly reduced, for particular computationally-intensive applications. Two classes of transformations, namely the Residue Number System (RNS) and the Logarithmic Number System (LNS). Several properties of the LNS and the RNS are reviewed, the exploitation of which reduces the data activity, the strength of the operators, or even the number of actual operations required to perform certain computational tasks. Since the reduction of these factors directly reduces power dissipation, the use of alternative arithmetic emerges as a low-power design methodology.

Keywords: Computer arithmetic, Signal processing, Low power design, Residue arithmetic, Logarithmic arithmetic.

6.1   Introduction

Computer arithmetic is investigated in this chapter as a means to minimize the power dissipated by digital signal processing systems.

Power dissipation has evolved into an instrumental design optimization objective due to the growing demand for portable electronics equipment as well as due to excessive heat generation in high-performance systems. In the former case, low-power techniques are employed to prolong battery life, while, in the latter case, low-power techniques are required to mitigate the potential reliability problems. The dominant component of power dissipation for well-designed

D. Soudris et al. (eds.), Designing CMOS Circuits for Low Power, 97–116.
© 2002 *Kluwer Academic Publishers. Printed in the Netherlands.*

CMOS circuits is *dynamic* power dissipation, given by [1]

$$P = aC_L f V_{dd}^2, \tag{6.1}$$

where a is the activity factor, C_L is the switching capacitance, f is the clock frequency, and V_{dd} is the supply voltage. A variety of design techniques are commonly employed to reduce the factors of product (6.1), without degrading system performance. As slower circuits tend to dissipate less power, the low-power design problem can be seen as an attempt to achieve a specified system performance by employing slow components.

The reduction of the various factors, which determine power dissipation, is sought at all levels of the design abstraction. In particular, techniques that reduce the power dissipation at higher design-abstraction levels aim to reduce the computational load and the number of memory accesses required to perform a certain task, and to introduce parallelism and pipelining in the system [2]. At the circuit and process levels, minimal feature size circuits are preferred, capable of operating at minimal supply voltages, while leakage currents and device threshold voltages are minimized.

The application of the LNS and the RNS can reduce the power dissipation by minimizing the particular factors; however the goal is sought by means of an innovative way, namely the application of computer arithmetic techniques. The proper choice of the *number system*—i.e., the way numbers are represented in a digital system—can reduce power dissipation. The effect is significant as the number system has an effect on several levels of the design abstraction. In particular power dissipation reduction due to the appropriate selection of the number system stems from:

- The reduction of the number of the operations;
- The reduction of the strength of the operators; and
- The reduction of the activity of the data.

A judicious choice of number system can reduce the number of the actual operations required to accomplish certain computational tasks; therefore it can reduce the computational load of an application. Furthermore, both data activity and the strength of the operators are influenced by the choice of the number system. Finally, power dissipation can be reduced by using low-power arithmetic circuit architectures. Again, the possible architectures are determined by the number system.

Several authors address the issue of low-power arithmetic; a lot of work in this field is on the definition of new low-power circuit-level architectures [3] and the identification among existing architectures of those that dissipate minimal power for the basic operations, such as addition and multiplication [4]. In addition, comparisons of widely used number representations, such as sign-magnitude and two's-complement systems [5], in terms of underlying bit activity, have been reported [6].

In this chapter, the properties of the constituent transformations, namely the Residue Number System (RNS) and the Logarithmic Number System (LNS) are studied, with the objective to reveal the mechanisms by means of which power dissipation reduction is achieved.

The remainder of this chapter is organized as follows: Section 6.2 briefly addresses aspects of the conventional arithmetic representations, Section 6.3 addresses the impact of logarithmic arithmetic on power dissipation, while Section 6.4 discusses the low-power aspects of residue arithmetic. Finally, conclusions are discussed in Section 6.5.

6.2   Conventional Arithmetic and Low-power design

A particular alternative technique is not expected to offer efficient implementations for all basic operations and algorithms. Therefore, the conventional arithmetic circuits are of particular interest for general-purpose computations. In addition conventional arithmetic circuits can be used as components in certain alternative arithmetic structures.

Parhami [5] offers an overview of low-power techniques for arithmetic circuits. Common techniques for low-power logic design can be applied to arithmetic circuits as well [5]. Such techniques are based on the following guidelines:

- Avoid wasted power: glitching minimization, not clocking idle modules;
- Barely meet performance requirements, since slower circuits dissipate less power; and
- Minimize signal activity by properly encoding data.

In some cases, the power required for a particular task can be reduced by several times, by minimizing the corresponding computational load. In Section 6.4, the appropriate selection of the number system will be shown to reduce the computational load in certain tasks.

6.2.1   Basic Arithmetic Operations

Callaway and Swartzlander [7] have focused on low-power arithmetic at the gate level; they have characterized several adder and multiplier architectures in terms of power dissipation. They offer area, time, and power dissipation measures for various architectures and word lengths. In terms of minimal power dissipation for 16-bit adders, the constant-width carry-skip adder emerges as the optimal choice. However, minimal absolute power dissipation may not be the optimization objective in a design. In most cases, a more complex criterion, the power-delay product, is more applicable, because it describes the combined effect of reducing power dissipation at the cost of increasing circuit delay. Returning to the 16-bit adder example, the utilization of the power-delay product criterion points out a different topology as an optimal solution, namely the variable-width carry-skip adder [7].

This example demonstrates that there is not an optimal choice of architecture that demonstrates optimal characteristics and is applicable to every design situation. Instead, the design specifications, expressed as area, time, and power dissipation constraints, should be met, while minimizing an appropriately defined cost function. A similar discussion for multipliers of word sizes between 8 and 32 bits reveals that Wallace and Dadda architectures outperform array multipliers for low-power operation [7].

Nannarelli and Lang [8] have studied the impact of the low-power design techniques on the power dissipation of division. While division is not as common as the other basic operations, the energy dissipation in the divider is comparable to that of the other basic arithmetic units, due to its relatively long latency. The presented results show that a 30% to 40% reduction on the energy dissipation is possible and an additional 20% is possible by using dual voltage. Energy dissipation is found to be the same for the radices 4, 8 and 16. However, power dissipation proportional to the average current, increases with the radix. The smallest energy dissipation is achieved for a particular latency by using a radix-16 divider architecture with dual voltages.

6.2.2   Number Representations

Bit assertion activity is another factor, which affects power dissipation and depends on the number system selection. It has been shown that the probabilistic distribution of the input signals largely affects the performance of the number representation in terms of bit activity. Landman and Rabaey demonstrate this effect by introducing the Dual-Bit Type (DBT) Method for modeling the bit activity in a data word [6], assuming two's complement and sign-magnitude representations. While the sign magnitude representation is found to exhibit less bit activity than two's complement coding, a general conclusion on the power dissipation behavior cannot be drawn, since the complexity of the corresponding processing circuitry is different. Since sign-magnitude arithmetic requires more complicated adders and subtractors than two's complement arithmetic, the increased activity of the latter can be compensated from a power dissipation viewpoint.

6.3   The Logarithmic Number System

Conventional arithmetic circuits and low-power design techniques can be applied to realize components of VLSI architectures that implement LNS circuits. The application of LNS aims to reducing the strength of particular operations and to reduce the switching activity.

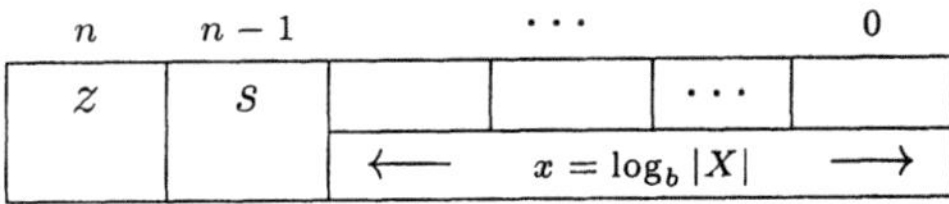

Figure 6.1.   The organization of a $(n + 1)$-bit LNS digital word.

The LNS [9] has been employed in the design of low-power DSP devices, such as a digital hearing aid by Morley *et al.* [10]. More recently, Sacha and Irwin report that LNS can reduce power dissipation in adaptive filtering processors [11].

6.3.1   LNS Basics

The LNS maps a linear number X to a triplet as follows

$$X \xrightarrow{\text{LNS}} (z, s, x = \log_b |X|), \tag{6.2}$$

where z is a single-bit flag which, when asserted, denotes that X is zero, s is the sign of X, and b is the base of the logarithmic representation. The organization of an LNS word is shown in Fig. 6.1. The inverse mapping of a logarithmic triple (z, s, x) to a linear number X is defined by

$$(z, s, x) \xrightarrow{\text{LNS}^{-1}} X: X = (1 - z)(-1)^s b^x. \tag{6.3}$$

Mapping (6.2) is of practical interest because it can simplify certain arithmetic operations, i.e., it can reduce the implementation complexity (also called *strength*) of several operators. For example, due to the properties of the logarithm function, the multiplication of two linear numbers $X = b^x$ and $Y = b^y$, is reduced to the addition of their logarithmic images, x and y. The basic arithmetic operations and their LNS counterparts are summarized in Table 6.1. The zero flag z and the sign flag s are omitted for simplicity. Table 6.1 reveals that while the complexity of most operations is reduced, the complexity of LNS addition and LNS subtraction is significant. In particular, LNS addition requires the computation of the nonlinear function

$$s_a(d) = \log_b(1 + b^{-d}), \tag{6.4}$$

which substantially limits the data word lengths for which LNS can offer efficient VLSI implementations.

LNS arithmetic example: Let $X = 3.25$, $Y = 6.72$ and $b = 2$. Perform the operations $X \cdot Y$, $X + Y$, $\sqrt{X}$ and Y^2 using the LNS. Initially the

Table 6.1.   Basic linear arithmetic operations and their LNS counterparts.

Linear Operation	Logarithmic Operation
$Z = XY = b^x b^y = b^{x+y}$	$z = \log_b Z = x + y$
$Z = X/Y = b^x/b^y = b^{x-y}$	$z = x - y$
$Z = \sqrt[m]{X} = \sqrt[m]{b^x} = b^{\frac{x}{m}}$	$z = x/m, \quad m, \text{integer}$
$Z = X^m = (b^x)^m$	$z = mx, \quad m, \text{integer}$
$Z = X + Y = b^x + b^y = b^x(1 + b^{y-x})$	$z = x + \log_b(1 + b^{y-x})$
$Z = X - Y = b^x - b^y = b^x(1 - b^{y-x})$	$z = x + \log_b(1 - b^{y-x})$

data are transferred to the logarithmic domain, as implied by (6.2):

$$X \xrightarrow{\text{LNS}} (z_x, s_x, x = \log_2 |X|) = (0, 0, x = \log_2 3.25)$$
$$= (0, 0, 1.70044) \qquad (6.5)$$
$$Y \xrightarrow{\text{LNS}} (z_y, s_y, y = \log_2 |Y|) = (0, 0, y = \log_2 6.72)$$
$$= (0, 0, 2.74846) . \qquad (6.6)$$

Using the LNS images (6.5) and (6.6), the required arithmetic operations are performed as follows: The logarithmic image z of the product $Z = X \cdot Y$ is given by

$$z = x + y = 1.70044 + 2.74846 = 4.44890. \qquad (6.7)$$

As both operands are of the same sign, i.e., $s_x = s_y = 0$, the sign of the product is $s_z = 0$. Also since, $z_x \neq 1$ and $z_y \neq 1$, the result is non-zero, i.e., $z_z = 0$.

To retrieve the actual result Z from (6.7), inverse conversion (6.3) is used as follows:

$$Z = (1 - z_z)(-1)^{s_z} 2^z = 2^{4.44890} = 21.83998. \qquad (6.8)$$

By directly multiplying X by Y it is found that $Z = 21.84$. The difference is due to round-off error during the conversion from linear to the LNS domain.

The calculation of the logarithmic image z of $Z = \sqrt{X}$ is performed as follows:

$$z = \frac{1}{2}x = \frac{1}{2}1.70044 = 0.85022. \qquad (6.9)$$

The actual result is retrieved as follows:

$$Z = 2^{0.85022} = 1.80278. \qquad (6.10)$$

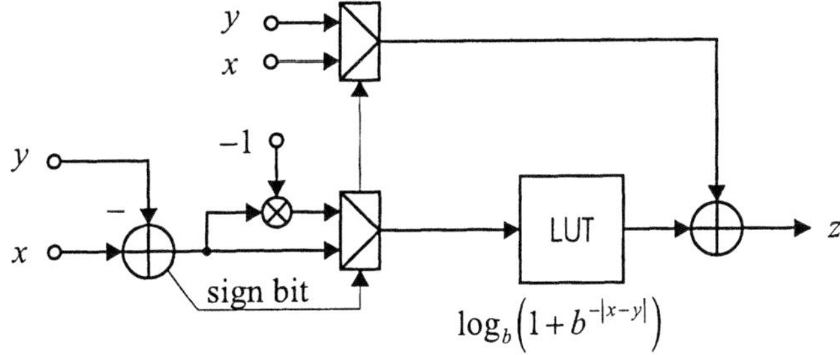

Figure 6.2. The organization of an LNS adder, composed of a subtractor, two multiplexers, a sign-inversion unit, a look-up table and a final adder.

The calculation of the logarithmic image z of $Z = X^2$ can be done as:

$$z = 2 \cdot 1.70044 = 3.40088. \tag{6.11}$$

Again, the actual result is obtained as

$$Z = 2^{3.40088} = 10.56250. \tag{6.12}$$

The operation of logarithmic addition is rather awkward and its realization is usually based on a memory look-up table operation. The logarithmic image z of the sum $Z = X + Y$ is

$$
\begin{aligned}
z &= \max(x, y) + \log_2(1 + 2^{\min(x,y)-\max(x,y)}) & \text{(6.13)}\\
&= 2.74846 + \log_2(1 + 2^{-1.04802}) & \text{(6.14)}\\
&= 3.31759. & \text{(6.15)}
\end{aligned}
$$

The actual value of the sum $Z = X + Y$, is obtained as

$$z = 2^{3.31759} = 9.96999. \tag{6.16}$$

The organization of the realization of an LNS adder is shown in Fig. 6.2. It is noted that in order to implement LNS subtraction, i.e., the addition of two quantities of opposite sign, a different memory Look-Up Table (LUT) is required. The LNS subtraction LUT contains samples of the function

$$s_s(d) = \log_b(1 - b^{-d}). \tag{6.17}$$

The main complexity of an LNS processor is the implementation of the LUTs for storing the values of the functions $s_a(d)$ and $s_s(d)$. A straight-forward implementation is only feasible for small word lengths. A different technique can be used for larger word lengths, based on the partitioning of a LUT into an assortment of smaller LUTs. The particular partitioning becomes possible due

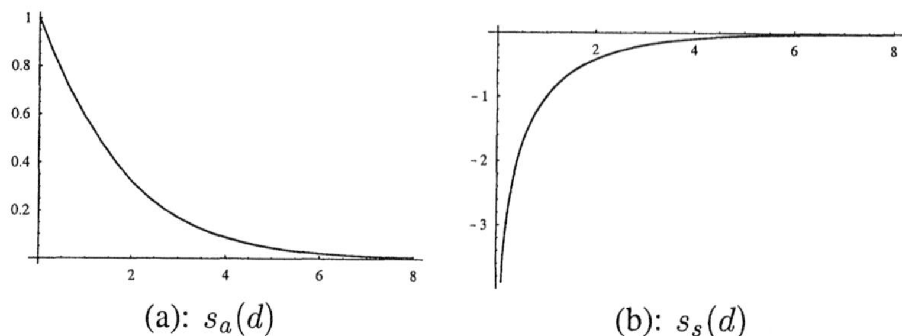

(a): $s_a(d)$ (b): $s_s(d)$

Figure 6.3. The functions $s_a(d)$ and $s_s(d)$, the approximation of which is required for LNS addition and subtraction.

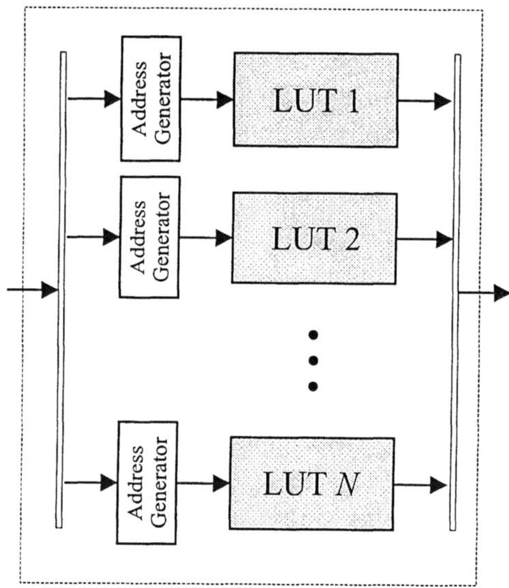

Figure 6.4 The partitioning of the LUT that stores the addition and subtraction function into a set of smaller LUTs leads to memory compression.

to the nonlinear behavior of the functions $\log_b(1 + b^{-d})$ and $\log_b(1 - b^{-d})$, depicted in Fig. 6.3 for $b = 2$. By exploiting the different minimal word length required by groups of function samples, the overall size of the LUT is compressed, leading to the organization of Fig. 6.4. In order to utilize the benefits of LNS, a conversion overhead is required in most cases, to perform the forward LNS mapping defined by (6.2). It is noted that conversions (6.2) and (6.3) are required in the case that an LNS processor receives as input and transmits as output linear data in digital format. Since all arithmetic operations can be performed in the logarithmic domain, only an initial conversion is imposed; therefore, as the amount of processing implemented in LNS grows, the contribution of the conversion overhead to power dissipation and to area-time complexity becomes negligible since it remains constant.

In stand-alone DSP systems the adoption of a different solution to the conversion problem is possible. In particular, the LNS forward and inverse mapping

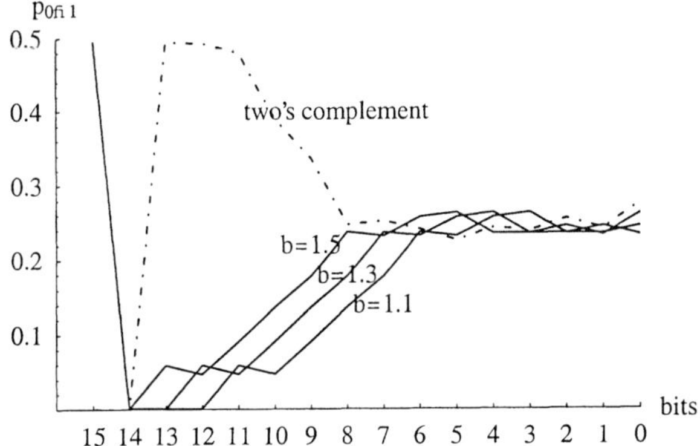

Figure 6.5.   Probability $p_{0\rightarrow1}$ per bit for two's complement and LNS encoding for $\rho = -0.99$.

overhead can be mitigated by employing logarithmic A/D and D/A converters, instead of linear converters, followed by corresponding digital conversion circuitry. Such an approach has been adopted by Morley *et al.* in the design of a digital hearing-aid processor [10].

6.3.2   LNS and Power Dissipation

LNS is applicable for low-power design because it reduces the strength of certain arithmetic operators and the bit activity. The operator strength reduction.by LNS reduces the switching capacitance, i.e., it reduces the C_L factor of (6.1). Sacha and Irwin have studied the impact of the number system choice on the QRD-RLS algorithm [11]. They have compared the amount of switched capacitance per algorithm iteration for several implementations of QRD-RLS, each using a particular arithmetic, namely CORDIC, floating-point, fixed-point, and LNS. A performance comparison of the various implementations, reveals that LNS offers accuracy comparable to floating-point, but only at a fraction of switched capacitance per iteration of the algorithm.

The reduction of average switched capacitance due to LNS stems from the simplification of basic arithmetic operations, shown in Table 6.1. However, LNS can affect power dissipation in an additional way, the bit activity, i.e., the a factor of (6.1). A design parameter, which is often neglected, although it plays a key role in an LNS-based processor performance, is the base of the logarithm, b [12][13], as demonstrated in Fig. 6.5. The choice of base has a substantial impact on the average bit activity. Fig. 6.5 shows activity per bit position, i.e., the probability of a transition from "low" to "high" in a particular bit position, for a two's complement word and several LNS words, each of a different base b. It can be seen that departing from the traditional choice $b = 2$ can substantially reduce the signal activity in comparison to the two's-complement representation. The input data are sampled from a zero-mean Gaussian process with a correlation factor $\rho = -0.99$, similarly to the derivation of the DBT model.

Since multiplication-additions are important in DSP applications, the power requirements of an LNS and a linear fixed-point adder-multiplier have been compared. Paliouras and Stouraitis report that approximately two times reduction in power dissipation is possible for operations with word size of 8 to 14 bits. Given a sufficient number of multiplication-additions, the LNS implementation becomes more efficient from the low-power dissipation viewpoint, even when a constant conversion overhead is taken into consideration.

6.4   The Residue Number System

A different concept than the nonlinear transformation is followed by mapping of data to appropriately selected finite-fields.

The Residue Number System (RNS) [14] has recently been shown to offer significant power-dissipation savings in the design of signal processing architectures for FIR filters [15] and frequency synthesizers [16]. It is shown that RNS can even reduce the computation load in complex-number processing, thus providing savings at the algorithmic level of the design abstraction.

6.4.1   RNS Basics

The RNS maps an integer X to a N-tuple of *residues* x_i,

$$X \xrightarrow{\text{RNS}} \{x_1, x_2, \ldots, x_N\}, \tag{6.18}$$

where $x_i = \langle X \rangle_{m_i}$, $\langle \cdot \rangle_{m_i}$ denotes the mod m_i operation, and m_i is a member of the set of the co-prime integers $B = \{m_1, m_2, \ldots, m_M\}$, called *moduli*. Co-prime integers have the property that $\gcd(m_i, m_j) = 1$, $i \neq j$. The modulo operation $\langle X \rangle_m$ returns the integer remainder of the integer division x div m, i.e., an integer k such that $x = m \cdot l + k$, where l is an integer.

RNS is of interest because basic arithmetic operations can be performed in a digit-parallel carry-free manner, i.e.,

$$z_i = \langle x_i \circ y_i \rangle_{m_i}, \tag{6.19}$$

where $i = 1, 2, \ldots, M$, and the symbol $\circ$ stands for addition, subtraction, or multiplication. Every integer in the range $0 \leq X < \prod_{i=1}^{N} m_i$ has a unique RNS representation. Inverse conversion is accomplished by means of the Chinese Remainder Theorem (CRT) or mixed-radix conversion [17]. The CRT retrieves an integer from its RNS representation as follows:

$$X = \left\langle \sum_{i=1}^{N} \overline{m_i} \langle \overline{m_i}^{-1} x_i \rangle_{m_i} \right\rangle_M, \tag{6.20}$$

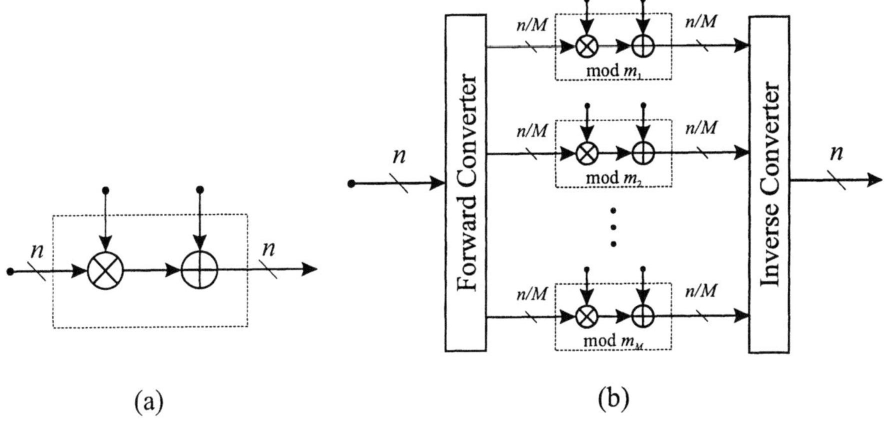

$$(a) \qquad\qquad\qquad\qquad (b)$$

Figure 6.6. (a) Structure of a binary architecture and (b) the corresponding RNS processor.

where $\overline{m_i} = \frac{M}{m_i}$, $M = \prod_{i=1}^{N} m_i$, and $\overline{m_i}^{-1}$ is the multiplicative inverse of $\overline{m_i}$ modulo m_i, i.e., an integer such that $\langle \overline{m_i} \cdot \overline{m_i}^{-1} \rangle_{m_i} = 1$.

RNS arithmetic example: Consider the base $B = \{3, 5, 7\}$ and two integers $X = 11$ and $Y = 6$. The RNS images of X and Y are

$$X \xrightarrow{\text{RNS}} \{\langle 11 \rangle_3, \langle 11 \rangle_5, \langle 11 \rangle_7\} = \{2, 1, 4\} \qquad (6.21)$$

$$Y \xrightarrow{\text{RNS}} \{\langle 6 \rangle_3, \langle 6 \rangle_5, \langle 6 \rangle_7\} = \{0, 1, 6\}. \qquad (6.22)$$

The RNS image of the sum $Z = X + Y$ is obtained as follows:

$$Z \xrightarrow{\text{RNS}} \{\langle 2 + 0 \rangle_3, \langle 1 + 1 \rangle_5, \langle 4 + 6 \rangle_7\} = \{2, 2, 3\}. \qquad (6.23)$$

To retrieve the integer that corresponds to the RNS representation $\{2, 2, 3\}$ by applying (6.20), the following quantities are pre-computed: $M = 3 \cdot 5 \cdot 7 = 105$, $\overline{m_1} = \frac{105}{3} = 35$, $\overline{m_2} = \frac{105}{5} = 21$, $\overline{m_3} = \frac{105}{7} = 15$, $\overline{m_1}^{-1} = 2$, $\overline{m_2}^{-1} = 1$, $\overline{m_3}^{-1} = 1$. The value of the sum in integer form, is obtained by applying (6.20):

$$\begin{aligned} Z = X + Y &= \langle 35 \langle 2 \cdot 2 \rangle_3 + 21 \langle 1 \cdot 2 \rangle_5 + 15 \langle 1 \cdot 3 \rangle_7 \rangle_{105} \\ &= \langle 122 \rangle_{105} = 17. \end{aligned} \qquad (6.24)$$

To verify the result of (6.24), notice that $X + Y = 11 + 6 = 17$ and that

$$17 \xrightarrow{\text{RNS}} \{\langle 17 \rangle_3, \langle 17 \rangle_5, \langle 17 \rangle_7\} = \{2, 2, 3\}, \qquad (6.25)$$

i.e., the result obtained in (6.24).

The basic architecture of an RNS processor in comparison to a binary counterpart, is depicted in Fig. 6.6. Fig. 6.6 shows that the word length n of the binary counterpart is partitioned into M sub-words, the residues, which can be processed independently and are of word length significantly smaller than n. The architecture in Fig 6.6 assumes that the moduli are of equal word length. The ith residue channel performs arithmetic modulo m_i. Conceptually, RNS introduces a subword-level parallelism into an algorithm; therefore its hardware implementation can enjoy the low-power benefits of parallel architectures [2].

6.4.2   RNS and Power Dissipation

Freking and Parhi have studied the power dissipation of FIR filter architectures that employ RNS. They report that RNS can reduce power dissipation since it reduces the hardware cost, the switching activity, and the supply voltage [15]. By employing the binary-like RNS filter structures by Ibrahim [18], Freking and Parhi report that RNS reduces the bit activity up to 38% in (4×4)-bit multipliers. As the critical path in an RNS architecture increases logarithmically with the equivalent binary word length, RNS can tolerate a larger reduction in the supply voltage than the corresponding binary architecture, while achieving a particular delay specification. To demonstrate the overall impact of the RNS on the power budget of an FIR filter, Freking and Parhi report that a filter unit with 16-bit coefficients and 32-bit dynamic range, operating at 50 MHz, dissipates 26.2 mW on average for a two's complement implementation, while the RNS equivalent architecture dissipates 3.8 mW. Hence, power dissipation reduction becomes more significant as the number of filter taps increases, and a three times reduction is possible for filters with more than 100 taps.

A different approach to low-power RNS is proposed by Chren. Chren suggests to one-hot encode the residues in an RNS-based architecture, thus defining *One-Hot RNS* (OHR) [16]. Instead of encoding a residue value x_i in a conventional positional notation, an $(m - 1)$-bit word is employed. In this word, the assertion of the ith bit denotes the residue value x_i. The one-hot approach allows for further reducing bit activity and power-delay products using residue arithmetic. OHR is found to require simple circuits for processing. The power reduction is rendered possible since all basic operations, i.e., addition/subtraction and multiplication, and the RNS-specific operations of scaling, (i.e., division by constant), modulus conversion, and index computation are performed using transposition of bit lines and barrel shifters. The performance of the obtained residue architectures is demonstrated through the design of a Direct Digital Frequency Synthesizer, which exhibits a power-delay product reduction of 85% over the conventional approach [16].

6.4.3   RNS Signal Activity for Gaussian Input

In the following, the bit activity in an RNS architecture with positionally-encoded residues, is experimentally studied for the encoding of 8-bit data using

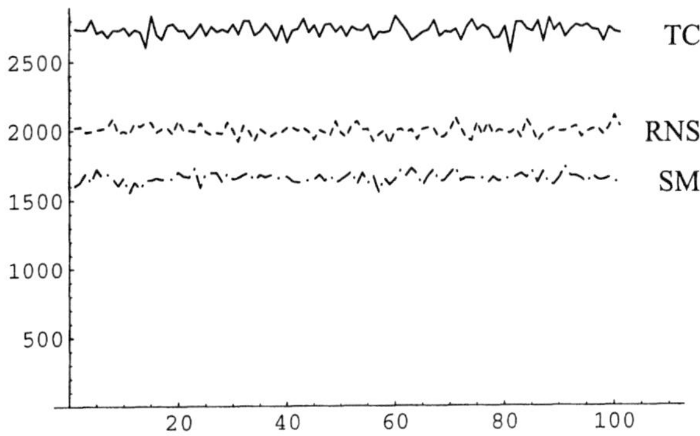

Figure 6.7. Number of low-to-high transitions, assuming strongly anti-correlated ($\rho = -0.99$) Gaussian data, for two's complement, RNS, and sign-magnitude number systems, for 100 Monte Carlo runs.

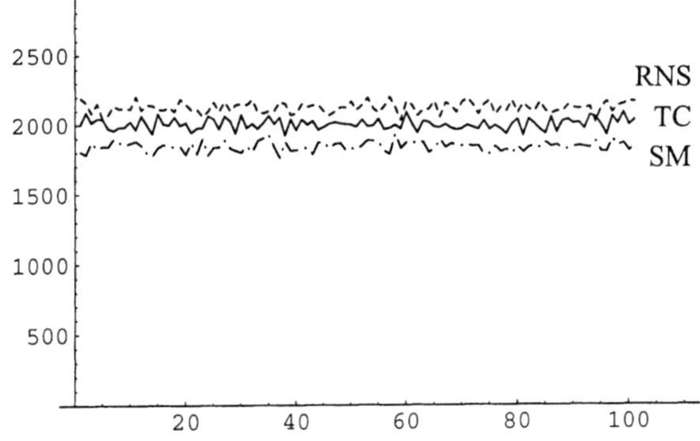

Figure 6.8. Number of low-to-high transitions, assuming uncorrelated ($\rho = 0$) Gaussian data.

the base $\{2, 151\}$, which provides a linear fixed-point dynamic range of approximately 8.24 bits. Assuming data sampled from a Gaussian process, the bit assertion activities of the particular RNS, an 8-bit sign-magnitude, and an 8-bit two's-complement system are measured and compared. The results are depicted in Figs. 6.7–6.9 for 100 Monte Carlo runs. It is observed that RNS performs better than two's complement representation for anti-correlated data and slightly worse than sign-magnitude and two's-complement representations, for uncorrelated and correlated sequences.

6.4.4 Algebraic extensions: The Quadratic RNS

Residue arithmetic can be exploited to reduce the number of real operations required to perform complex-number multiplication. This is achieved by em-

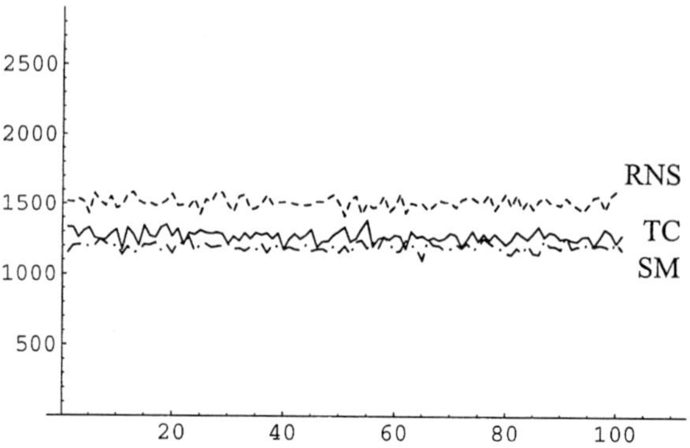

Figure 6.9.   Number of low-to-high transitions, assuming strongly correlated ($\rho = 0.99$) Gaussian data.

ploying an extension of RNS, the *Quadratic RNS* (QRNS) [17]. The direct complex-number multiplication can be performed as

$$
\begin{aligned}
p &= (a + jb)(c + jd) & \text{(6.26)}\\
&= (ac - bd) + j(bc + da), & \text{(6.27)}
\end{aligned}
$$

where j is the imaginary unit, i.e., $\sqrt{-1}$, and a, b, c, and d are real numbers. Parhami [5] shows a different technique to reduce the number of multiplications to three, by performing five additions or subtractions, with an extra computational step. According to this technique, the complex product is computed as

$$
p = [c(a + b) - b(c + d)] + j[c(a + b) - a(c - d)], \qquad \text{(6.28)}
$$

where the common term $c(a + b)$ is initially computed.

In case that the moduli are primes of the form $m_i = 4k + 1$, a QRNS mapping can be established, such that the residue pair of the real and imaginary part modulo m_i a_i and b_i respectively, can be mapped to a quadratic residue as

$$
(a_i, b_i) \xrightarrow{\text{QRNS}} (q_i, q_i^*), \qquad \text{(6.29)}
$$

where q_i and q_i^* are the quadratic images of a_i and b_i, respectively. The quadratic images are obtained as

$$
\begin{aligned}
q_i &= \langle a_i + \bar{\jmath} b_i \rangle_{m_i} & \text{(6.30)}\\
q_i^* &= \langle a_i - \bar{\jmath} b_i \rangle_{m_i}, & \text{(6.31)}
\end{aligned}
$$

while the mapping is inversed as follows

$$
\begin{aligned}
a_i &= \langle 2^{-1}(q_i + q_i^*) \rangle_{m_i} & \text{(6.32)}\\
b_i &= \langle 2^{-1}\bar{\jmath}^{-1}(q_i - q_i^*) \rangle_{m_i}, & \text{(6.33)}
\end{aligned}
$$

where $\bar{\jmath}$ is the solution of

$$\left\langle \bar{\jmath}^2 + 1 \right\rangle_{m_i} = 0. \tag{6.34}$$

QRNS mapping example: Assume that $m = 17$. This is a valid QRNS modulo choice, as 17 is prime of the form $4k + 1$. It can be noted that roots of the equation

$$\left\langle x^2 + 1 \right\rangle_{17} = 0 \tag{6.35}$$

are $x_0 = 4$ and $x_1 = 13$. Notice that $\langle -x_0 \rangle_{17} = x_1$. The multiplicative inverse of 2 mod 17 is 9, since

$$\langle 2 \cdot 9 \rangle_{17} = \langle 18 \rangle_{17} = 1, \tag{6.36}$$

while the multiplicative inverse of 4 is 13, since $\langle 4 \cdot 13 \rangle_{17} = 1$. Hence, the complex RNS number $X = a + jb = 8 + j7$ is mapped to its QRNS image (q_I, q_I^*), as follows:

$$q_X = \langle 8 + 4 \cdot 7 \rangle_{17} = 2 \tag{6.37}$$
$$q_X^* = \langle 8 - 4 \cdot 7 \rangle_{17} = 14. \tag{6.38}$$

The inverse QRNS mapping is

$$a = \langle 9(2 + 14) \rangle_{17} = 8 \tag{6.39}$$
$$b = \langle 15(2 - 14) \rangle_{17} = 7, \tag{6.40}$$

where it has been used that $\langle \bar{\jmath}^{-1} 2^{-1} \rangle_{17} = \langle 9 \cdot 13 \rangle_{17} = 15$.

The quadratic mapping is of practical importance because is alleviates the dependency of the real and imaginary parts of a complex product from both the real and imaginary parts of both the operands, as shown by (6.27). In other words, it eliminates the cross-product terms. Therefore, by exploiting the QRNS, the complex product $\{(qp_i, qp_i^*)|i = 1, 2, \ldots, N\}$ of two QRNS-encoded complex numbers, $\{(qa_i, qa_i^*)|i = 1, 2, \ldots, N\}$ and $\{(qb_i, qb_i^*)|i = 1, 2, \ldots, N\}$, can be evaluated as the direct product of the corresponding quadratic images, i.e.,

$$qp_i = \langle qa_i \cdot qb_i \rangle_{m_i} \tag{6.41}$$
$$qp_i^* = \langle qa_i^* \cdot qb_i^* \rangle_{m_i}. \tag{6.42}$$

Eqs. (6.41) and (6.42) show that a complex multiplication requires only two residue multiplications instead of four multiplications, an addition, and a subtraction. Therefore, by paying an initial cost for conversion, a significant computational complexity reduction can be achieved by the QRNS mapping, which is directly translated to power savings.

QRNS complex multiplication example: Continuing the QRNS mapping example, the complex integer $Y = 3 + j4$, is mapped to the QRNS image

$$q_Y = \langle 3 + 4 \cdot 4 \rangle_{17} = 2 \tag{6.43}$$

$$q_Y^* = \langle 3 - 4 \cdot 4 \rangle_{17} = 4. \tag{6.44}$$

In order to compute the complex product $P = X \cdot Y$, it suffices to compute its QRNS image as

$$q_P = \langle q_X \cdot q_Y \rangle_{17} = \langle 2 \cdot 2 \rangle_{17} = 4 \tag{6.45}$$

$$q_P^* = \langle q_X^* \cdot q_Y^* \rangle_{17} = \langle 14 \cdot 4 \rangle_{17} = 5. \tag{6.46}$$

To retrieve the actual complex residue, the inverse QRNS mapping is applied, as follows:

$$a = \langle 9(4 + 5) \rangle_{17} = 13 \tag{6.47}$$

$$b = \langle 15(4 - 5) \rangle_{17} = 2. \tag{6.48}$$

Hence the complex residue product is $P = 13 + j2$. The direct computation of the product is

$$\langle (8 + j7)(3 + j4) \rangle_{17} = \langle (8 \cdot 3 - 4 \cdot 7) + j(3 \cdot 7 + 4 \cdot 8) \rangle_{17} \tag{6.49}$$

$$= \langle -4 + j53 \rangle_{17} = 13 + j2. \tag{6.50}$$

Consider the Monte Carlo runs of the following experiment: Assuming that the real and imaginary parts of the factors of a complex product are taken from two Gaussian random processes, the total bit activity in the intermediate results is measured for the complex product evaluation. Specifically, 10-bit sign-magnitude and 10-bit two's-complement operations are compared to QRNS operations that cover a dynamic range in excess of 20 bits. Ten Monte-Carlo runs, each of 1000 samples, compose the experiment, which is repeated for uncorrelated ($\rho = 0$), correlated ($\rho = 0.99$), and anti-correlated ($\rho = -0.99$) Gaussian data; results are shown in Figs. 6.10–6.12, respectively. Even in the case that QRNS provides significantly larger dynamic range, it can be seen that the bit activity is reduced approximately two times.

D'Amora *et al.* have compared the implementation of a direct-form complex FIR filter with its QRNS counterpart [19]. They report that, for a particular throughput rate, the QRNS-based implementation requires half the area and a third of the power dissipation of the conventional implementation. The conventional implementation is assumed to utilize the four-multiplication scheme for complex-number multiplication, while the QRNS implementation exploits the *index transform.*

The index transform reduces a modulo-m multiplication to a modulo-$(m-1)$ addition, for m prime, resembling the reduction of multiplication to addition

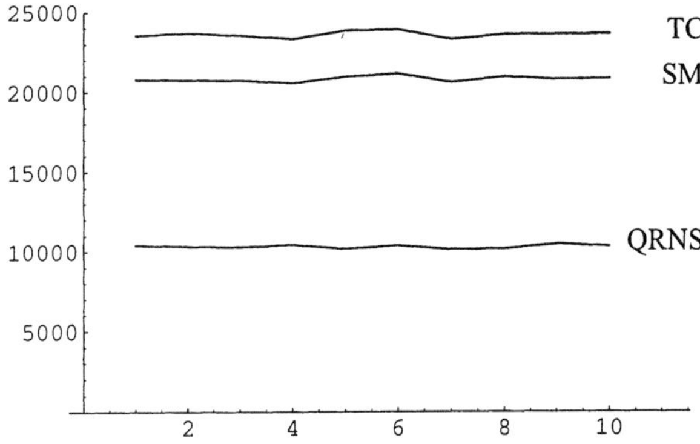

Figure 6.10. Number of low-to-high transitions for complex-number multiplication, assuming uncorrelated ($\rho = 0$) Gaussian operands.

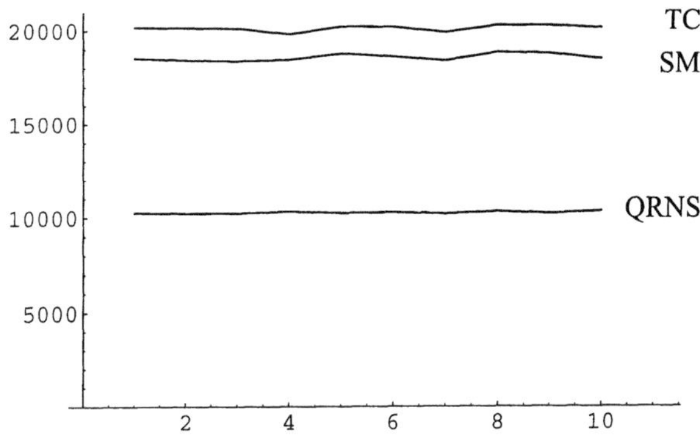

Figure 6.11. Number of low-to-high transitions for complex-number multiplication, assuming strongly correlated ($\rho = 0.99$) Gaussian operands.

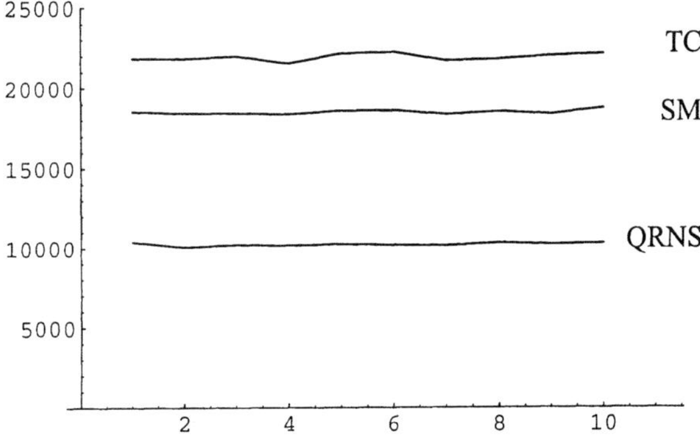

Figure 6.12. Number of low-to-high transitions for complex-number multiplication, assuming strongly anti-correlated ($\rho = -0.99$) Gaussian operands.

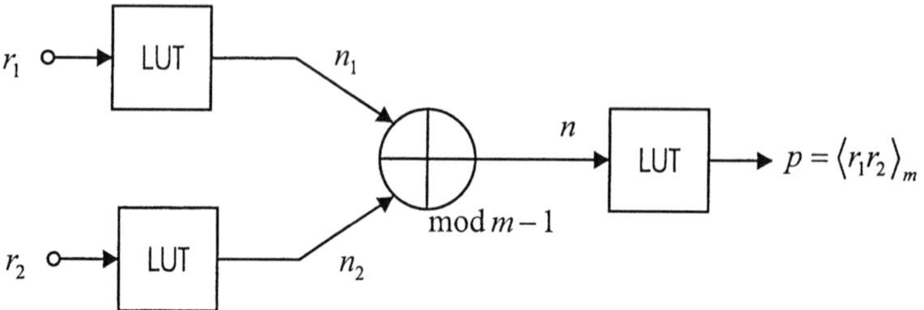

Figure 6.13.   The architecture of a residue multiplier based on the index transform.

r	$\langle 2^n \rangle_5$	n
1	$\langle 2^0 \rangle_5$	0
2	$\langle 2^1 \rangle_5$	1
3	$\langle 2^3 \rangle_5$	3
4	$\langle 2^2 \rangle_5$	2

Table 6.2   The primitive root $\rho = 2$ of the field of residues modulo 5, generates residues $[1, 5)$.

by LNS. An integer root ρ can be determined, such that the residues $r \in [1, m)$ can be written as

$$r = \langle \rho^n \rangle_m, \tag{6.51}$$

and the multiplication of the residues can be reduced to addition modulo $(m-1)$ of the indices, which correspond to the residues to be multiplied. Therefore, the product p modulo m of two residues, r_1 and r_2 is

$$p = \langle r_1 r_2 \rangle_m = \langle \rho^{n_1} \rho^{n_2} \rangle_m = \langle \rho^{n_1 + n_2} \rangle_m = \rho^n, \tag{6.52}$$

where

$$n = \langle n_1 + n_2 \rangle_{m-1} . \tag{6.53}$$

Hence, residue multiplication modulo m can be performed as residue addition modulo $(m - 1)$, preceded and followed by a mapping of the operands to their indices and of the result to the residue as shown in Fig. 6.13. These mappings are commonly implemented as table look-ups [17].

Index transform example Assume the operation $\langle r_1 r_2 \rangle_5$, where $m = 5$. It can be noted that $\rho = 2$ is a primitive root of the corresponding field. All residues in $[1, 5)$ can be written in the form $\langle 2^n \rangle_5$, as shown in Table 6.2. Hence, in order to perform the multiplication $\langle 3 \cdot 4 \rangle_5$, the corresponding indices are obtained from Table 6.2 and added modulo $5 - 1 = 4$, as $\langle 3 + 2 \rangle_4 = 1$. The obtained index value 1 corresponds to the residue 2, as it can be seen in Table 6.2, since $\langle 2^1 \rangle_5 = 2$. It is noted that addition modulo a power of two can be realized as a carry-ignore operation.

The QRNS can exploit the index transform because the utilized moduli need to be prime. In addition in the case of DSP architectures such as FIR filters, the coefficients can be directly stored in index-residue form, thus the strength of each multiplication can be further reduced, since the determination of the corresponding indices is not repeated for every residue multiplication. The significant power dissipation savings reported by D'Amora *et al.*, assume the utilization of the index transform for residue multiplication [19].

6.5   Conclusions

Recent advances in computer arithmetic offer interesting alternative solutions for low-power design. Depending on an assortment of factors that need to be considered, such as signal statistics, computational load, type of arithmetic operations, accuracy and dynamic range, it is worth evaluating the LNS or the RNS and the various information-based transformations for hardware implementations of computationally-intensive tasks.

The choice of arithmetic can lead to substantial power savings, because it affects several levels of the design abstraction, as it can reduce the number of operations, the signal activity, and the strength of the operators; the impact of the arithmetic in a digital system is not limited to the definition of the architecture of arithmetic circuits.

References

[1] A. P. Chandrakasan, S. Sheng, and R. Brodersen, "Low-power CMOS digital design," *IEEE Journal of Solid-State Circuits*, vol. 27, pp. 473–484, Apr. 1992.

[2] J. M. Rabaey and M. Pedram, *Low Power Design Methodologies*. Boston: Kluwer Academic Publishers, 1996.

[3] K. K. Parhi, "Low-energy CSMT carry generators and binary adders," *IEEE Transactions on VLSI Systems*, vol. 7, pp. 450–462, Dec. 1999.

[4] T. K. Callaway and E. E. Swartzlander, Jr., "Power-delay characteristics of CMOS multipliers," in *Proceedings of the 13th Symposium on Computer Arithmetic (ARITH13)*, (Asilomar, USA), pp. 26–32, July 1997.

[5] B. Parhami, *Computer Arithmetic - Algorithms and Hardware Designs*. New York: Oxford University Press, 2000.

[6] P. E. Landman and J. M. Rabaey, "Architectural power analysis: The Dual Bit Type method," *IEEE Transactions on VLSI Systems*, vol. 3, pp. 173–187, June 1995.

[7] T. K. Callaway and E. E. Swartzlander, "Low power arithmetic components," in *Low Power Design Methodologies* (J. M. Rabaey and M. Pedram, eds.), Boston: Kluwer Academic Publishers, 1996.

[8] A. Nannarelli and T. Lang, "Low-power division: Comparison among implementations of radix 4, 8 and 16," in *Proceedings of 14th Symposium on Computer Arithmetic*, (Adelaide, Australia), Apr. 1999.

[9] E. Swartzlander and A. Alexopoulos, "The sign/logarithm number system," *IEEE Transactions on Computers*, vol. 24, pp. 1238–1242, Dec. 1975.

[10] R. E. Morley, Jr., G. L. Engel, T. J. Sullivan, and S. M. Natarajan, "VLSI based design of a battery-operated digital hearing aid," in *Proceedings of the IEEE International Conference on Acoustics, Speech and Signal Processing*, pp. 2512–2515, 1988.

[11] J. R. Sacha and M. J. Irwin, "The Logarithmic Number System for strength reduction in adaptive filtering," in *Proceedings of International Symposium on Low-Power Electronics and Design, (ISLPED '98)*, (Monterey, CA), pp. 256–261, 1998.

[12] V. Paliouras and T. Stouraitis, "Signal activity and power consumption reduction using the Logarithmic Number System," in *Proceedings of the 2001 IEEE International Symposium on Circuits and Systems (ISCAS)*, vol. 2, (Sydney), pp. II–653–II–656, 2001.

[13] V. Paliouras and T. Stouraitis, "Low-power properties of the Logarithmic Number System," in *Proceedings of 15th Symposium on Computer Arithmetic (ARITH15)*, (Vail, CO), pp. 229–236, June 2001.

[14] N. Szabó and R. Tanaka, *Residue Arithmetic and its Applications to Computer Technology*. New York, NY: McGraw-Hill, 1967.

[15] W. L. Freking and K. K. Parhi, "Low-power FIR digital filters using residue arithmetic," in *Proceedings of Thirty-first Asilomar Conference on Signals, Systems, and Computers*, vol. 1, pp. 739 – 743, 1997.

[16] W. A. Chren, Jr., "One-hot residue coding for low delay-power product CMOS design," *IEEE Transactions on Circuits and Systems – Part II*, vol. 45, pp. 303–313, Mar. 1998.

[17] M. A. Soderstrand, W. K. Jenkins, G. A. Jullien, and F. J. Taylor, *Residue Number System Arithmetic: Modern Applications in Digital Signal Processing*. IEEE Press, 1986.

[18] M. K. Ibrahim, "Novel digital filter implementations using hybrid RNS-binary arithmetic," *Signal Processing*, vol. 40, no. 2–3, pp. 287–294, 1994.

[19] A. D'Amora, A. Nannarelli, M. Re, and G. C. Cardarilli, "Reducing power dissipation in complex digital filters by using the Quadratic Residue Number System," in *Proceedings of the 34th Asilomar Conference on Signals, Systems, and Computers*, 2000.

Chapter 7

REDUCING POWER CONSUMPTION IN MEMORIES

Alexander Chatzigeorgiou
University of Macedonia, Thessaloniki, Greece
achat@uom.gr

Spiridon Nikolaidis
Aristotle University of Thessaloniki, Thessaloniki, Greece
snikolaid@physics.auth.gr

Abstract   Memory circuits which form an integral part of every digital system, consume often significant power compared to processing units, especially in embedded systems for data-intensive applications. Power reduction techniques for SRAM and DRAM memories, address three key issues that affect energy consumption: static current, supply voltage and charging/discharging capacitance. Past efforts have resulted in continuous power reduction for each memory generation; however, future research has to deal with tradeoffs arising from performance degradation for lower supply voltage and standby power dissipation due to subthreshold leakage currents.

Keywords: Low-Power memory, SRAM, DRAM

7.1   Introduction

Power consumption due to memory accesses in a computing system, often constitutes the dominant portion of the total power consumption [1]. As a result, power dissipation of memory circuits has been the focus of many research efforts during the recent years. This chapter examines high-performance Random Access Memory (RAM) circuits with emphasis on the sources of power consumption and techniques for low-power operation. As memory capacity and chip size are continuously increasing, it has become clear that investigating new means of reducing both active and standby power consumption in mem-

117

D. Soudris et al. (eds.), Designing CMOS Circuits for Low Power, 117–140.
© 2002 *Kluwer Academic Publishers. Printed in the Netherlands.*

ory circuits is a critical issue in the design of efficient, low-power computing devices.

As the sources of power dissipation in SRAMs are slightly different than those in DRAMs, appropriate low-power design techniques for each type have been developed. In the first half of this chapter the sources of power consumption in SRAMs will be described briefly, while several well-known methods for active power reduction in SRAMs are considered. Most of these methods concentrate on the reduction of static currents and on power optimizations of circuits whose power consumption relies heavily on static current, such as sense amplifiers. In DRAMs however, the dominant component of the power consumption comes from the memory array, especially during charging/discharging of the bit-lines. Therefore, techniques that allow the reduction of this source of power are described in the remainder of this chapter.

7.2   Static Random Access Memories

Static Random Access Memories are read/write (R/W) memory circuits which permit the modification of data bits stored in memory cells as well as their retrieval on demand. The term *static* stems from the fact that as long as sufficient power supply voltage is provided the stored data is retained indefinitely. *Random access* indicates that the access time is independent of the physical location of the data to be read/stored in the memory array. SRAM features high speed and therefore is used for the main memory in supercomputers or cache memory of mainframe computers [2].

The basic data storage cell which consists of a simple latch circuit with two stable operating points (states) is shown in Figure 7.1 and requires six transistors per bit (also called *full SRAM cell*). Access to the data being held in the memory cell is enabled by the word line (*WL*), which controls pass transistors M_5 and M_6. Connection to the 1-bit SRAM cell is implemented by two complementary bit lines (both polarities are used to improve noise margins during read/write operations).

For example, when the word line is not selected (*WL*="0") pass transistors M_5 and M_6 are turned off preventing the modification of the cell value, which is preserved by the latch consisting of the two cross coupled inverters. At this point, the bit lines are charged-up by column pull-up transistors.

If the memory cell is selected by raising the voltage of the word line to "1", pass transistors M_5 and M_6 are turned on. Once the SRAM cell is selected either a read or write operation can be performed: To store a logic "1" to the cell, bit line ($\overline{BL}$) is forced to logic "0" by the data-write circuitry. This action turns driver transistor M_3 off and transistor M_4 on, forcing Q to logic "1". When a logic "1" is stored in the cell and the cell is selected in order to read its value, the voltage of column ($\overline{BL}$) is slightly pulled down by transistors M_1 and M_5.

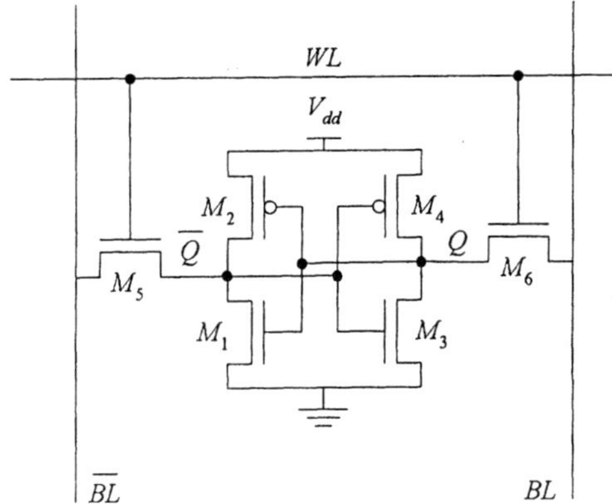

Figure 7.1.   Full SRAM cell

This small voltage difference at the bit line is sensed by the data-read circuitry and amplified as a logic "1" to the output. Writing and reading a logic "0" is performed in a symmetrical manner.

The most significant advantages of SRAMs over other memory types (such as DRAMs) is that the memory cells do not need to be refreshed, fast reading/writing of data due to differential bit-lines and simple operation because row and column address signals are simultaneously loaded.

The full SRAM cell is advantageous from the power consumption point of view, since it does not present any steady-state currents during standby operation. However, it is expensive in terms of area since it requires 6 transistors and both p and n well regions, thus making the layout less compact. Modified configurations of the presented SRAM cell exist, depending on the load devices, which in Figure 7.1 are realized by transistors M_2 and M_4. Other designs are based on resistive loads or depletion-loads using nMOS devices, which reduce the number of transistors by two, but suffer from standby power consumption. Schottky-barrier diode (SBD) loads employing a cross-coupled pair of bipolar inverters have also been reported with improved speed performance compared to CMOS SRAM cells [3].

Figure 7.2 shows technology trends concerning SRAM chip capacity and die size with respect to the minimum feature size for each generation (Source: SIA Technology Roadmap [4]). It is evident that memory capacity quadruples every three years exploiting increased die size and shrinkage in cell area.

7.2.1   Sources of power dissipation in SRAMs

The total power dissipated in a typical SRAM architecture is the sum of two components: active and standby power [2]. Standby power consumption (which takes place even when no operation is performed) is due to leakage

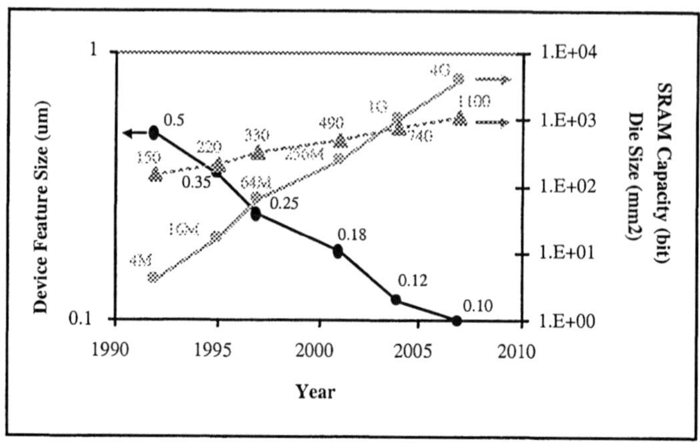

Figure 7.2.   Trends in SRAM chip capacity and size

currents of the cross-coupled CMOS inverters within each cell. This power component for state-of-the-art SRAMs is usually negligible. However, standby power increases significantly if resistive load devices are used: In such circuits, for every value stored in the cell, one of the load devices will always conduct a non-zero current causing steady-state power consumption. If i_{hld} is the steady-state current flowing through each of the unselected memory cells, the standby power dissipated in an SRAM with an external supply voltage V_{dd} is given by:

$$P_{s\tan dby} = mnV_{dd}i_{hld} \qquad (7.1)$$

for a memory array of n rows and m columns. Even if full SRAM cells are used, threshold voltage scaling imposed by ultra-low voltage operation modes (less than 2V) can lead to a situation where standby power becomes significant. Sub-threshold currents generated from an extremely large number of inactive cells can be higher than the currents generated from the remaining active cells and this can dominate the total memory power.

Considering a typical SRAM architecture (Figure 7.3) the following sources of active power consumption can be identified [5] :

- The row and column decoders where power is consumed for driving lines with large parasitic capacitances.

- The memory array.

- The sense amplifiers, which consume the largest portion of power, and

- The remaining peripheral circuits, such as I/O buffers, write circuitry, etc

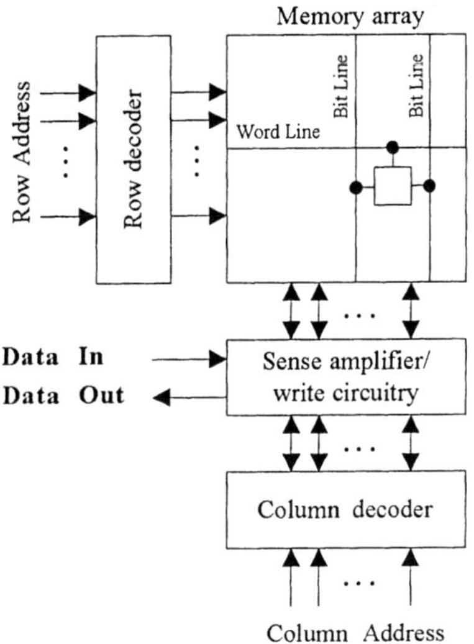

Figure 7.3. Typical SRAM architecture

The above components of active power, for a normal read cycle, can be summarized by the formula [6], [7]:

$$P_{active} = V_{dd}I_{dd} = V_{dd}\left[(mI_{act}\Delta t + C_{PT}V_{INT})f + I_{DCP}\right] \qquad (7.2)$$

where I_{act} is the effective active current flowing during word line activation (i.e. during charging/discharging of bit-line capacitances), Δt is the word-line activation time, C_{PT} is the total capacitance of the peripheral circuits, V_{INT} the internal supply voltage and I_{DCP} the total static current (due to column circuitry and differential amplifiers on the I/O lines), while f is the operating frequency. In the above equation, power consumption due to leakage currents in the $m(n\text{-}1)$ unselected cells and due to decoder charging currents is considered negligible [6].

7.2.2 Low-Power SRAM Circuit Techniques

Acknowledging the intense requirements for low power memories in current high-performance computing devices, circuit designers have developed a number of power optimizing techniques which target several sources of energy dissipation in an SRAM. Figure 7.4 shows the evolution of the average power consumption for past SRAM generations versus the memory capacity. SRAM low power techniques are described in the following sections.

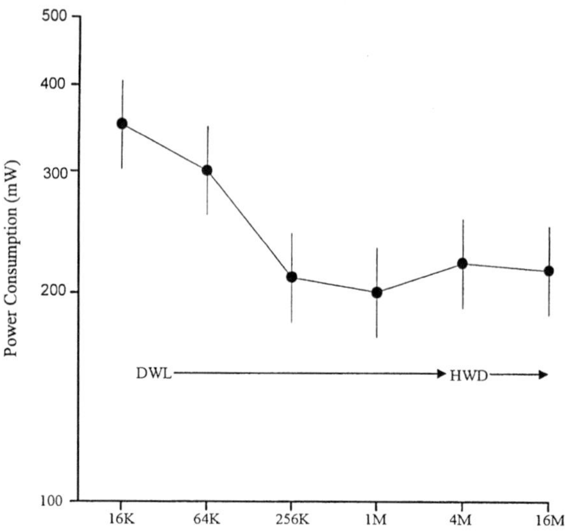

Figure 7.4.   Trends in SRAM power consumption

7.2.2.1   Divided Word-Line Approach.   The increase in memory size
in recent RAM technologies causes an unavoidable increase in word-line length,
which eventually increases the physical capacitance and power consumption,
and thus leads to speed degradation [2]. To solve this problem, the Divided
Word-Line (DWL) approach has been proposed in [8] and is based on a two-
stage hierarchical row decoder structure. The DWL concept is schematically
shown in Figure 7.5 and greatly reduces the static current of SRAMs (I_{DCP}
in the equation for active power). The memory array is divided into n sub-
arrays and each one is driven by a divided word-line which is activated if the
corresponding Word Line and Block Select Line are selected. Consequently,
during an access cycle, only the memory cells which are connected to one
divided word-line are activated. In other words, the memory array driven by
a word line is partitioned into smaller segments so that only the addressed
sub-array is activated. This hierarchical structure leads to reduced *DC* column
currents, since only the selected columns switch, providing significant power
reduction. Furthermore, the word-line selection delay (i.e. the delay from the
address input to the divided word-line) is reduced, improving memory speed.

Although the DWL scheme has been widely applied in high-density SRAM
circuits, for a capacity of more than 4Mb the number of sub-arrays becomes
prohibitive due to the increased capacitance of the global word-lines. To cope
with this issue, the Hierarchical Word Decoding (HWD) scheme was proposed
in [9]. The innovation of HWD is that each word-line is divided into multiple
levels, forming a word-line hierarchy, which significantly reduces the total
word-line capacitance. Hence, the delay and power are reduced. For memory

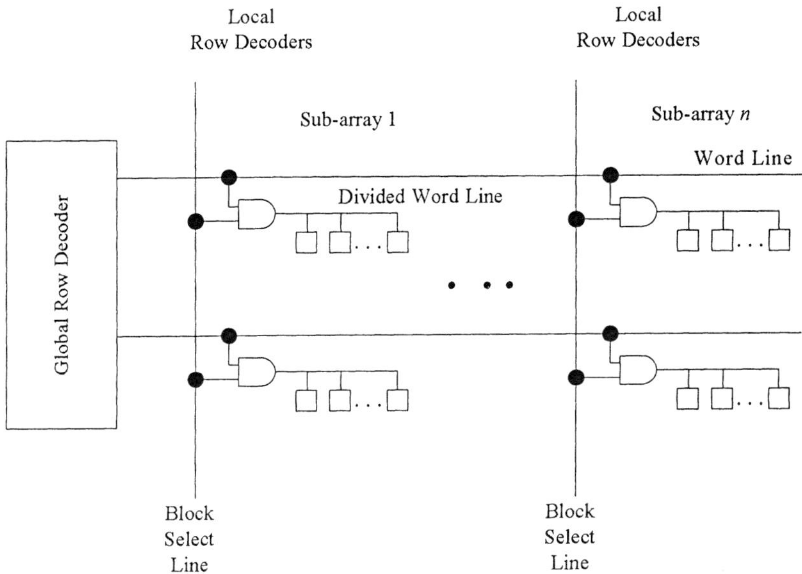

Figure 7.5.   Divided Word-Line Scheme for SRAM

sizes around 4Mb, a 20%-30% reduction in the total load capacitance is achieved by HWD over DWL.

7.2.2.2   Divided Bit-Line Approach.

The *Divided Bit-Line* (DBL) approach proposed in [7] aims at reducing the power which is dissipated during read or write operation by reducing the capacitance of bit-lines. The bit-line capacitance is mainly composed of the drain capacitance of the pass transistors in each SRAM cell and the metal capacitance of bit-lines. DBL attempts to combine two or more SRAM cells so that the number of transistors connected to a bit-line is reduced. Figure 7.6 shows four SRAM cells combined together and connected via one pass transistor to the bit-line. Thus, the number of pass transistors connected to the bit-line is reduced by four.

In the general case, if the number of rows in the memory array is n, and the number of combined cells in DBL denoted by M, the capacitance C_1 at the terminal of the pass transistor connected to the bit line, and capacitance C_2 at the terminal connected to the sub-bit line are [7] $C_1 = \frac{C}{M} + 0.1C$, $C_2 = C\frac{M}{n}$, where C is the original drain capacitance of n rows.

This approach can be extended in a similar manner to the Divided Word-Line Approach, employing an *hierarchical divided bit-line* architecture, where each bit-line is divided into more than two levels. Using these techniques, the active power can be reduced by as much as 50-60%, while the access time is reduced by 20-30% [7]. However, the fact that in this approach the sub-bitlines are

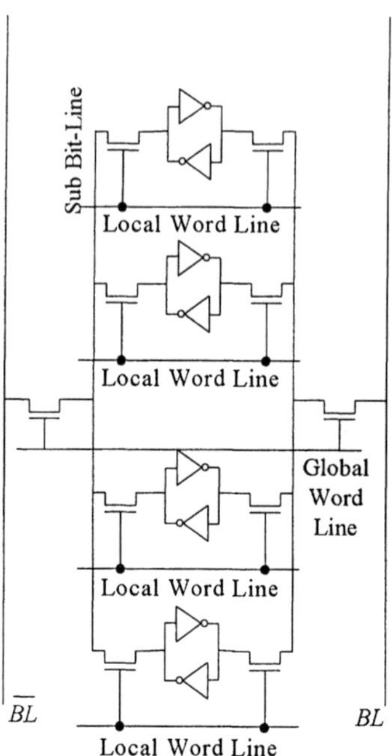

Figure 7.6.   Divided Bit-Line Approach

connected to the main bitlines via several pass-transistors, can have a negative impact on speed, if the number of levels becomes large. A technique to avoid pass-transistors in series by "physically" split bitlines has been proposed in [10].

7.2.2.3   Pulse operation of Word-Line Circuitry.

Another effective approach to reduce active power consumption in SRAM circuits is to shorten the duration of the active duty cycle, which is the period when read/write operation can be performed after the arrival of the address signals. By pulsing the word-lines only for the minimum time required for accessing the contents of a memory cell, significant power savings are achieved, while the reduction ratio is equal to the ratio of the pulse width to the cycle time [6], [11]. Short pulses are generated from the *Address Transition Detector* (ATD) shown in Figure 7.7 (a) for each address change from logic "0" to logic "1" and vice-versa (the two inverters in sequence act as a delay element). The corresponding waveform for an address change in one address bit is shown in Figure 7.7 (b). The ATD-generated pulses for all address line transitions are summed up through an OR gate to a single pulse ϕ_{ATD}. This summation pulse is then stretched out with a delay circuit and used to reduce power or speed up signal propagation by turning memory elements on and off as it passes through.

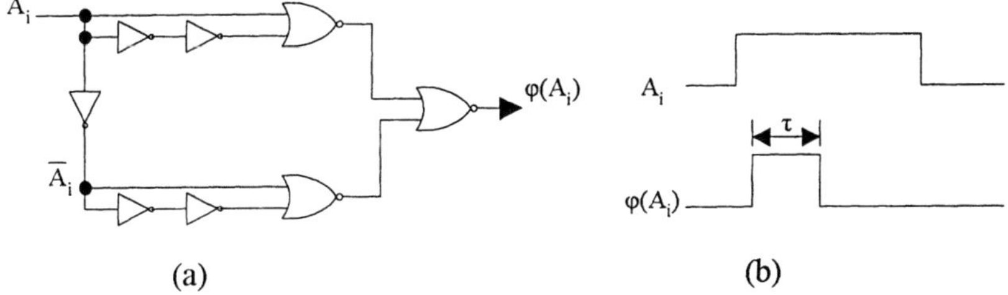

Figure 7.7. a) Address Transition Decoder, (b) corresponding waveforms

Pulsed operation techniques are also used to reduce power consumption by reducing the signal swing on the bit lines [11] or as a front-end of the sense amplifiers. For example, power hungry memory sense amplifiers are powered up only for that portion of cycle time when they are needed to sense the signals.

Significant power savings come from operating the bit-lines at half-supply voltage than at V_{dd} [11]. This approach is based on half-swing pulse mode gate families (Figure 7.8 shows an nMOS half-swing pulse mode AND gate [12]). For the example shown in Figure 7.8, positive half-swing (transitions from a rest state $V_{dd}/2$ to V_{dd} and back to $V_{dd}/2$) on the two inputs and a negative half-swing (transitions from a rest state $V_{dd}/2$ to GND and back to $V_{dd}/2$) combined with the received-gate logic style, result in an output signal with full-swing with negligible effects of the half-swing inputs on the performance of the receiver. Considering two wire-load-dominated, high-capacitance lines, which for a conventional pulsed design present a full-rail swing [12], the power dissipation to drive this lines is $2CV_{dd}^2f$ (where C is the capacitance being switched and f the pulse frequency). If the lines are reduced to half-swing the power becomes $2CV_{dd}(V_{dd}/2)f$, a 50% power saving. If half-swing pulse mode techniques are combined with charge recycling (exploiting the charge used to generate the assert transition of a positive pulse to generate the reset transition of a negative pulse), then the power drawn from the $V_{dd}/2$ power supply can be zeroed, leading to power savings of 75% on high-capacitance lines. In the context of SRAM design, the half-swing pulse-mode gates can be used in row and column decoders leading to power savings of 18% in read power and 46% in the write power for a 2-K x 16-b SRAM in 0.25um technology.

7.2.2.4   Low power design techniques for sense amplifiers.   During the read cycle for an SRAM memory cell, the differential bit lines are initially precharged and then one of the bit lines goes low, while the other remains at its initial logic level. Unfortunately, the small size of the discharging transistor in the memory cell (transistors M_1 or M_3 in Figure 7.1) and the large bit-capacitance causes the bit-line discharge to be very slow, leading in turn to a

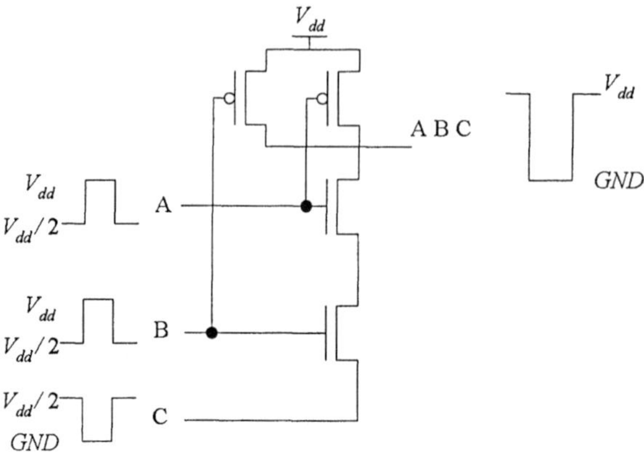

Figure 7.8.   Half-swing pulse-mode *n*MOS AND gate

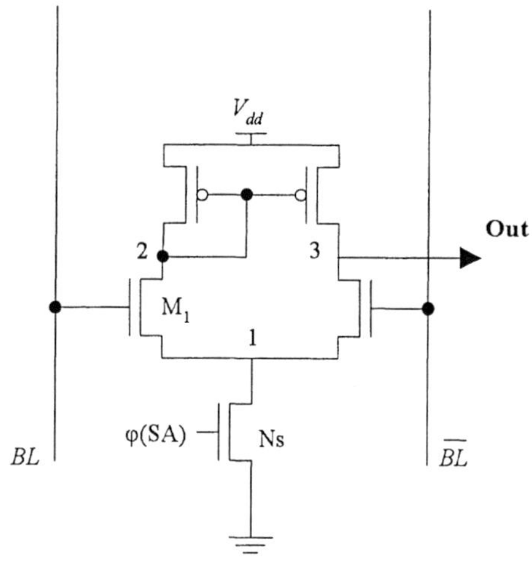

Figure 7.9.   Single-ended Differential Sense Amplifier

slow read operation. The role of the *Sense Amplifier* (SA) is to detect the small differential signal from the bit-lines when one line starts going low, and amplify it. It becomes apparent that sense amplification of the bit lines is the key to fast memory operation (the delay of a sense amplifier represents 30% to 40% of the whole read access time [5]). More important to low power design, is the fact that sense amplifiers can contribute to overall power reduction, because reducing the voltage swing on the bit lines can eliminate a substantial part of the power, which is dissipated to charge and discharge the bit lines. On the other hand, the circuit of the sense amplifier itself, consumes often the largest portion of the total power in an SRAM.

The basic circuit used in most sense amplifiers shown in Figure 7.9 is called a single-ended sense amplifier. The circuit has two differential inputs (BL and $\overline{BL}$) and the noise affects both inputs and only their difference is detected. Before signal ϕ_{SA} is asserted, bit-lines BL and $\overline{BL}$ are high, and internal voltages at nodes 1, 2 and 3 are also high. Once a memory cell is selected for the "read operation" one of the bit-line voltages starts to drop slightly, and at the same time signal ϕ_{SA} is asserted (by an ATD pulse) turning transistor Ns on. If we assume that the stored data in the memory cell causes BL to drop, transistor M_1 starts to turn off. This causes the output voltage at node 3 to drop immediately (principle of differential inverter operation [13]). Thus, the small differential input is amplified by the gain of the differential amplifier.

In many SRAM circuits, multi-stage sense amplifiers are widely used to achieve large voltage gains and high-speed "read" memory operation. These circuits employ the double-ended sense amplifier shown in Figure 7.10. This structure consists of two complementary differential sense amplifiers both of which receive the bit-line voltages at their inputs. Here, the output is also differential and output voltages change in opposite directions when a differential input is detected. The complementary outputs of the first-stage are fed to the input terminals of a second-stage, generating the output signal. The output delay of the two-stage sense amplifier can be as much as an order of magnitude smaller than the delay of a one-stage sense amplifier [14]. Unfortunately, the major disadvantage of this topology is the increased power dissipation for high-density memories. [6] reports a DC current of 1 to 5 mA flowing in each sense amplifier on the I/O line. The power dissipation becomes the largest portion of the total chip power as the number of I/O lines increases to obtain higher data throughput for high-speed processors.

Several circuits have been proposed to reduce the power consumption of sense amplifiers, while improving their speed. One of them is the *n*MOS *cross-coupled* structure shown in Figure 7.11 proposed in [15], which resembles the DCVSL structure. Unlike the basic single-ended sense amplifier where the *p*MOS devices form a current mirror, in Figure 7.11 the *n*MOS transistors are cross-coupled, while the differential outputs are connected to their gates. To outline its operation, it is assumed that the bit lines are initially precharged and that the voltage at line BL starts to drop slightly when the memory cell is accessed. When signal ϕ_{SA} is asserted, the voltage at the gate of the M_1 transistor is slightly higher than the gate voltage of transistor M_2 and consequently transistor M_1 turns on first, pulling the voltage of bit-line BL down, which in turn reduces the gate-to-source voltage of transistor M_2 making it more difficult for M_2 to conduct. Eventually, M_2 turns off completely and M_1 fully turns on discharging the bit line BL. The operation of this positive feedback is fundamentally different that the principle of the differential amplifier. Here the output voltage level does not correspond to the polarity of the voltage difference

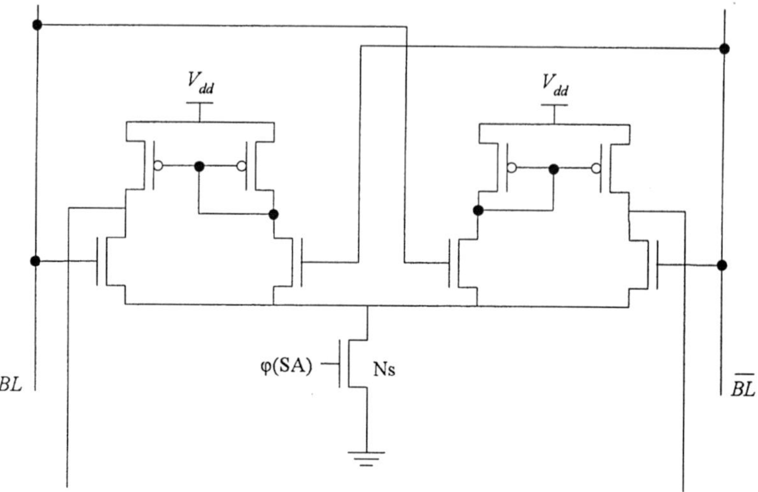

Figure 7.10.   Double-ended Differential Sense Amplifier

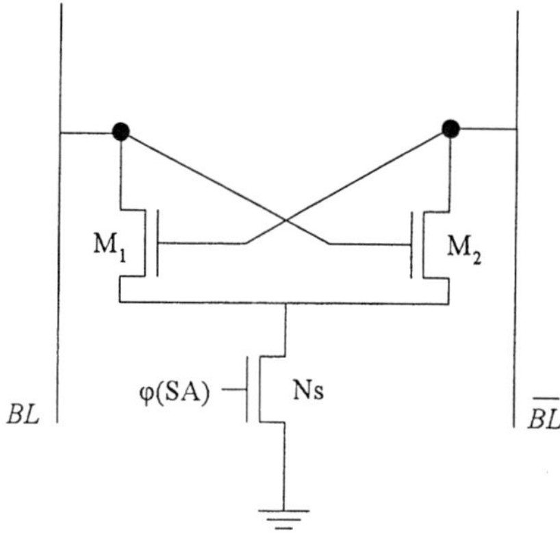

Figure 7.11.   Cross-coupled nMOS sense amplifier [15]

between the two bit lines, but rather it amplifies the small voltage difference that already exists [14]. This scheme achieves faster sensing and power reduction: Simulation results have shown that the delay time of this scheme is approximately one half of that of the conventional single-ended amplifier, while its power consumption is reduced to 20% of the power of the original circuit [15].

7.3   Dynamic Random Access Memories

The "heart" of all SRAM cells, discussed in section 7.2, consists of a two-inverter latch circuit, which holds the memory value and is accessed via two pass transistors for "read" and "write" operations. The only function of the load

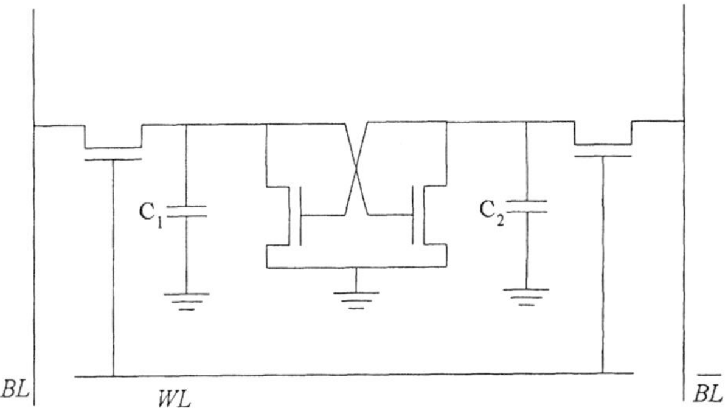

Figure 7.12.   Four-transistor DRAM cell

devices in the inverters, whether transistors or resistors, was to replenish the charge, which is lost by leakage currents. As a result, an SRAM cell requires four to six transistors per bit, four to five lines connecting each cell, including power and ground supplies. These unavoidable elements lead to considerable area requirements, which is often a prohibitive factor for high-density random access memories. Moreover, most SRAM cells, except for the full 6 transistor cell, have non-negligible standby power dissipation.

In a *Dynamic Random Access Memory* (DRAM), the principle is to eliminate the load devices in each cell by simply storing binary data as a charge in a capacitor, whose state is periodically *refreshed.* The refresh operation, which consists of a read followed by a write operation, is necessary, since the data stored as a charge in a capacitor, cannot be retained indefinitely because of leakage currents removing the stored charged.

Consequently, the obvious advantages of DRAM over SRAM cells are the much smaller silicon area required and the absence of any static power dissipation within the cell. Due to its lower cost compared to SRAM, DRAM is extensively used for the main memory in personal computers or mainframes [2]. By removing the load devices from a six-transistor SRAM cell, the simplest and earliest version of the DRAM cell is derived, namely the four-transistor (4T) DRAM cell (Figure 7.12). The binary data is stored in the parasitic diffusion capacitances C_1 and C_2 shown in the diagram, while it is apparent that the area advantage is marginal compared to the full SRAM cell or vanishes at all compared to four-transistor SRAM cells.

By observing the redundancy that was present in the first DRAM cell, i.e. that the cell stores both the data value and its complement, one more transistor has been removed leading to the three-transistor (3T) DRAM cell shown in Figure 7.13, which was the first widely used dynamic memory cell. Here, the binary information is stored in the parasitic capacitance C_1. The cell is written by placing the appropriate data value on *BL/write* and asserting the *write select*

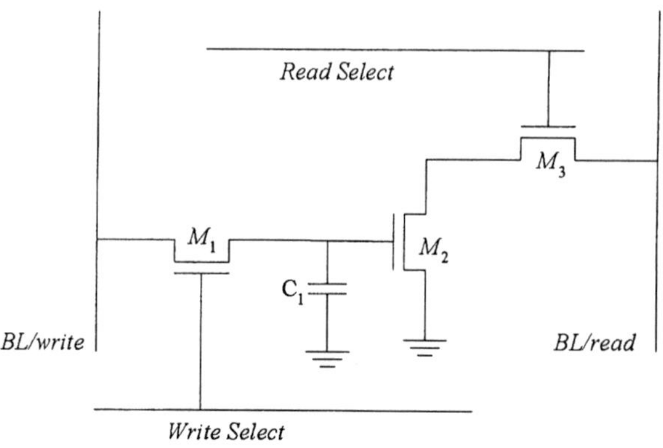

Figure 7.13.   Three-transistor DRAM cell

line. For reading the cell, the *read select line* is raised and the storage transistor M_2 is either on or off depending on the charge stored in C_1. Bit Line *BL/read* is either clamped to V_{dd} with the aid of a load device or is precharged to V_{dd} (or V_{dd}-V_T). When a logic "1" is stored, *BL/read* is pulled low, while if a logic "0" is stored, transistor M_2 is off and *BL/read* remains high. It is important to note that when a logic "1" is written in the cell, the voltage on the storage node equals $V_{Write\_Select\_Line} - V_{TN}$. This threshold voltage drop reduces the current flowing through transistor M_2 increasing the read access time. Some DRAM designs employ a technique called *bootstrapping* to raise the voltage of write select line higher than V_{dd} [14].

To refresh the contents of this 3T DRAM cell, the most common approach is to read the stored data, place its inverse value on *BL/write* and assert *write select* line. It can be seen that this memory cell does not dissipate any static power, while the additional peripheral circuitry required for scheduling the precharge and read-write events and the refresh cycles does not significantly overshadow the area advantages gained by the small cell size [14]. The following two important properties of the 3T cell are worth mentioning [3]: a) In contrast to the SRAM cell, no constraints exist on the device ratios, b) In contrast to other DRAM cells, the data value which is stored in the cell is not affected by a read (*nondestructive reading*).

By sacrificing some of the properties of the 3T DRAM cell it is possible to obtain a memory cell with even lower complexity, namely the one-transistor (1T) DRAM cell, shown in Figure 7.14. It consists of one explicit storage capacitor (C_1) and one access transistor (M_1). Binary information is stored as the presence or absence of charge in the storage capacitor, while the much larger

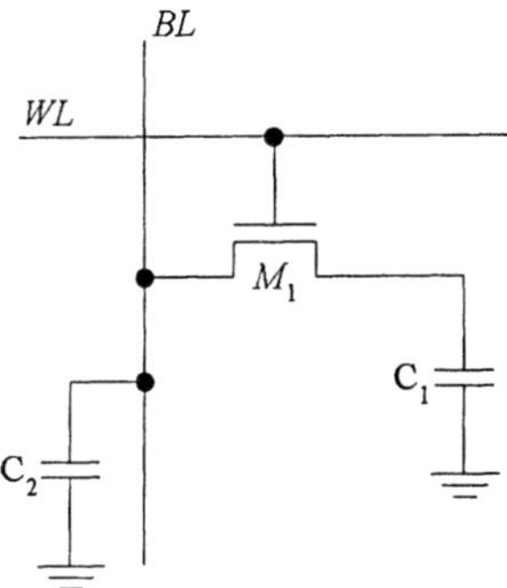

Figure 7.14.   One-transistor DRAM cell

parasitic column capacitance associated with the word line has to be taken into account in the operation of the 1T cell, due to charge sharing. During a write cycle, the data to be stored is placed on the bit line (*BL*) raising word-line (*WL*) at the same time. For a logic "1" the cell capacitor is being charged, otherwise it is being discharged through access transistor M_1.

Because reading a 1T DRAM cell is destructive, a read-refresh circuit has to be employed. Before a read operation is performed, the bit-line is precharged to a voltage level, precharging also the column capacitance C_2. When the word line is pulled high in order to activate the access transistor M_1 charge redistribution between the column and storage capacitance takes place, resulting in a voltage change on the bit line. The direction of this voltage change (increase or decrease) determines the value of the data stored in the cell. Charge redistribution inevitably alters the stored charge on the storage capacitor destroying the contents of the cell. Hence, a refreshing of the cell contents has to be performed during each read cycle. As the storage capacitance is normally one or two orders of magnitude smaller than the column capacitance, the voltage change is very small, making amplification of this voltage change to a full voltage swing necessary.

Figure 7.15 shows DRAM chip capacity/die size versus year of production for each DRAM generation together with the minimum feature size used. Similar to SRAM, memory capacity quadruples every 3 years while die size grows by a factor close to 1.5 in the same period.

7.3.1   Sources of power dissipation in DRAMs

Similar to SRAMs, total power consumption in a DRAM is the sum of standby and active power. Active power dissipation in dynamic random access memory is due to the following components:

- The row and column decoders

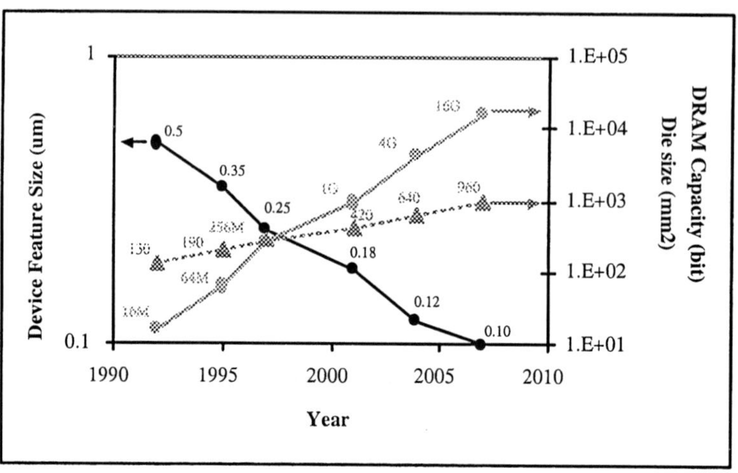

Figure 7.15. Trends in DRAM chip capacity and size

- The memory array, which is the dominant source of power consumption in DRAMs

- The sense amplifiers

- Circuit blocks such as the refresh circuit, the substrate back-bias generator, the boosted level generator, the voltage reference circuit and the half-voltage generator

- Peripheral circuit blocks such as the main sense amplifier, the I/O buffers, the write circuitry etc.

The power sources can be summarized in the following equation [6]:

$$P_{active} = V_{dd}I_{dd} = V_{dd}\left[(mC_D\Delta V_D + C_{PT}V_{INT})f + I_{DCP}\right] \qquad (7.3)$$

where C_D is the bit-line capacitance and ΔV_D the bit-line voltage swing. From equation 7.3, it can be concluded that the following issues are key to reducing the active power dissipation in DRAMs [2]:

- Reducing the bit-line capacitance C_D and the total periphery capacitance C_{PT}. This can be achieved by techniques such as partial activation of multi-divided bit-lines and word-lines and shared I/O resources.

- Scaling down of the internal supply voltage by precharging the bit-lines using the half-voltage generator or by reducing the external supply voltage

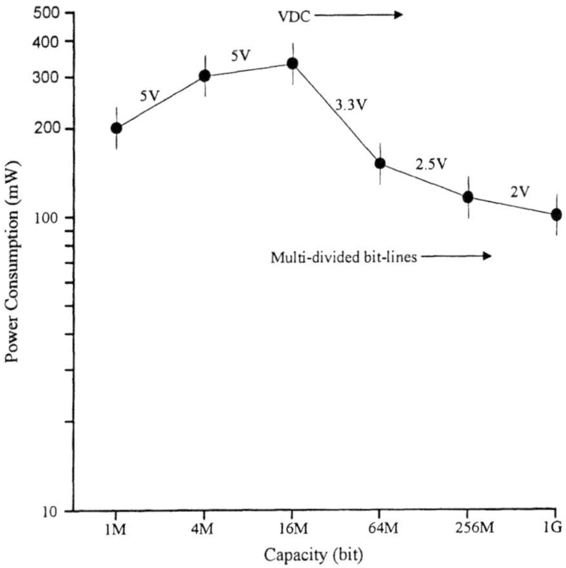

Figure 7.16.   Trends in DRAM Power Consumption

- Reducing the static current flowing through the peripheral circuitry using static CMOS decoders or by applying pulsed operation techniques employing an ATD.

The main source of standby power dissipation in a DRAM is the need for refreshing the contents of the cells. Other sources of standby power are circuits such as the back-bias generator, the half-voltage generator etc.

7.3.2   Low-Power DRAM Circuit Techniques

The low-power DRAM techniques which will be described next, aim at reducing either the charging capacitance of the bit-lines (multi-divided bit-lines and word-lines, refresh time increase) or the operating voltage (half-voltage precharging, on-chip voltage converters). Figure 7.16 shows the average power evolution for past DRAM generations versus memory capacity, as a result of the applied low-power techniques.

7.3.2.1   Multi-divided bit-lines.   With increasing memory capacity, a major concern of DRAM designers is to reduce the bit-line capacitance, which eventually not only reduces power dissipation but it also improves the signal-to-noise ratio of the memory cell [6]. An effective technique to achieve this is to reduce the number of memory cells per bit-line using a multi-divided bit-line scheme. In other words, one bit-line is divided into several sections and each

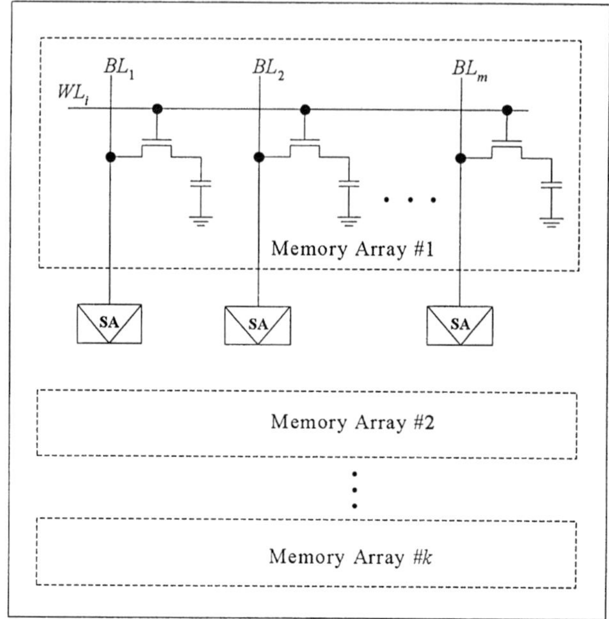

Figure 7.17.   Multi-divided DRAM architecture

time only one section is activated. Figure 7.17 shows a DRAM architecture with a memory array consisting of m columns by n rows, where each bit line is divided into k sections to obtain a smaller C_D. In this scheme, the number of amplifiers, which sense the small variations in the bit-lines, is given by mk while the number of cells physically connected to one amplifier is n/k.

The detailed concepts behind the multi-divided bit-line approach are shown in Figure 7.18. In order to achieve noise cancellation an actual bit-line structure consists of a pair of complementary bit lines [16], which differs from the conceptual one shown in Figure 7.17. One pair of bit-lines, BL and $\overline{BL}$, is multi-divided by repeating the basic unit as shown in Figure 7.18. In the resulting scheme, if one sub-bit-line is activated, the remaining sub-bit-lines in the same column remain inactive, reducing significantly the switched capacitance and thus the active power dissipation. Figure 7.18 (a) shows the conventional scheme, which was used in 64Kb and 256Kb DRAM generations. A Y decoder controls an Y switch (named YS in [16]), which in turn is responsible for connecting each of the sub-bit-lines to the corresponding I/O lines, for accessing the cell contents.

Figure 7.18 (b) shows the evolution of the conventional scheme using a shared amplifier. In this structure, which has been widely used in the 256Kb to 4Mb DRAM generations, two sub-bit-lines share one amplifier. The read signal from

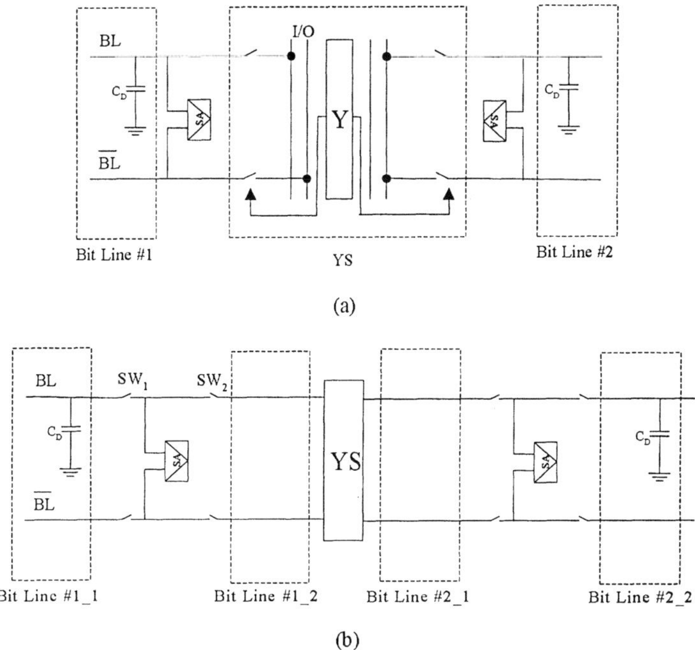

Figure 7.18. Multi-divided bit-line schemes: (a) conventional, (b) shared amplifier

a cell connected to the bit-line #1_1, is fed to the amplifier by turning switch SW_1 on and then amplified having switch SW_2 turned off. Then, the amplified signal is transferred to sub-bit-line #1_2 and to the I/O line by successively turning switches SW_2 and YS on. Since the bit-line capacitance contributing to each cell signal is half that of the conventional scheme, this structure can offer a doubled signal voltage [16]. It is obvious that this scheme requires complicated timing, which results in slower speed. Moreover, although chip area can be reduced using fewer amplifiers and YS switches, the bit-line capacitance contributing to power dissipation remains unchanged, ending up in double power dissipation for actual designs [16]. For 1Mb to 16Mb DRAM generations schemes employing a combination of shared sense amplifiers, shared I/O and shared Y decoder have been proposed [16], [6], which, coupled with other low power techniques have continuously scaled power consumption down.

7.3.2.2 Multi-divided word-lines.

A hierarchical word-line topology has been proposed in [17] for an 256Mb DRAM. This scheme is similar to the one used in SRAM architectures: The memory array is divided into several blocks and each block is divided into sub-arrays. The Sub-Word-Line (SWL) circuitry embedded in each sub-array is responsible for selecting one sub-word-line at a time, while the SWL is selected by both the Main-Word-Line (MWL) and the row-select line (RX), allowing partial activation of word lines. The SWL circuit is shared between sub-arrays being aside, meaning that only two sub-arrays from the whole number of sub-arrays are activated each time, leading to almost half power consumption. However, the additional selections required introduce a rather not acceptable speed penalty.

7.3.2.3   Refresh Time Increase.

As shown in the previous paragraphs, power dissipation is closely related to the active current (I_{act}), which in turn depends on the total bit-line capacitance (C_D) being charged and discharged. Apart from multi-divided data lines, C_D can be significantly reduced by increasing the refresh cycle, during which a refresh operation is performed for all world lines of the memory array. During a refresh operation all m cells on one selected word-line are simultaneously refreshed.

The maximum refresh cycle time is given by $t_{REF_{\max}}/n$ where $t_{REF_{\max}}$ is the maximum refresh time, during which each cell should be refreshed to ensure correct operation and is determined by the cell leakage current. The ratio β of the maximum refresh cycle time to the minimum normal "read" cycle time $t_{RC_{\min}}$ expressing the busy-refresh rate [16] is:

$$\beta = \frac{t_{REF_{\max}}/n}{t_{RC_{\min}}} \tag{7.4}$$

A larger value for β is preferable since it involves less conflict between refresh and normal operation. Expressing the active current I_{act} as [16]:

$$I_{act} = mnC_D V_p \frac{\beta}{t_{REF_{\max}}} \tag{7.5}$$

where V_P is the bit-line precharge voltage, and observing that $t_{REF_{\max}}$ doubles at each DRAM generation, it is possible to achieve a smaller C_D by doubling the refresh cycle time: Assuming that by moving to the next DRAM generation it is required to quadruple the original memory capacity, three alternatives exist for realizing a fixed busy-refresh rate β:

a) Quadrupling the number of rows (n), resulting in quadrupled refresh cycle and the same bit-line capacitance (mC_D). However, this option requires too difficult fabrication processes to control the leakage current so that the maximum refresh time can be increased by a factor of 4.

b) Quadrupling the number of columns (m), which results in quadrupling the bit-line capacitance even though the requirements for refresh time are preserved.

c) Doubling both m and n which only doubles the capacitance and the refresh time, ending up in the optimum solution for each successive generation of DRAMs achieving a double maximum refresh time. Indeed, this scheme has been used for DRAMs from 256Kb to 16Mb.

For recent generations, it seems difficult to keep the pace of doubling $t_{REF_{\max}}$ at each generation and new refreshing schemes have been proposed [17]. These techniques use a reduced m for normal operation, thus reducing the maximum power, while m used for refresh operations remains unchanged, in order to preserve β.

7.3.2.4 Half-voltage bit-line precharging.

An excellent technique to reduce the memory array power consumption in a DRAM is to precharge the bit-lines to half-V_{dd}, instead of a full-V_{dd} voltage, employing a half-voltage generator [6]. In half-V_{dd} precharging, during the sensing operation, one bit-line switches from $V_{dd}/2$ to V_{dd} and the other switches from $V_{dd}/2$ to zero. Ideally, this results in a reduction of the bit-line power consumption by a factor of 4 compared to the full-V_{dd} precharge case as the voltage swing is halved. However, from comparisons of the half-V_{dd} precharging scheme with an nMOS sense amplifier and the full-V_{dd} precharging with a CMOS sense amplifier, it follows that a power reduction of about 50% can be achieved together with area savings of about 30% [6]. In addition, the peak current is significantly reduced, resulting in noise reduction. Moreover, the precharge time is reduced and the cycle time is shortened. Full-voltage bit-line precharging was popular until the 256-Kb DRAM generation while half-voltage precharging is still popular in recent commercial multi-Mb architectures.

7.3.2.5 On-chip Voltage-Down Converter.

Low-voltage operation is essential for low-power, high-performance memory circuits. Some of the parts of a memory, however, still need to operate at the standard external supply voltage V_{dd}. Therefore, an on-chip Voltage-Down Converter (VDC) is used in recent DRAM architectures in order to generate an internal supply voltage (3.3V or lower) from the external power supply [5], [18], [19], [20]. A simple voltage conversion approach is shown in Figure 7.19. In this architecture, the memory array, the sense amplifiers and all peripheral circuits are fed by the internal power supply, which is generated by the VDC circuit, while the I/O buffers are supplied by the external one. Assuming that the internal voltage V_{INT} is scaled down by a factor of k and employing scaled devices for the memory core circuits, leads to power consumption which is also reduced by a factor of k [21].

The schematic for a typical VDC is also shown in Figure 7.19, which provides from an external 5V supply an internal 3.3V V_{INT}. It is composed of a Reference Voltage Generator, a driver circuit and a time-dependent load. The driver consists of a differential amplifier and a common-source drive pMOS transistor. The peak current for a DRAM array can be more than 100 mA with a peak width of around 20ns [21], and in order to be able to deliver such a large current, the width of the pMOS transistor should be sufficiently large. More-

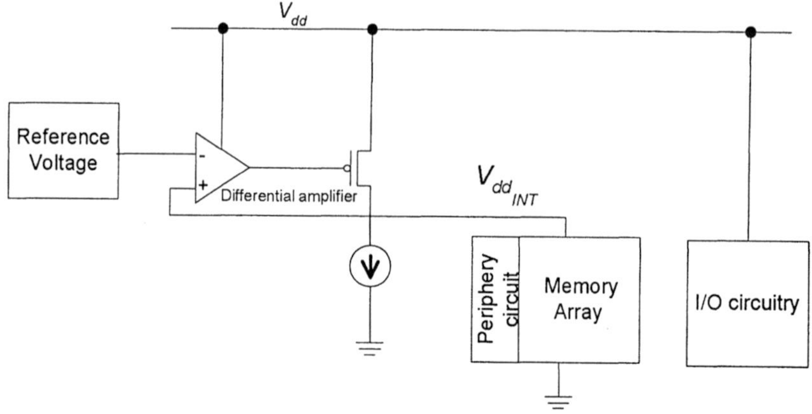

Figure 7.19.   On-Chip Voltage-Down Converter scheme

over, when the output current changes rapidly, the output voltage V_{dd} decreases by ΔV_{dd}. To minimize this output voltage drop, the gate voltage of the pMOS transistor has to respond quickly when the output goes low. This is achieved by increasing the tail current through the differential amplifier [5]. The current source is required to clamp the output voltage to its nominal V_{dd} value when the load current becomes almost zero.

7.4   Conclusions

The investigation of several low power techniques for static and dynamic Random Access Memory circuits reveals that significant power savings can be obtained by addressing the three dominant issues of low-power operation: The reduction of the capacitance which is associated with the memory array bit and word lines, the operating voltage scaling for each memory component and the reduction of the static current.

In static RAMs, techniques that allow partial activation of multi-divided word-lines offer significant power savings by reducing the DC current, while sophisticated sense amplifier design and pulsed operation of word-line circuitry can contribute to the overall power reduction. On the other hand, the main aim in dynamic RAMs is to reduce the charging capacitance employing multi-divided memory arrays or by increasing the refresh time. Lowering the operating voltage through on-chip voltage conversion or half-voltage precharging are additional techniques leading to reduced power dissipation.

The continuous progress in RAM chip scaling and the emergence of ultra-low power operation, which inevitably leads to subthreshold current problems and extreme operating frequencies, call for rapid advances in both memory core as well as in peripheral circuit and sense amplifier designs. Combined with the emergence of portable and high-performance computing devices with huge storage needs, these issues impose intense requirements and many challenges for low power memory innovations.

References

[1] F. Catthoor, S. Wuytack, E. De Greef, F. Balasa, L. Nachtergaele, A. Vandecappelle, Custom Memory Management Methodology: Exploration of Memory Organization for Embedded Multimedia System Design, Kluwer Academic Publishers, Boston, 1998.

[2] K. Itoh, VLSI Memory Chip Design, Springer Series in Advanced Microelectronics, Springer Verlag, 2001.

[3] J. M. Rabaey, Digital Integrated Circuits: A Design Perspective, Prentice Hall, Upper Saddle River, 1996.

[4] The international technology roadmap for semiconductors, Semiconductor Industry Association, 2001.

[5] A. Bellaouar, and M. I. Elmasry, Low-Power Digital VLSI Design: Circuits and Systems, Kluwer Academic Publishers, Boston, 1995.

[6] K. Itoh, K. Sasaki, and Y. Nakagome, Trends in Low-Power RAM Circuit Technologies, Proceedings of the IEEE, vol. 83, no. 4, pp. 524-543, April 1995.

[7] A. Karandikar and K. K. Parhi, Low Power SRAM Design using Hierarchical Divided Bit-Line Approach, in Proc. of IEEE Int. Conference on Computer Design (ICCD), Austin, October 1998.

[8] M. Yoshimito, K. Anami, H. Shinohara. T. Yoshihara, H. Takagi, S. Nagao, S. Kayano, and T. Nakano, A Divided Word-Line Structure in the Static RAM and its Application to a 64K Full CMOS RAM, IEEE Journal of Solid-State Circuits, vol. SC-18, no. 5, pp. 479-485, October 1983.

[9] T. Hirose, H. Kuriyama, S. Murakami, K. Yuzuriha, T. Mukai, K. Tsutsumi, Y. Nishimura, Y. Kohno, and K. Anami, A 20-ns 4-Mb CMOS SRAM with Hierarchical Word Decoding Architecture, IEEE Journal of Solid-State Circuits, vol. 25, no. 5, pp. 1068-1074, October 1990.

[10] J.-M. Masgonty, S. Cserveny, C. Piguet, Low-Power SRAM and ROM Memories, in Proc. of 11th Int. Workshop on Power and Timing Modeling, Optimization and Simulation (PATMOS), Yverdon, Switzerland, September 2001.

[11] M. Margala, Low-Power SRAM Circuit Design, in Proc. of 7th IEEE International Workshop on Memory Technology, Design and Testing (MTDT'99), San Jose, August 1999.

[12] K. W. Mai, T. Mori, B. S. Amrutur, R. Ho, B. Wilburn, M. A. Horowitz, I. Fukushi, T. Izawa, and S. Mitarai, Low-Power SRAM Design Using Half-Swing Pulse-Mode Techniques, IEEE Journal of Solid-State Circuits, vol. 33, no. 11, pp. 1659-1671, November 1998.

[13] N. H. E. Weste, and K. Eshraghian, Principles of CMOS VLSI Design, Addison Wesley, Reading, 1994.

[14] S.-M. Kang, and Y. Leblebici, CMOS Digital Integrated Circuits: Analysis and Design, McGraw-Hill, New York, 1996.

[15] K. Sasaki et al, A 9-ns 1-Mbit CMOS SRAM, IEEE Journal of Solid State Circuits, vol. 24, no. 5, pp. 1219-1225, October 1989.

[16] K. Itoh, Trends in Megabit DRAM Circuit Design, IEEE Journal of Solid State Circuits, vol. 25, no. 3, pp. 778-789, June 1990.

[17] T. Sugibayashi et al., A 30-ns 256-Mb DRAM with a multidivided array structure, IEEE Journal of Solid-State Circuits, vol. 28, no. 11 , pp. 1092-1098, November 1993.

[18] T. Furuyama et al, A New On-Chip Voltage Converter for Submicrometer High-Density DRAMs, IEEE Journal of Solid-State Circuits, vol. 22, no. 3, pp. 437-441, June 1987.

[19] M. Horiguchi et al, Dual-Operating-Voltage Scheme for a Single 5-V, 16-Mbit DRAM, IEEE Journal of Solid-State Circuits, vol. 23, no. 5, pp. 1128-1132, October 1988.

[20] M. Horiguchi et al, A Tunable CMOS-DRAM Voltage Limiter with Stabilized Feedback Amplifier, IEEE Journal of Solid-State Circuits, vol. 25, no. 5, pp. 1129-1135, October 1990.

[21] K. Itoh, Low Power Memory Design, in Low Power Design Methodologies, eds. J. M. Rabaey and M. Pedram, Kluwer Academic Publishers, Boston, 1996.

Chapter 8

LOW-POWER CLOCK, INTERCONNECT AND LAYOUT DESIGNS

Christian Piguet

CSEM: Centre Suisse d'Electronique et de Microtechnique, Neuchâtel, Switzerland

& LAP-EPFL, Lausanne, Switzerland

christian.piguet@csem.ch

Abstract   This chapter describes low power issues at the low level down to layout. The first very important issue is the clocking scheme, as clock trees consume more and more power due to clock skew minimization. Some techniques are presented, such as two loaded and unloaded parallel clock trees and latch-based clocking schemes including gated clocks. Larger and larger interconnect delays and crosstalk effects in deep submicron technologies are the second issue, resulting in new routing strategies with hierarchical metal layers. The third issue is layout design for low-power standard cell libraries.

Keywords: Clock Power, clock distribution, latch-based, interconnect delays, crosstalk, layout design.

8.1   Introduction

Good digital or computer design is presented in [1] with four basic rules:

- rule 1: Simplicity favors regularity

- rule 2: Smaller is faster

- rule 3: Good design demands compromise

- rule 4: Make the common case fast

However, most of the errors in digital circuits are timing errors. Timing design problems are rarely presented in literature. Design teams are secretive about their clock systems: either because they believe they are doing something new, or because they think they are doing nothing new [2]. Consequently, this

D. Soudris et al. (eds.), Designing CMOS Circuits for Low Power, 141–168.
© 2002 *Kluwer Academic Publishers. Printed in the Netherlands.*

chapter will present the three most difficult elements to design for low-power in deep submicron technologies, i.e. the clocking system, the interconnections and the layout of digital blocks.

The key element, one is tempted to say the heart, of most digital systems is the clock. It is well-known today that the clock could consume 1/3 or more of the total power of a chip. It takes into account the clock generation and distribution as well as the structures of latches and flip-flops. Several clocking schemes have been and are used today such as a single clock or phased clocks. Emphasis is put on safe, reliable and finally low-power clocked designs. Problems such as reliability, clock skew, metastability and clock power are more and more crucial in the design of low-power chips in deep submicron technologies.

Interconnect delays are well known to be increased in deep submicron technologies, requiring new interconnection strategies, but crosstalk could also be a stopping point. These problems are strongly related to the design of drivers, transistor sizing and finally to layout. This is a difficult problem as reliability in deep submicron technologies is mandatory, but these techniques have also to be capable of reducing the power consumption.

8.2 Low-Power Clock Design

8.2.1 Clock Skew

Microprocessors are clocked by a global clock signal. The largest difficulty is to have a high frequency global clock and to design the clock tree for such a frequency. This is due to the clock skew that occurs in the clock tree [3]. Proposed solutions are to balance the clock tree delays [4] while using various tricks (wire snaking, balance of the number of buffers, of the number of crossovers, etc). These solutions generally increase the power consumption. Furthermore, they are not acceptable for very low voltage circuits, as slightly V_{th} variations can change significantly the delay values of the clock tree buffers.

Example: A DSP core synthesized with the CSEM low-power library in TSMC 0.25 µm. The test bench A contains only a few multiplication operations, while the test bench B performs a large number of MAC operations. Table 8.1 shows that the power is sensitive to the application program as well as to the required skew: 100% of power increase from 10 ns to 3 ns skew.

Table 8.1.   Power consumption of the same core with various test benches and skew.

Skew	Test bench A	Test bench B
10 ns	0.44 mW/MHz	0.76 mW/MHz
3 ns	0.82 mW/MHz	1.15 mW/MHz

Large hierarchical buffers are used for clock distribution. The local clock distribution is performed by a grid (Sparc III clock grid is connected to 80,000 flip-flops). Furthermore, clock gating is performed in such clock trees. It reduces the power consumption but introduces some extra clock skew. In most microprocessors, delayed clock are created by analog delays to have more clock edges to perform all the work in a single pipeline stage. These clocks are also delayed to provide time borrowing [5]. These delays are programmable in the Intel Itanium mainly for debug purposes, but it could also be used in working mode [6].

Figure 8.1 shows how low skew is implemented in the old DEC Alpha 21164 [7]: skew delays (90 ps) on the clock inputs of Flip-flops or latches are always shorter than register or logic delays. This can be achieved by putting always some logic between latches. The total capacitance on the clock wire of this 21164 µP is 3.75 nF at 300 MHz (power = 12 Watts) and the clock driver of this clock wire contains several transistors with a total W=58cm!

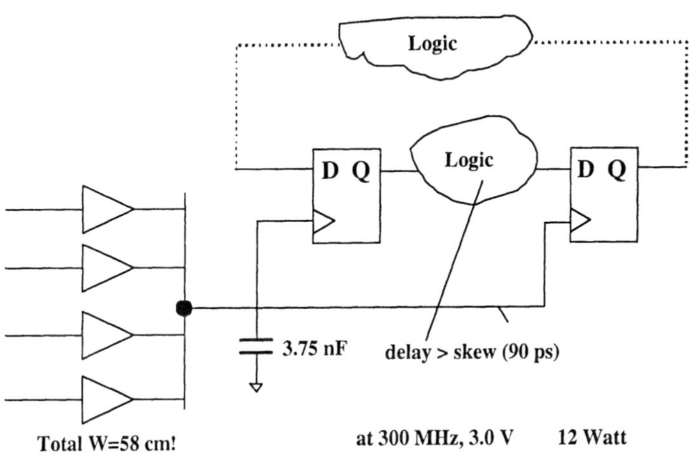

Figure 8.1.   Clock Drivers in the DEC Alpha 21164.

Clock power is a dramatic problem. In some very fast chips, the switched capacitance of the clock node can be a significant part (i.e. 25%) of the total switched capacitance. In other words, the clock power consumption can be as large as the logic power [8]. Self-timed and asynchronous techniques are probably a way to think, but most probably, complex System-on-Chip(SoC) will be globally asynchronous but locally synchronous (GALS). In ISSCC 2000, it was proposed to have $N \times N$ regions, each one clocked by its own Phase Locked Loop (PLL) that have to search for synchronization with the adjacent PLL. Phase detectors located at the region borders are used to synchronize these PLL [6].

The clock distribution is designed as a hierarchical clock tree, according to the decomposition principle [8]. Another important point is the clock input capacitances of the D flip-flops or latches connected to the clock node. Several flip-flops types can be considered such as gated flip-flops, race-free static flip-flops with six transistors connected to the clock input or dynamic flip-flops [9]. For the StrongArm, DIGITAL has proposed flip-flops with only three transistors connected to the clock line [10].

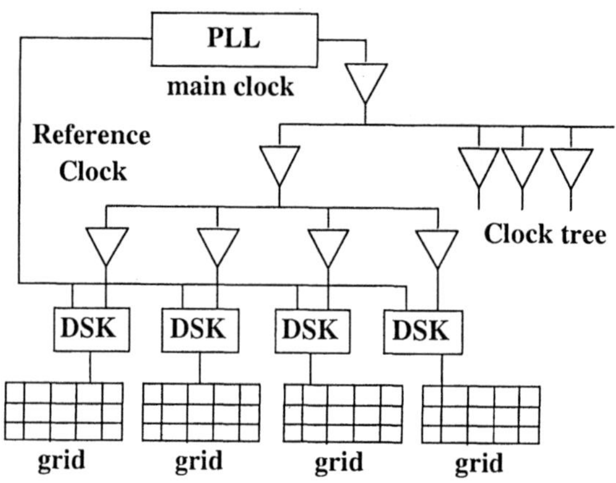

Figure 8.2.   Clock Trees of the Itanium microprocessor.

A very good idea has been proposed by Intel for the Itanium processor [6]. The PLL produces 1 GHz after the divider by 2. This frequency is distributed twice with 2 separated clock trees. The first clock tree is the normal clock tree loaded with x00,000 flip-flops or latches. The second clock tree is not loaded, therefore its skew is quite low. It is the reference clock. So the working clock is synchronized to the reference clock by several local PLLs called Deskew Clusters (DSK). Each DSK provides a clock to a grid of M4 and M5 clocks (Figure 8.2).

Many papers have proposed techniques capable of reducing clock power consumption, such as:

- hierarchical clock tree (decomposition principle [11])

- careful design of the clock drivers to avoid the short circuit current [12]

- minimum number of MOS connected to the clock (careful design of D-flip-flops and latches)

- frequency reduction by using double edge flip-flops or parallelized logic modules [11]

- multirate clocks, i.e. to clock some blocks at $f/2$, $f/4$,....

- reduced swing for the clock distribution [8]. Such a technique could be promising, but it is not easy to implement due to noise problems.

8.2.2 Latch-based Clocking Scheme

The design methodology [5] using latches and two non-overlapping clocks has many advantages over the use of D-flip-flops methodology with a single-phase clock (Figure 8.3). Due to the non overlapping of the clocks as long as half period of the master frequency and the additional time barrier caused by having two latches in a loop instead of one D-flip-flop, latch based designs support greater clock skew before failing than a similar D-flip-flops design (each targeting the same throughput). This allows the synthesizer and router to use smaller clock buffers and to simplify the clock tree generation, which will reduce the power consumption of the clock tree.

With latch-based designs, the clock skew becomes relevant only when its value is close to the non-overlapping of the clocks. When working at lower frequency and thus increasing the non-overlapping of clocks, the clock skew is never a problem. It can even be safely ignored when designing circuits at low frequency. A shift register made with D-flip-flops can have clock skew problems at any frequency.

Furthermore, if the chip has clock skew problems at the targeted frequency after integration, the clock frequency of the latch-based design can be reduced. It results that the clock skew problem will disappear, allowing the designer to test the chip functionality and eventually to detect other bugs or to validate the design functionality. This can reduce the number of test integration needed to validate the chip. With a D-flip-flops design, when a clock skew problem appears, you have to reroute and integrate again.

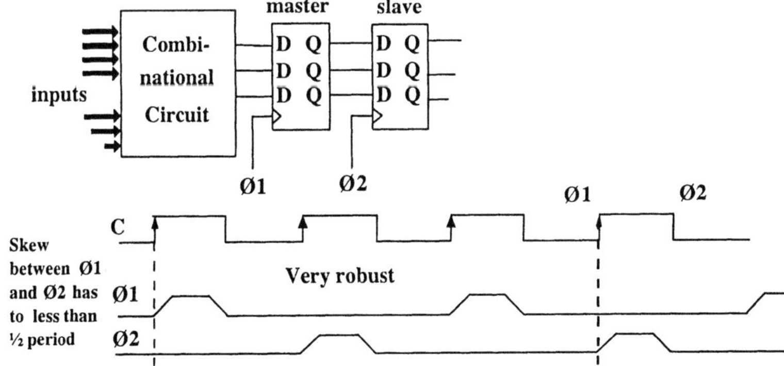

Figure 8.3. Non-overlapping clocks in double-latch scheme.

Using latches for pipeline structure is also very good for power consumption when using such a scheme in conjunction with clock gating. The latch design has additional time barriers, which stop the transitions and avoid unneeded propagation of signal and thus reduce power consumption. The clock gating of each stage (latch register) of the pipeline with individual enable signals can also reduce the number of transition in the design compared to the equivalent D-flip-flops design.

Another advantage with a latch design is the time borrowing (Figure 8.4). It allows a natural repartition of computation time when using pipeline structures. With D-flip-flops, each stage of logic of the pipeline should ideally use the same computation time, which is difficult to achieve, and in the end, the design will be limited by the slowest stage (plus a margin for the clock skew). With latches, the slowest pipeline stage can borrow time from either or both the previous and next pipeline stage. And the clock skew only reduces the time that can be borrowed.

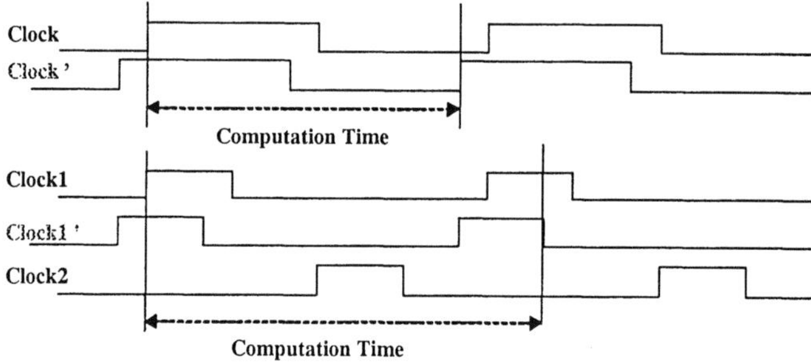

Figure 8.4.   Time Borrowing.

In the Itanium and Motorola PowerPC processors, clocks (or precharge signals) are delayed by self-timed delays to increase performances, while applying the "Time borrowing" technique [5]. This technique allows some pipeline stages to have a larger delay than the previous or next stages by delaying the clock. It can be useful depending on the amount of work that is performed by the various pipeline stages. Figure 8.5 shows another implementation for time borrowing with latches [5]. The available computation time of a given pipeline stage can be longer if the next stage requires less computation time.

8.2.3   Gated Clock with Latch-based Designs

The latch-based design also allows a very natural and safe clock gating methodology [5]. Figure 8.6 shows a simple and safe way of generating enable signals for clock gating. This method gives glitch free clock signals without the adding of memory elements, as it is needed with D-flip-flops clock gating.

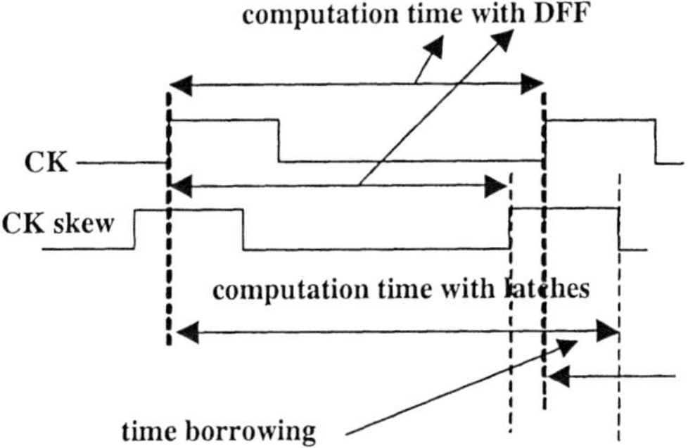

Figure 8.5. Time Borrowing with latches.

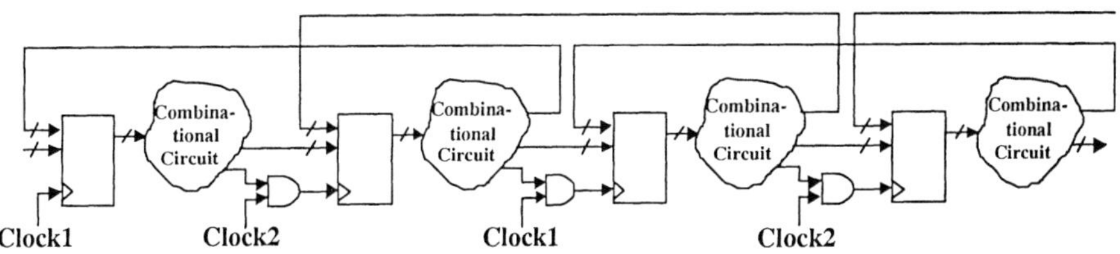

Figure 8.6. Latch-based Clock Gating.

According to the synthesis performed at CSEM, Synopsys handles very nicely the latch-based design methodology. It performs nicely the time borrowing and seems to analyze correctly the clocks for speed optimization. So it is possible to use this design methodology with Synopsys, although there are a few points of discussion linked with the clock gating.

This clock gating methodology cannot be inserted automatically by Synopsys. The designer has to write the description of the clock gating in his VHDL code. This statement can be generalized to all designs using the above latch-based design methodology. We believe Synopsys can do automatic clock gating for pure double latch design (in which there is no combinational logic between the master and slave latch, Figure 8.3), but such a design causes a loss of speed over similar D-flip-flops design.

The most critical problem is to prevent the synthesizer from optimizing the clock gating AND gate with the rest of the combinational logic. To ensure a glitch free clock, this AND gate has to be placed as shown in Figure 8.6. This can be easily done manually by the designer by placing these AND gates in a separate level of hierarchy of his design or placing a *"don't touch"* attribute on them.

Forcing a *"don't touch"* on these gates presents the drawback that this part of the clock tree will not be optimized for speed or clock buffering. Remark that the AND gate shown in Figure 8.6 represents a NAND gate followed by an inverting clock buffer. It would be interesting that the tool handles this gate in a special way to keep it in front of the latch clock input. Maybe by placing a specific attribute on it in such a way that it can recognize it as a clock gating gate, which forbid the optimizer to move logic between it and the latch, but still allows it to size the NAND and the clock buffer.

8.2.4   Results

A synthesizable by Synopsys CoolRISC-DL 816 core [5] with 16 registers has been designed according to the proposed Double Latch (DL) scheme (clocks O1 and O2) and provides the estimated (by Synopsys) following performances (only the core, about 20,000 transistors) in TSMC 0.25 µm: - 2.5 Volt, about 60 MIPS (but 120 MHz single clock). It is the case with the core only, if a program memory with 2 ns of access time is chosen, as the access time is included in the first pipeline stage, the achieved performance is reduced to 50 MIPS - 1.05 Volt, about 10 mW/MIPS, about 100,000 MIPS/watt

The core "DFF+Scan" is a previous CoolRISC core designed with flip-flops [13]. The CoolRISC-DL "double latch" cores with or without special scan logic provide better performances.

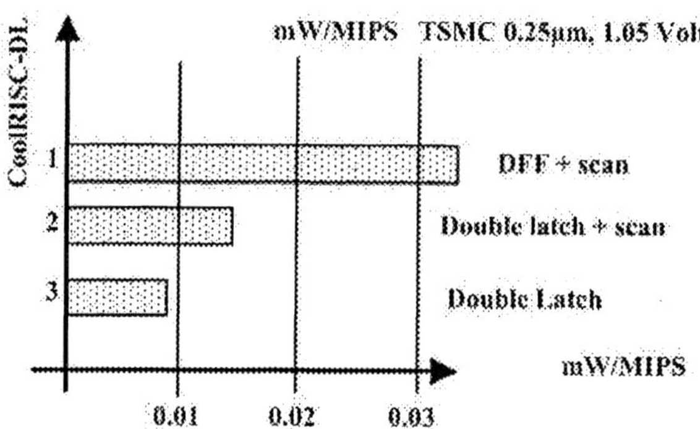

Figure 8.7.   Power consumption comparison of "soft" CoolRISC cores.

8.2.5   Asynchronous Design

If clock trees are more and more difficult to design and consume a large part of the total power, why not to remove completely the clock and design [13] asynchronous architectures. The 2001 SIA Roadmap [14] recognizes that, in 2007, asynchronous design will be used in many designs. If ϑ is the distance traveled by a signal in one clock cycle, due to higher frequencies and increasing interconnect delays, a chip will contain several time zones of sizes $\vartheta \times \vartheta$. Due also to the increasing die size, this number of time zones will grow very rapidly with the new technologies, up to 10,000 zones in 2012. To synchronize 10,000 time zones is a true asynchronous problem [14]. It is why asynchronous design is strongly required for chip architectures in the future [15].

Asynchronous architectures are often presented as being capable of reducing significantly the power consumption [15]. Looking at the results from many papers, it turns out that the largest power savings are obtained with circuits or applications that present a very irregular behavior [16]. An irregular behavior is, for instance, a processor that presents a quite tricky instruction set with multi-bytes instructions executed in a various number of cycles and phases. Or an application for which the controller has to very often stop and restart depending on the application. A very regular behavior is a 32-bit RISC core for which all instructions are always executed in one clock. Figure 8.8 illustrates this basic law. Basically, SoCs will present more an irregular behavior than a regular one.

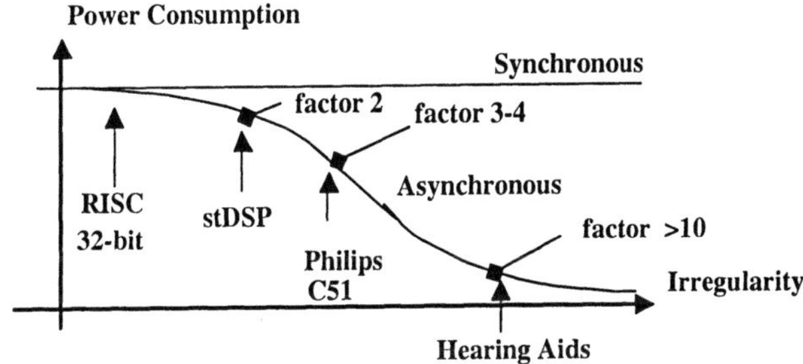

Figure 8.8. Power Consumption of Asynchronous versus Synchronous Architectures.

At the architecture level, power and V_{dd} management with behavior prediction of the user will be used extensively, as well as low-power communication protocols between the various processors on a single chip. These protocols have to be kept simple and will be asynchronous due to the fact that the various cores will be clocked (if not asynchronous cores) with many different frequencies.

8.3   Interconnect Delays

In deep submicron technologies, the wire delays are a very important issue, as they are larger and larger than the gate delays, as shown in the SIA Roadmap [14]. The gain of copper metal lines for interconnection is only one technology generation. So other design techniques have to be used, such as large metal lines and interconnection hierarchy.

8.3.1   The increased delays of wires

New generations of technologies can be modelized while using a reduction factor S (Fig. 8.9). It is a theoretical approach [17], [18] that provides relatively good trends. The gate capacitance $C_g = (\varepsilon \times W \times L)/D$ is reduced by a factor of S (ε=electric field). Transistor width W, length L and oxide thickness D are reduced by the same factor of S. The current $I_{ds} = 1/2 \times \beta \times (V_{gs} - V_t)^2$ is reduced by a factor of S, because $(V_{gs} - V_t)^2$ is S^2 times smaller, but $\beta = (\mu \times \varepsilon \times W)/(L \times D)$ is S times larger. The gate delay $T_d = (C \times V_{dd})/I_{ds}$ is reduced by a factor of S, as C, V_{dd} and I_{ds} are reduced by a factor of S.

Table 8.2.   Reduction Factor S.

Parameter	Reduction factor
Dimensions W, L, D	$1/S (D=$ oxide thickness)
Substrate doping $N sub$	S
Voltages V_{dd} V_T	$1/S$
Current I_{ds}	$1/S$
Gate Capacitance C_g	$1/S$
MOS Resistance	1
Gate Delay	$1/S$
Power $P = I * V_{dd}$	$1/S^2$
Density MOS/mm^2	S^2

It is very important to see that the voltages have been reduced by a factor of S. Therefore, the electric field in the transistor is kept constant. If not, with a smaller transistor, with a constant V_{dd}, the electric field is increased and reliability becomes a main issue. The advantage of a constant V_{dd}, as it was in the past, is the speed (the delay is reduced by a factor of S^2). But today, for these reliability reasons, V_{dd} has to be scaled down.

Using this theoretical approach, Figure 8.9 shows that interconnection delays are increased of a factor S^2 for a scaled technology, assuming that the length of the interconnection is kept constant. This assumption is right if existing chips are considered, as their sizes are not reduced in scaled technologies, due to the fact that they are more complex. So interconnection delays are a main issue. Figure 8.9 shows that RC delays are increased by a factor of S^2 due to the fact that the resistance R of an interconnection is increased by a factor of S^2 and the capacitance is unchanged. This can be easily understood if one considers the section $W \times H$ of a wire, which is reduced by a factor of S^2.

From Table 8.2 and Figure 8.9, one can note that gate delays are reduced by a factor of S and that interconnection delays are increased by a factor of S^2. The conclusion is faster and faster cells and slower and slower interconnection

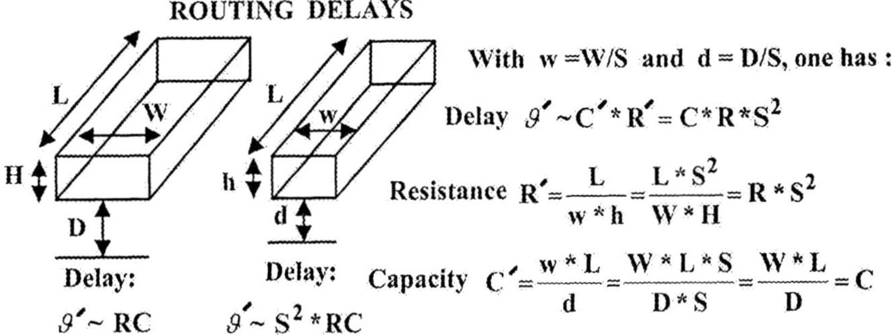

Figure 8.9. Interconnection Delays

lines! One has to consider the strength of the gate that drives the interconnect line. For fast gates (large buffers), with a short delay, the proportion of the interconnect delay to the total delay is increasing [19]. In other words, a large buffer is more sensitive to the interconnect delay.

Predictions [20] on wire resistances and RC delays are shown in Table 8.3, demonstrating that theory and reality are quite in accordance. However, other predictions show that interconnection delays could be increased less than S^2: for a 1mm wire, one has a resistance of 94 ohm (factor 1) in 0.5µm technology and 1340 ohm (factor 14.2 instead of 25) in a 0.10 µm technology.

Table 8.3. Wire resistance

Year	1995	1997	2001	2007
L_{min} (µm)	0.35	0.25	0.18	0.10
Metal Resistance (ohm/µm)	0.15	0.19	0.29	1.34
Wire Capacitance (fF/µm)	0.17	0.19	0.21	0.27
RC Wire Delay	1	1.4	2.4	14
S Factor	1	1.4	1.95	3.5
Theory S^2	1	1.9	3.8	12.2

Furthermore, if one considers a macrocell, a core or a synthesizable VHDL Intellectual Property block (IP), when integrated in a new process, the length of the interconnections will also be reduced by a factor of S. In this case, the wire resistance is not increased by a factor of S^2, but only S. Furthermore, wire capacitance as well as gate capacitance are reduced by a factor of S, so the RC delay is quite constant.

Figure 8.10 shows the decrease of gates delays and the increase of wire delays with advanced technologies [20], as explained in the SIA Roadmap [14]. As expected, gate delays are significantly reduced for deeper submicron technologies. For a reasonable wire length (about 43 mm), while using aluminium, the wire delay is increased very dramatically for 0.13 μm and 0.10 μm technologies. Copper and low k material with significantly reduced resistance can be used as well, with a smaller increase in resistance. However, it has to be noticed that gate delay and wire delay curves cross each other at 0.25 μm technology with aluminum wires, while this crossing occurs for a 0.18 μm technology if copper and low k are used. The conclusion is that the gain of copper metal lines for interconnection is only one technology generation.

8.3.2   New Materials for wires and dielectric

New materials can improve the interconnect layers, i.e. the use of gold, copper or silver [22] instead of aluminium. Copper reduces the resistance of about 40% [23]. IBM is using copper in its 0.20 μm CMOS-7S process (1998). It is quite difficult to avoid transistor copper contamination. It is a main challenge to introduce other material interconnection (copper, silver, etc). The complete fabrication process has to be adapted and reliability and yield issues are very important.

Another technology improvement is the introduction of a low-k insulator between metal lines that decreases the capacitance. Traditionally, SiO_2 was used, with a dielectric constant of about 4.0. IBM introduced in 2000 a low-k insulator SiLK with a dielectric constant of 3.0. The result from a AL/SiO_2 process to a Copper/SiLK is 37% delay reduction for a 200 mm M3 metal wire [23].

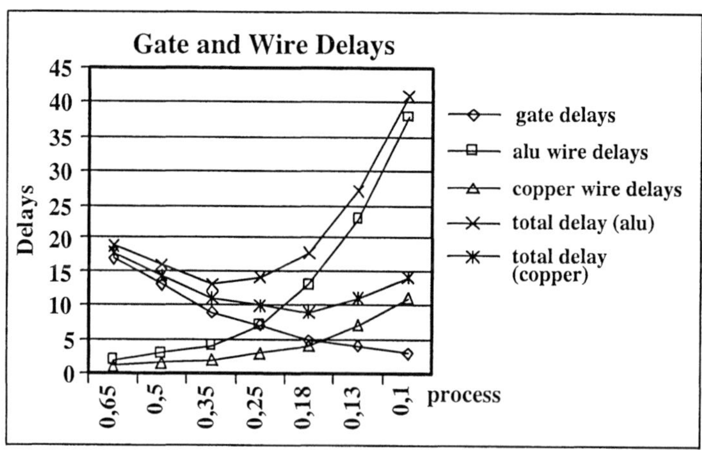

Figure 8.10.   SIA 2001: Gate delays versus interconnect delays.

8.3.3 Design methods taking into account interconnection delays

Designing small logic blocks, well-designed clock trees, buffers for long interconnections, are some of the methods that can be used. One has also to use a hierarchy of interconnect levels [24]. Short wires with a few hundred microns as orthogonal metal 1 and 2, sometimes metal 3 and 4 for medium length metal wires, and long wires in metal 5 and 6 (Figure 8.11). These long wires cannot shrink with the rest of the chip dimensions and are called "fat" wires. The resistance R of these fat wires is very low. For instance, in 0.18 μm technology, metal pitch is not scaled at the minimum 0.18 μm feature fro the PA-RISC microprocessor:

- 0.64 μm pitch for metal 1

- 0.93 μm for metal 2 and 3

- 1.60 μm for metal 4

- 2.56 μm for metal 5.

The 2001 Roadmap [14] stresses such a hierarchy, as copper short interconnections delays within basic cells also scale down with the technologies, as the basic cells are smaller and smaller. Therefore, the interconnection lengths are smaller as well as the load capacitances. Consequently, the interconnection delay problem is mainly at the upper level with long-range interconnections. The number of metal levels will increase up to 11 in 2016. Even with quite "fat" wires for these long-range interconnections, the problem still remains.

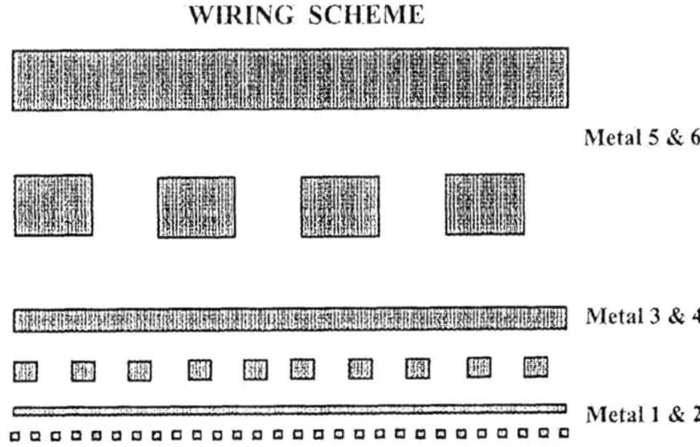

Figure 8.11. Hierarchy of interconnect levels

Figure 8.12 shows a photograph of metal lines in a 4 metal layer technology. The vias (contacts between metal lines) can be seen quite easily. If Figure 8.12

shows metal line widths that are larger than the height, it will be not the case in 0.10 μm. The height of the six metal levels will quite large compared to the width, in such a way to get a reasonable section $W \times H$ not too resistive. However, crosstalk will be increased (§8.3.4).

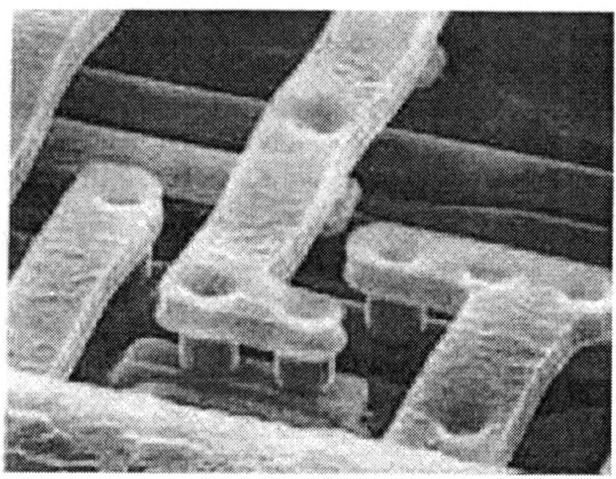

Figure 8.12. Photograph of metal lines in a 4 metal layer technology

Figure 8.13 shows a cut of an integrated circuit in 0.18 μm technology. One can see the 6 metal layers with quite large 5 and 6 metal lines on the top. Quite impressive, a transistor is so small compared to metal interconnections that it can be easily understood that more and more, delay problems come from interconnections and not from transistors.

One has to note that the delay propagation of signals in one clock period becomes so large that less and less logic gates are reached in this clock period. It means that the delay propagation of a signal from one side to the other of a single chip is much larger than one clock period (very soon tens of clock periods). It is why the SIA Roadmap predicts that asynchronous design will be more and more important by defining timing zones in which delay propagation is less than one clock period (§8.2.5).

8.3.4   Crosstalk

Shrinking feature sizes implies a large increase in crosstalk between closely spaced wires (Figure 8.14). Furthermore, as the wire section has to be as large as possible to reduce the wire resistance, the height of the wire is already bigger than the width. As a result, the coupling capacitance and consequently crosstalk is significantly increased. The increasing number of metal layers also decreases

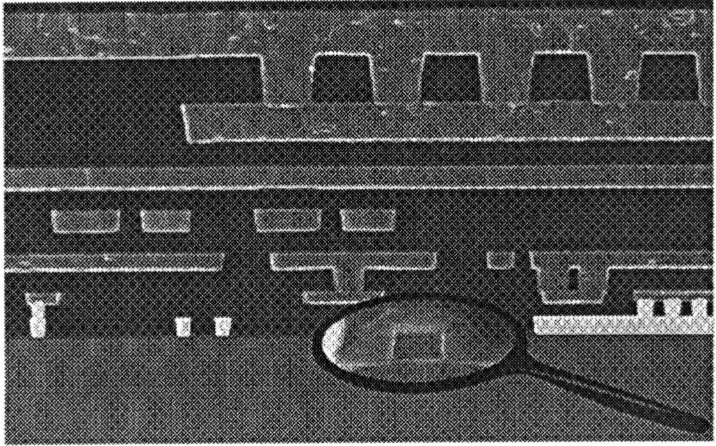

Figure 8.13. Copper wires and very small transistors.

the capacitance to the ground and consequently increases the relative coupling capacitance [25].

Crosstalk between parallel metal wires, mainly in memories and busses, can result in incorrect switching of an adjacent victim wire due to the switching of an aggressor wire. Delays can also be affected if the 0 to 1 switching of an aggressor slows down the 1 to 0 switching of a victim line.

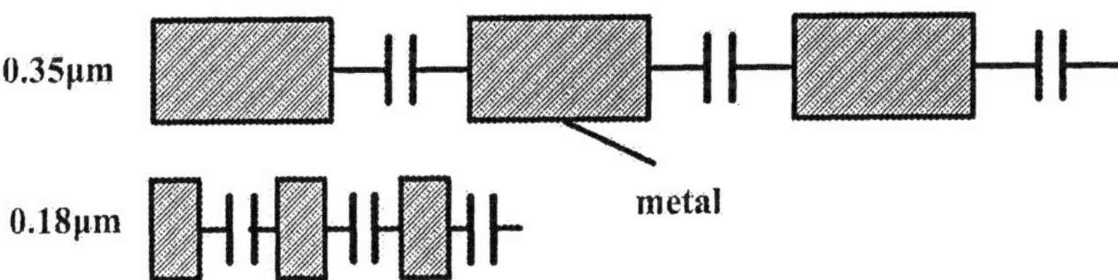

Figure 8.14. Crosstalk Effect.

In 0.13 and 0.10 μm technologies, crosstalk will be a dramatic problem if the tools to analyze it adequately as well as the design tools to cope with this problem (place and route tools) are not available. Some companies acknowledge that some chips in very deep submicron never work due to crosstalk. Today, one can use only limited tools that are capable of avoiding parallel metal lines on a too long distance. These tools consider only one victim node and one aggressor node, which is far from reality in which several aggressors and several victims node have to be considered. However, on average, the situation with one aggressor and several victim nodes is less dramatic than the former situation, as the aggressor has to share its crosstalk effect over several victims [25]. Considering several aggressor nodes could be more dangerous, but on average several aggressor nodes do not switch at the same time in the same direction.

There are several ways to reduce the crosstalk:

- very large drivers or repeaters to be less sensitive to aggressor nodes

- to space the metal wires if they run parallel on a too long distance

- routers that are conscious of the crosstalk effect and that are capable of twisting wires - reduction of the height of the metal wires (but results in an increased R)

- use of an intra-metal dielectric with a low k (to reduce C)

- copper metal lines (to reduce R), so the metal height can be reduced

- to design larger metal lines that increase the ground capacitance and consequently decreases the relative coupling capacitance

- to work at very low V_{dd}, as the voltage induced by crosstalk on victim nodes is depending on V_{dd}

One can notice that some of these techniques do increase the power consumption (larger buffers, larger metal lines, spaced metal lines), but some other techniques can also reduce power (lower V_{dd}, low k, twisted wires).

8.4   Low-Power Optimization for Layout Design

Twenty-five years ago, a 1000 MOS integrated circuit was considered as a very big circuit. Development time was over one year. The circuit was designed, "simulated", lay-out and verified by hand. Today, many CAD tools allow the designer to design a digital chip without considering layout issues. However, one has to notice that the layout of standard cells and memories (also analog circuits) are always designed by hand and that layout of several parts of very high performance microprocessors are also designed by hand. Hand layout is always more performing than automatic layout and is still performed by specialists.

Today, the layout work has been divided in two very different tasks. For digital logic blocks, the designer has to use CAD tools (VHDL, logic synthesizers and place & route tools) that produce very large pieces of layout for any logic block. These CAD tools are based on standard cell libraries that has been designed by specialists and very well characterized in a given technology. The second task in layout design is precisely the design of such standard cell libraries as well as memories (ROM, RAM). This work is still performed by hand by layout specialists, and deep submicron technologies implies a much more complex work.

8.4.1 Masks and Layout

The design of layout masks (or layout) consists in defining all the geometries that define an integrated circuit. A layout is therefore a colored drawing with a very large number of rectangles and polygons. This layout is then used to fabricate the masks that are used in the fabrication process of the integrated circuit itself. Layout specialists have to master the various layout rules that define the minimum distances between geometries. Productivities in manual layout design are about 4 to 6 transistors per day and per operator. The technological process is a sequence of steps. Each major step requires a given mask that defines the geometries of the corresponding layer.

8.4.2 Industrial Layout Rules

The fabrication process of integrated circuits has to align very precisely the successive masks on the chip itself. There are several alignment figures on the chip that are used for this alignment, but some misalignment is unavoidable. There are also some imperfections due to the process itself. It is therefore necessary to design the layout with these problems in mind. It is why any foundry provides layout rules for its fabrication process that take into account all these imperfections as well as the performances of the production machines. The goal of industrial layout rules is to have the better packing density as well as the best yield (two contradictory goals). It explains why layout rules are more or less different from foundry to foundry even for the common 0.25 μm process. It results that it is not possible to use the layout for a given process for another process. Furthermore, the conversion process from one layout to another is tricky.

Figure 8.15 shows a single MOS transistor designed in a very old micronic process in which one can still see transistors. One can see the diffusion defining a transistor with a polysilicon gate in the middle as well as two contacts and associated metal lines at both extremities. Figure 8.16 shows the layout of the nMOS transistor shown in Figure 8.15. This example requires the application of 22 over 32 layout rules of this very old micronic process (including rules to well and to doping masks not shown on Figure 8.16). The diffusion rectangle is crossed by the poly-silicon gate that defines a transistor of width W and length L.

The poly-silicon gate overlap is quite large. In case of misalignment between diffusion and poly-silicon masks, it is therefore impossible that the transistor diffusion will not be completely covered by the transistor gate. A direct conducting path (short-circuit) between drain and source of the transistor could therefore be avoided Drain and source of the transistor are connected to metal wires through contacts as well as the poly-silicon gate. Metal overlaps around

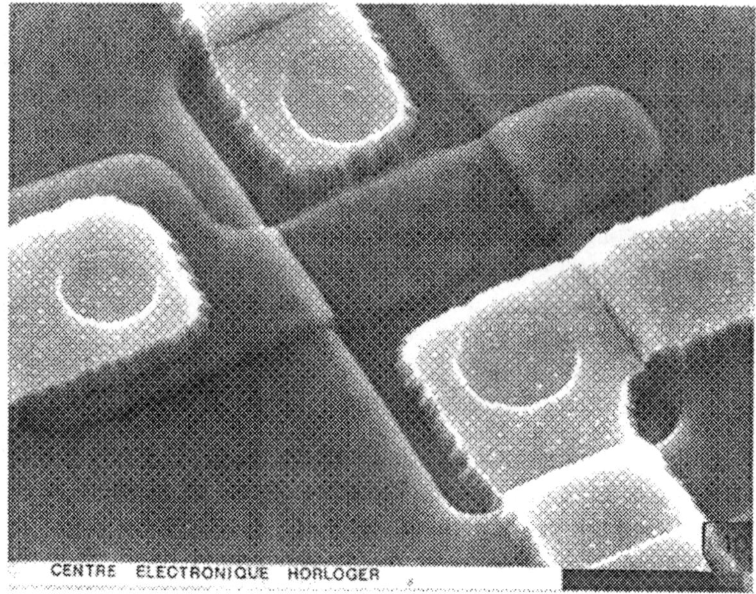

Figure 8.15.   Photograph of a MOS transistor.

contacts are different depending on the direction, in order to increase the layout density.

8.4.3   Layout Optimization for Low-Power

At the layout level (the lowest level in the design abstraction), power reduction is quite difficult to achieve, contrary to the high levels [30]. A low-power standard library generally achieves a factor of 2 of power reduction compared to a general library. The design of low-power circuits at layout level is strongly related to the circuit level, i.e. detailed transistor schematics and transistor sizing. However, what is generally of interest in the capacitance reduction. The latter can be used simultaneously to reduce the power consumption and to increase the speed. Basically, this can be achieved by keeping as small and simple as possible the basic elements that have to be designed (circuits, logic cells, layouts) [31], [32], [33]. The capacitance reduction is very effective at the cell level, i.e. in the design of low-power standard cell libraries.

Among several logic styles, the most simple and robust is static logic [9]. It also fits to low-voltage standard libraries. Furthermore, a very good regularity of both cell schematics and layouts is provided by a static logic style called "branch-based logic" [9], [32], [33]. The logic cells are designed exclusively with branches as shown in Figure 8.17. In such cells, branches are made of

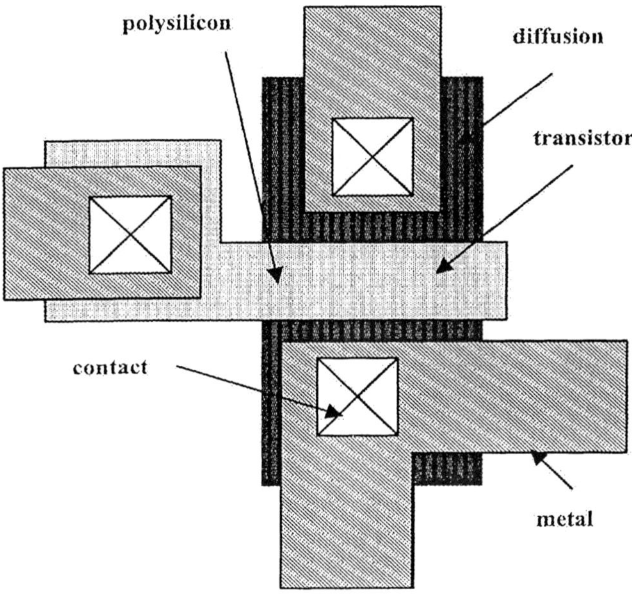

Figure 8.16.   Layout of a nMOS transistor

transistors in series (limited to 3 MOS) connected between a power line and a gate output. Therefore, the pMOS network of a logic gate cannot be the topological dual of the nMOS network. As shown in Figure 8.17, a particular simplification method using Karnaugh maps [34] can produce Boolean sums of products for both networks. "0" cubes are chosen for the nMOS network and "1" cubes of the same logic function for the pMOS network (each variable in the resulting expression has to be inverted, as a pMOS is on with a "0" on its gate). Only three pMOS branches are necessary for the branch-based logic gate of Figure 8.17, although the pMOS network of the original gate presents four paths between Vdd and the gate output. Any combinatorial and sequential cell can be designed with branches.

An obvious drawback of this technique is the increasing number of transistors as shown in Figure 8.17 for the nMOS network. However, one can note that the branch-based pMOS network contains the same number of transistors. However, this problem is not so serious as a standard cell library does not provide so many complex gates. The most used cells are simple gates, flip-flops, latches and multiplexers. Fortunately, as explained in §8.4.4, the new CSEM library contains only a very limited number of very simple cells.

As shown in Figure 8.18, the symbolic layout of the non branch-based P-ch network contains two supplementary contacts with two drain parasitic capaci-

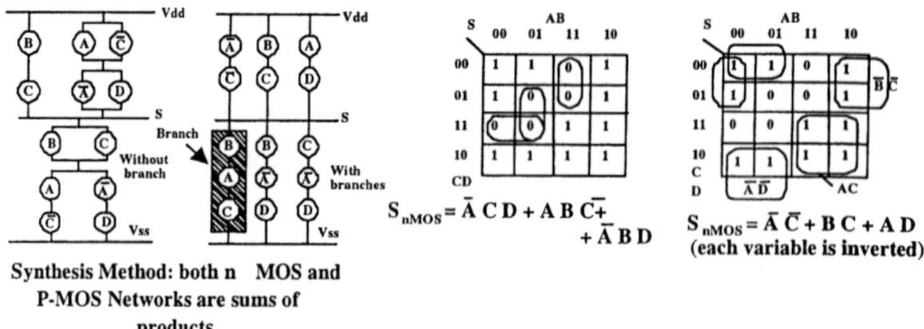

Figure 8.17.   Branch Based Logic.

tances that can be removed in the more compact branch-based implementation, reducing the parasitic capacitance. Figure 8.18 describes also a very regular geometrical branch-based layout of three branches. It provides no diffusion interruption, common drain of two branches, a minimal number of contacts and few metal connections [35]. It is not the case with serial-parallel and parallel-serial transistor configurations, as shown in Figure 8.19, in which the branch-based implementation contains one more transistor but a more compact layout with less contacts. This branch-based style has been used successfully for adders in Silicon-on-Insulator (SOI) technologies [36].

8.4.4   Low-power Standard Cell Libraries

To compile VHDL models into low power implementations, a very low-power standard cell library is required. It is sometimes assumed that a library with 300 to 500 cells is better than a library with a very limited number of cells. However, experience with a library based on a limited set of standard cells [37], [38], [39], shows that this approach provides many advantages. The number of functions into the new library has been reduced to 22 and the number of layouts to 92. It can be seen that the ratio between the number of layouts and the number of functions is larger (92/22=4.2 instead of 220/60=3.6 for the previous library). It means that the number of cell sizing and buffering is larger. For speed and power optimization achieved by the logic synthesizer, this enlarged ratio is beneficial.

It seems obvious that the logic synthesizer could do a better job if the number of cells in the library is large. With a larger choice, one can think that a better solution could be provided. But this is not the case. Experiments of Table 8.4 and 8.5 show that the delay of some operators is significantly reduced with the new library resulting in a quite small increase in silicon area (Table 8.4) and that the silicon area is reduced at the same speed with the new library (Figure 8.5). These results show that the logic synthesizer is more efficient because it

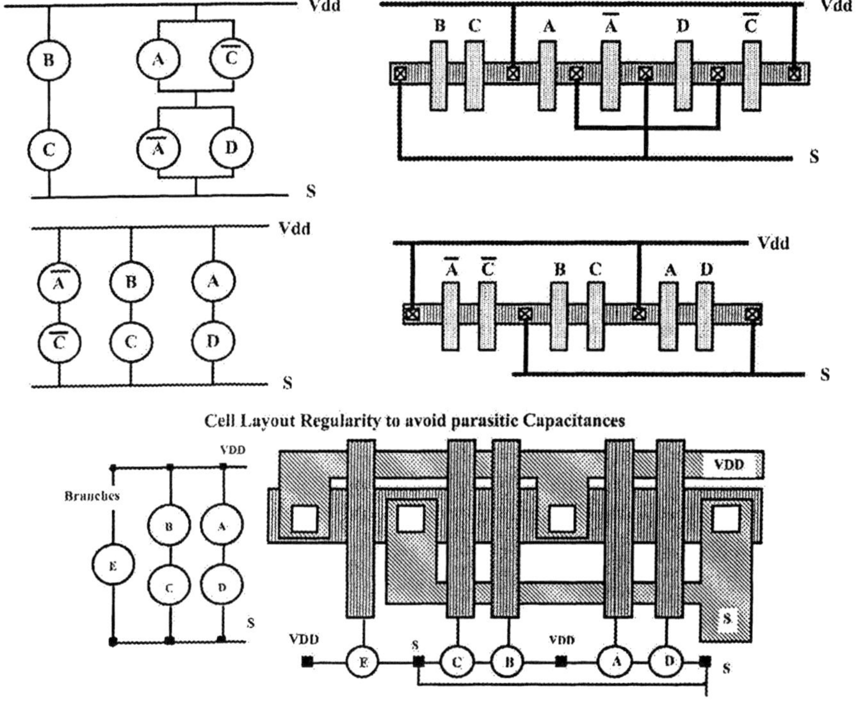

Figure 8.18. Branch-Based Layouts.

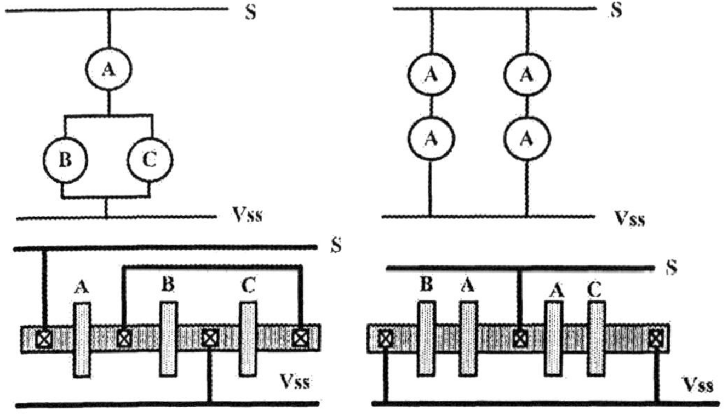

Figure 8.19. AOI & OAI Gates.

has a limited set of well-chosen cells and cell sizing adapted to the considered logic synthesizer. With significantly less cells than conventional libraries, the synthesizer is not lost in some optimizations due to a too large choice of cells.

The applied strategy for the design of the library was the following:

- to have only very fast cells, i.e. to remove all the cells with 3 pMOS transistors in series and to have a very limited number of cells with 2 pMOS transistors in series.

- to increase the number of cell sizing for the same function. However, it is not a simple increase from, for instance, sizing D1 (small transistors), D2 and D3 (medium-sized transistors) to D1, D2, D3, D4 and D5 (very large transistors). The cell sizing performed takes into account how the synthesizer uses the considered cells.

- to consider the combination of a given cell and of a buffer (or sized simple gate) to replace complex gates.

Such a strategy has to be checked through many experiments. The choice of the 22 functions was performed with a large number of experiments with and without a specific cell, and then the decision was made to insert or not this cell in the library. Similar experiments were performed with various sizing and buffering of the same cell. At the end, only 22 functions and 92 layouts were kept in the new library.

Table 8.4.   Delay Comparison (synthesis for maximum speed, 0.5µm process

	Old Library		New Library	
	Delay [ns]	μm^2	Delay [ns]	μm^2
32-bit multiply	16.4	907 K	12.1	999K
fp adder	27.7	510K	21.1	548 K
CoolRISC ALU	10.8	140K	7.7	170K

Table 8.5.   Silicon Area Comparison (synthesis for a given delay, 0.5µm process

	Old Library		New Library	
	Delay [ns]	μm^2	Delay [ns]	μm^2
32-bit multiply	17.1	868K	17.0	830K
fp adder	28.1	484K	28.0	472K
CoolRISC ALU	11.0	139K	11.0	118K

Furthermore, as the number of layouts is drastically reduced, it takes less time to design a new library for a more advanced process. A lot of time can also be saved for the library characterization, which is known to be often the most time consuming activity in library design. To reduce the number of layouts from 220 to 92 is definitively a great advantage.

It will also be a crucial point in future libraries for which more versions of the same function will be required while considering static power problems. A same function could be realized, for instance, with low or high V_{th} for double V_{th} technologies, or with several cells such as a generic cell with typical V_{th}, a low-power cell with high V_{th} and a fast cell with low V_{th}.

8.4.5 Automatic Layout Synthesis

Another approach to Standard Cell is a layout synthesis approach [40]. This approach consists in resynthesizing the layout each time a given cell is required. The major benefit is to allow the designer to choose exactly the complex cell that is needed and to size the transistors as required by the local constraints, such as timing specifications, interconnect lengths and loads. As the layout is automatically synthesized, a layout style has to be chosen, such as Gate Matrix style. The number of cells (called virtual) is therefore unlimited, and this approach is at the opposite from the cell library described in §8.4.4. As a result, the total number of transistors is reduced compared to a Standard Cell Library, but the density in transistors/mm^2 is still inferior. Furthermore, this method requires a on the fly cell characterization using very good timing and power models.

8.4.6 Transistor Sizing

Transistor sizing is very important for low-power cells. Over-sizing is generally the result provided by automatic sizing procedures. Branch-based logic with a sizing at the branch level (not at the gate level) seems better than other techniques. Figure 8.20 shows a simple NOR3 gate in which the three pMOS transistors in series are as large as possible, only limited by the standard cell height. If the three parallel nMOS transistors are also largely sized, they add parasitic capacitance to the gate output. The result is a slower critical rise time (3 pMOS in series) and a faster fall time.

Transistor sizing could provide bad results depending on the chosen assumptions. If the gate output capacitance is considered as a (large) fixed load, the result will be very large transistors and large input capacitances. However, in a design core, most of these gates will be loaded by other similar gates and therefore not by a fixed load. Reducing the transistor sizes will result in decreasing input capacitances that are also load capacitances for other gates. Such a sizing procedure will save much power while maintaining the speed. On the other hand, large fixed loads have to be driven by buffers.

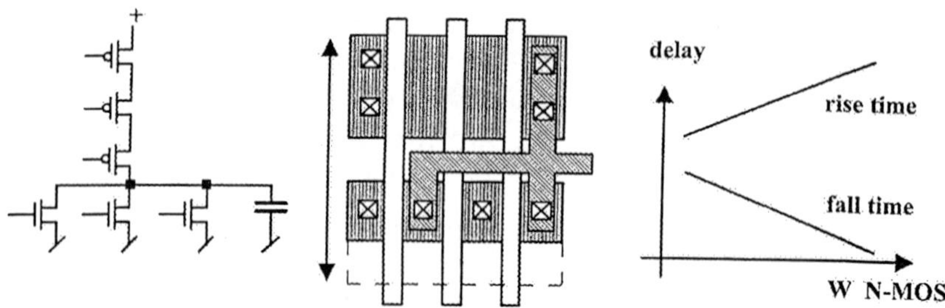

Figure 8.20.   Transistor Sizing.

Transistor sizing has also to take into account short circuit dissipation [12]. Minimal transistor sizing, for instance, resulting in reduced capacitive dynamic power, could increase the short-circuit dissipation due to slow slope of input signals. To reduce the short-circuit power, the input and output slopes have to be comparable, and at the gate level, buffers are generally introduced to avoid slow slopes of input signals.

8.4.7   Layout in Deep Submicron Technologies

The layout design of standard cells in very deep submicron technologies is not so different than in micronic technologies. It is mainly due to the fact that standard cells layouts are using one (sometimes two) metal levels even if the technology provides 6 to 7 metal layers (that are very useful for the routing). Regarding low-power, the same goal has to be achieved, i.e. to reduce the parasitic capacitances, to avoid to have too many contacts in series for internal interconnections, a careful examination of the internal layout topology to avoid misplaced elements that increase the cell area and parasitic capacitances. Mainly rules of thumb are applied to get good layouts.

However, as mentioned above, the transistor area becomes relatively smaller than interconnections and contacts that have large overlaps. It gives a strong push to have transistor schematics that reduce the number of contacts (as mentioned in §8.4.3). A main difference in deep submicron technologies is that cells can be connected inside the cells and not only on the top and bottom sides. Due to the large number of metal layers, cells can be abutted not only in rows, but the rows themselves are abutted, the V_{ss} and V_{dd} supply lines are shared between two rows. In that way, there no more routing channels and the density is significantly increased. Furthermore, it gives more freedom to design not too small cells, as some room is necessary for internal connectors and too small cells could imply some congestion for the cell router itself.

Figure 8.21 shows a layout in 0.25 μm provided by the PC tool Microwind largely used for education [41]. Three inverters in series are shown in a single

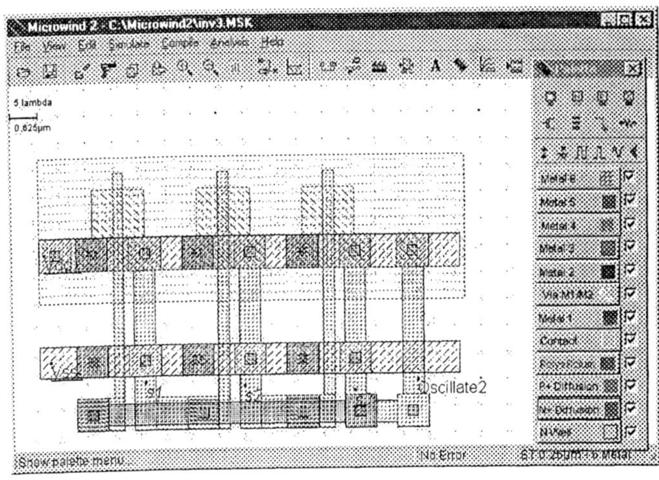

Figure 8.21.   Layout in 0.25 μm of a standard cell (Microwind).

layout. This tool provides design rules and transistor models down to 0.1 μm [42]. Figure 8.22 shows a 1-bit adder designed in 0.18 μm for the CSEM standard cell library. Only one metal level (M1) is used inside the cell.

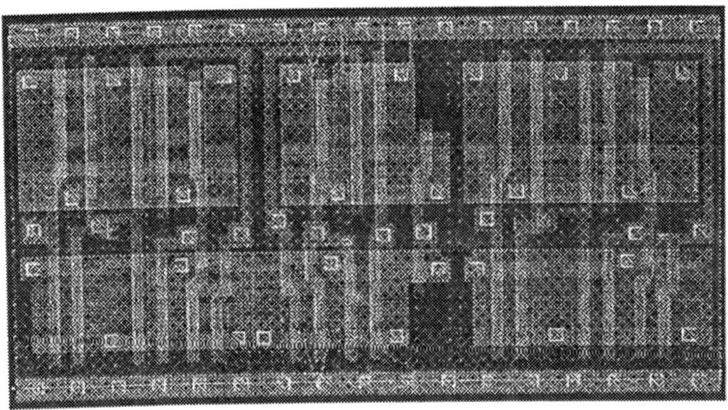

Figure 8.22.   Layout in 0.18 μm of a 1-bit adder (CSEM).

References

[1] D. A. Patterson, J- L. Hennessy, "Computer Organization & Design, The Hardware/Software Interface", Morgan Kaufmann Publishers Inc. San Francisco, 1994.

[2] K. D. Wagner, "Clock System Design" IEEE Design & Test of Computers, October 1988, pp. 9-27.

[3] D. Harris, M. Horowitz, "Skew-Tolerant Domino Circuits", IEEE JSSC Vol. 32 No 11, November 1997, pp. 1702-1711.

[4] R-S Tsay, "An Exact Zero-Skew Clock Routing Algorithm" IEEE Trans CAD, Vol 12, No 2, February 1993, pp. 242-249.

[5] C. Arm, J-M.Masgonty, C. Piguet, "Double-Latch Clocking Scheme for Low-Power I.P. Cores", PATMOS 2000, Goettingen, Germany, September 13-15, 2000.

[6] ISSCC 2000, February 2000, San Francisco, USA

[7] W. J. Bowhill et al. "A 300MHz 64b Quad-Issue CMOS RISC Microprocessor", Proc. ISSCC'95, February 15-17, 1995, San Francisco, CA, USA, pp.182-183.

[8] T. G. Noll, E. De Man, "Pushing the Performances Limits due to Power Dissipation of Future ULSI Chips", IEEE Intl. Symp. on Circuits & Systems, ISCAS'92, San Diego, May 1992, pp. 1652-1655.

[9] C. Piguet, et al. "Low-Power Design of a Standard Cell Library", Low Voltage - Low Power Workshop during ESSCIRC'95, Lille, Sept. 22, 1995, France.

[10] D. Dobberpuhl, "The Design of a High Performance Low Power Microprocessor", Invited Talk at ESSCIRC 1996, September 17-19, 1996, Neuchatel, Switzerland.

[11] C. Piguet et al. "Logic Design for Low-Voltage/Low-Power CMOS Circuits", 1995 International Symposium on Low Power Design, Dana Point, CA, USA, April 23-26, 1995.

[12] S. Turgis, "Conception Faible Puissance: Definition d'un Macro-Modele de Puissance Interne dans les Structures CMOS submicroniques", These Universite de Montpellier II, France, 30 Septembre 1996.

[13] C. Piguet et al,"Low-Power Design of 8-bit Embedded CoolRISC Microcontroller Cores", IEEE JSSC, Vol. 32, No 7, July 1997, pp. 1067-1078

[14] Sematech 2001 SIA roadmap. http://public.itrs.net/

[15] M. Renaudin. Asynchronous circuits and systems: a promising design alternative. In MIGAS Summer School on Microelectronics for Telecommunications: Managing High Complexity and Mobility, Autrans, France, June 28 - July 4, 2000.

[16] C. Piguet, "Low-Power Issues for SoCs",in proceedings of DATE 2001, Munich, 13-16 March 2001, pp. 488-490.

[17] H. B. Bakogu, "Circuits, Interconnections, and Packaging for VLSI", Addison-Wesley, 1990.

[18] C.Svensson, "High-Speed Single-Phase Dynamic Logic, Interconnects and Related Problems", CMOS & BiCMOS IC Design'96, EPFL, Lausanne, Switzerland, August 19-September 6, 1996.

[19] J. M. Daga, "Modelisation des Performances Temporelles des Circuits CMOS Submicroniques au niveau Porte Logique", These Universite Montpellier II, Septembre 1997.

[20] W. Rhines, Mentor Graphics, Keynote Address on the next century design environment at CICC'96.

[21] T. Nanya, "Asynchronous VLSI System Design" Research Center for Advanced Science and Technology, University of Tokyo, 1997.

[22] L. Gwennap, "IC Makers Confront RC Limitations", Microprocessor Report, August 4, 1997, pp. 14-18.

[23] K. Diefendorff, "IBM Paving the Way to 0.10 Micron", Microprocessor Report, May 01, 2000.

[24] B. Davari et al. "CMOS Scaling for High Performance and Low Power - The Next Ten Years", Proc. IEEE Vol. 83, No 4, April 1995, pp. 595-606.

[25] F. Ilponse, "Analyse du bruit du aux couplages capacitifs dans les circuits integres numeriques fortement submicroniques », Ph. D. Thesis, University Paris 6, February 28, 2002.

[26] M. Ishikawa et al. "Optically Interconnected Parallel Computing Systems", Computer Vol 31, No 2, February 1998, pp. 61-68.

[27] M.Koyanagi et al. "Multi-Chip Module with Optical Interconnection for Parallel Processor System", Proc. ISSCC'98, San Francisco, February 5-7, 1998, pp. 92-93

[28] P. K. Vasudev et al. "Si-ULSI with a Scaled-Down Future", Circuits and Devices, March 1998, pp. 19-29.

[29] J. A. Davis, J. D. Meindl, "Is Interconnect the Weak Link?", Circuits and Devices, March 1998, pp. 30-36.

[30] C. Piguet, "Low-Power Design of Systems on Chip", Chapter 18, pp. 18-1 to 18-18, in "The Computer Engineering Handbook", Edited by V. Oklobdzija, CRC Press, 2002.

[31] A. P. Chandrakasan, S.Sheng, R. W. Brodersen, "Low-Power CMOS Digital Design" IEEE J. of Solid-State Circuits, Vol. 27, No 4, April 1992, pp. 473-484.

[32] J-M. Masgonty et al. "Technology- and Power-Supply-Independent Cell Library" IEEE CICC'91, May 12-15, 1991, San Diego, CA, USA, Conf. 25.5

[33] J-M. Masgonty et al. "Branch-Based Digital Cell Libraries" EURO-ASIC'91, 27-31 May 1991, Palais des Congres, Paris.

[34] C. Piguet et al. "Metal-Oriented Layout Structure for CMOS Logic" JSSC Vol SC-19, No 3 June 1984, pp. 425-436.

[35] M. Robert et al. "Evaluation of VLSI Layout Style Implementation for Efficiency" Proc. EURO ASIC'91, May 27-31, 1991, Paris, pp. 362-365.

[36] A. Neve, D.Flandre, "Design of a Branch-Based Carry Select Adder IP portable in 0.25 µm Bulk and SOI CMOS Technologies", Proc. VLSI-SOC 2001, 11th IFIP International Conf. on VLSI, Le Corum, Montpellier, December 3-5, 2001, pp. 269-272.

[37] C.Piguet, " Low-Power Low-Voltage Library Cells and Memories ", ICECS 2001, Malte, September 3-5, 2001.

[38] Jean-Marc Masgonty, Stefan Cserveny, Claude Arm, Pierre-David Pfister, Christian Piguet, "Low-Power Low-Voltage Standard Library Cells with a Limited Number of Cells", PATMOS 2001, September 26-28, 2001, Yverdon, Switzerland.

[39] J-M. Masgonty et al. "Low-Power Low-Voltage Standard Cell and Gate-Array Libraries with a Limited Number of Cells", Proc. VLSI-SOC 2001, 11th IFIP International Conf. on VLSI Le Corum, Montpellier, December 3-5, 2001, pp. 279-284.

[40] A.Landrault et al. "A Design Approach of Reusable Cores based on Transistor Level Synthesis", Proc. VLSI-SOC 2001, 11th IFIP International Conf. on VLSI Le Corum, Montpellier, December 3-5, 2001, pp. 298-303.

[41] http://intrage.insa-tlse.fr/ etienne/Microwind/index.html

[42] E. Sicard, "An Illustration of 0.1 µm CMOS Layout on PC", EWME 2002, 4th European Workshop on Microelectronics Education, May 23-24, 2002 - "Parador de Baiona", Mancomunidad de Vigo, Spain.

Chapter 9

LOGIC LEVEL POWER ESTIMATION

George Theodoridis

Costas Goutis

University of Patras, Patras, Greece
{ theodor,goutis } @ee.upatras.gr

Abstract   Low-power design is based on two axes, the development of design techniques aiming at reducing the power consumption of a circuit, and the development of methods for estimating the dissipated power at every design step. In this chapter methods for estimating the power consumption at logic (or gate) level are presented and discussed. The aim of the chapter is to describe the main issues of the logic level power estimation and to present the basic concepts and principles of the most representative published methods.

Keywords: Power estimation, probabilistic power estimation, simulation-based power estimation, Markov models.

9.1   Introduction

Due to wide spread of portable electronic devices and the evolution of microelectronic technology, power dissipation has become a critical parameter in VLSI design in recent years. In the former case, the major requirement is to extend the operation time between two successive recharges of the battery-operated device. Considering that the progress of the battery technology is not so promising as the applications demand, the design of low-power devices is necessary to meet the above requirement. In the second case, thanks to the rapid development of microelectronic technology, the number of transistors and the operating frequency of the integrated circuits have been increased significantly. This results into an increase of the consumed energy and chip's heat, which in turn cause serious reliability problems. Taking into account the cost of

169

packaging and cooling systems, the design and implementation of low-power electronic systems is needed to cope with above problem.

Low-power design is taken place at all design levels of the design flow starting from system level and reaching down to layout level. By low-power design, it is meant that the application of techniques aiming at the reduction of the power consumption of a circuit, as well as the development of methods to estimate the dissipated power at every design step. The use of power estimators is very important, since it allows checking if the power specifications are met, avoiding the cost of the redesign process. Thus, a plethora of power estimators applied at all levels of the design flow have been presented in last years. Here, we will concentrate our attention on the power estimation methods applied at the gate level only.

The chapter is organized as follows: in Section 9.2 the power estimation problem is formulated, while the basic knowledge is given in Section 9.3. In Section 9.4 the estimation methods are categorized into two general groups and their advantages and disadvantages are discussed. The simulation-based power estimation methods are presented in Section 9.5, while in Section 9.6 the probabilistic methods are discussed. Finally, the conclusions are given in Section 9.7.

9.2   Problem Formulation

9.2.1   Sources of Power Dissipation

The total power dissipation of a CMOS circuit is the sum of the three components, which are: the dynamic power dissipation, P_{dyn}, the power consumption due to short circuit currents, P_{sc}, and the power dissipation due to leakage current, $P_{leakage}$. The dynamic power dissipation is by far the most dominant component, since its contribution ranges between 70%-80% of the total power consumption for non-submicron technologies [1], [2], [3]. Thus, we focus on methods, which estimate the dynamic power consumption.

Dynamic power dissipation is caused due to charging and discharging of the circuit nodes and is expressed for a node i by the following formula:

$$P_{dyn,i} = \frac{1}{2} C_i V_{dd}^2 f E_i(sw) \tag{9.1}$$

where C_i and $E_i(sw)$ are the capacitance and the switching activity (i.e. the average number of transitions per clock cycle) of node i, respectively, while V_{dd} and f are the supply voltage and clock frequency of the system.

9.2.2   Problem Statement

Considering that we are dealing with circuits at gate level, both the supply voltage and clock frequency have already been determined at previous design steps, while the exact capacitance value will be specified in following design steps. Thus, the power consumption in gate level can be estimated by calculating the switching activity, $E_i(sw)$, for each circuit node. A more accurate estimation can be achieved after the mapping of the circuit to a specific technology. Then, the gates capacitance is known and the only undetermined component is the interconnection capacitance.

In addition, power dissipation is strongly dependent on the applied input vectors. Each applied input vector acts as stimuli on the circuit causing the internal nodes to perform transitions according to the functionality and interconnection of the circuit gates. The same circuit under different input vectors (i.e. different stimulus) may have totally different behavior (i.e. different responses) in terms of switching activity and power consumption consequently. Hence, the applied input vectors must be taken into account.

Also, by power consumption estimation it is meant the computation of the average and worst case power consumption. Average power estimation is useful to determine the average lifetime of the battery, the average heat of the die and the required supply voltage. Worst case power dissipation refers to instantaneous power dissipation caused by the application of a specific input. It causes temporary malfunctions and delay faults and in many cases permanent damage in the die. In this chapter we are dealing with the estimation of average power dissipation.

Since we study static CMOS circuits, two input vectors must be considered to determine if a circuit node perform a transition. In contrary, in case of dynamic CMOS circuits only one vector is needed, since each node is evaluated in a steady value during the precharge phase. Also, we assume that the time between two successive input vectors is enough allowing the circuit to reach in a steady state. Finally, full ideal step transitions are assumed when a node switches.

In conclusion, the problem of power estimation at gate level is stated as follows:

Problem Statement: *"Given the Boolean description or the post mapping gate netlist of a synchronous static CMOS circuit and taking into account the applied input vector sequence, estimate the average power dissipation of the circuit by calculating the average switching activity of each circuit node"*

Therefore, the problem of estimating the power dissipation is transformed into the problem of deriving the switching behavior of a digital signal. In the following sections the basic concepts, the terminology, and the factors that affect the behavior of a digital signal and have to be considered during the switching activity estimation are presented.

9.3 Background

In general, the behavior of a digital signal, x, is modeled as time homogeneous, strict stationary, Markov stochastic process [4], [5]. The logic value, $x(t)$, of a signal, x, at time instance, t, is the random variable of a Markov stochastic process having two states, s, with $s \in S = \{0, 1\}$.

Definition 1. The **signal probability**, $p_x^1 = p(x = 1)$, of a signal, x, is defined as the average fraction of clock period where the signal stays at high logic level.

Definition 2. The **transition probability**, p_x^{ij}, is defined as the probability the signal, x, to perform a transition in two successive clock cycles and is expressed by the following formula.

$$p_x^{ij} = p(x(T) = j \land x(T-1) = i) \text{ with } i, j \in S \tag{9.2}$$

Note that $p_x^1 = p_x^{01} + p_x^{11}$, $p_x^0 = p_x^{10} + p_x^{00}$, and $p_x^{01} = p_x^{10}$ [4], [5].

Definition 3. The **switching activity**, $E_x(sw)$, of signal, x, is defined as the sum of p_x^{01} and p_x^{10} transition probabilities.

$$E_x(sw) = p_x^{01} + p_x^{10} \tag{9.3}$$

The switching behavior of a node depends on the correlations among the circuit signals, the gate structure of the circuit, and the delays of the gates. In the following sections these factors are presented and their influence in the switching activity calculation is discussed.

9.3.1 Signal Correlations

In general the value of a signal, depends on its history and the values of the remaining signals of the circuit as well. A signal is:

a) temporally correlated if its value depends on the values that signal has taken in the past,

b) spatially correlated if it is dependent on the values of other signals, and

c) spatiotemporal correlated if it is dependent both on its own and other signals' history.

The impact of the above correlations is critical in the switching activity calculation and it is presented in the circuit of Fig. 9.1.

Three different input vector sequences, V_1, V_2, and V_3, are applied at the inputs x, y, and z. A zero-gate delay model is assumed. Also, it assumed that the bits of the input vectors are changed simultaneously. The vectors of sequence V_1 are come from a random generator, where there is not any kind of dependencies. The signals of sequence V_2 are the outputs of a 3-bit counter, where the spatiotemporal dependencies are strong. Finally, the signals of V_3 are collected at the outputs of a Linear Feedback Shift Register (LFSR), where

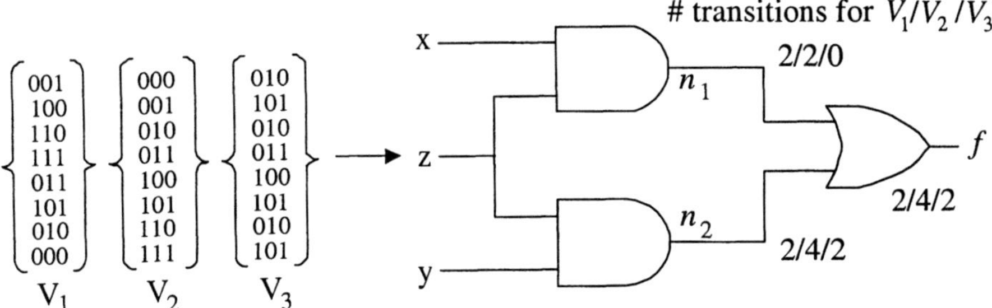

Figure 9.1. Impact of Signal Correlations at the Switching Activity [25]

the spatiotemporal correlations are weak. In Fig. 9.1 the number of transitions that perform the nodes n_1, n_2 and f are also presented.

When sequence V_1 is applied, the nodes n_1, n_2 and f perform 6 transitions, while in the case of V_2 application the number of transitions equals to10 (i.e. 66% increase). Finally, when the sequence V_3 is applied, the internal and output nodes perform 4 transitions resulting in a reduction of 33% compared with the corresponding number of transitions when sequence V_1 is applied. Hence, the same circuit under different applied input sequences has different behavior in terms of switching activity and power dissipation consequently. The reason of this phenomenon is the dependencies among the circuit signals.

The impact of all the types of correlations that appeared in logic circuits has been studied in [6]. Ignoring the spatiotemporal correlation at the primary inputs the error at the switching activity computation is 30% for the combinational circuits and 55% for sequential circuit. Also, if the structural dependencies are ignored the error is approximately 15%. Finally, ignoring the sequential correlation the error at the switching activity computation for mixed (combinational/sequential circuits) is 120%.

Definition 4. A set of signals, $(x_0(t), x_1(t), ..., x_n(t))$, are **uncorrelated** when their joint probability equals to the product of their individuals probabilities.

$$p\left(x_0(t) \cdot x_1(t) \cdot ... \cdot x_n(t)\right) = \prod_{i=1}^{n} p(x_i(t)) \quad \forall t \tag{9.4}$$

Definition 5. A signal, $x(t)$, **temporally correlated** when its probability, $p(x(t))$, in a time instance, t, depends on the values that the signal has been taken in the past.

$$p\left(x(t_0) \cdot x(t_1) \cdot ... \cdot x(t_n)\right) \neq \prod_{i=0}^{n} p(x(t_i)) \tag{9.5}$$

If there is not temporal correlation then $p_{ij}^x = p(x(t) = j)\, p\,(x(t - \delta) = i)$. The number of the time instances that are considered to evaluate the signal's value determines the order of correlation. In case of $n+1$ time instances are used, the correlation is of the n-th order.

Definition 6. A set of signals, $(x_0, \cdot x_1, ..., x_n)$, are **spatially correlated** when their joint probability is not equal to the product of their individual probabilities.

$$p(x_0 \cdot x_1 \cdot ... \cdot x_n) \neq \prod_{i=0}^{n} p(x_i) \tag{9.6}$$

Definition 7. Two or more signals, $(x_0(t), ..., x_i(t), ..., x_m(t))$, are **spatiotemporally correlated** when both the temporal and spatial correlation exist. Thus, the next formula holds.

$$p\left(x_0(t_0) \cdot ... \cdot x_m(t_0) \cdot ... \cdot x_0(t_n) \cdot ... \cdot x_m(t_n)\right) \neq \prod_{i=1}^{m} \prod_{j=1}^{n} p\left(x_i\left(t_j\right)\right)$$

$$\exists\, x_i\left(t_j\right) \in \{0.1\} \tag{9.7}$$

9.3.2   Structural Dependencies

Besides the previous dependencies there is an additional dependency that must be considered. It is introduced due to the reconvergent fanout regions.

Definition 8. Two paths of a circuit are **reconvergent** if they start at common node x and terminate at a node y in a successive circuit level. The paths are said to fanout at x and reconverge at y. The part of the circuit that is included between the reconvergent paths called **Reconvergent Fanout Region** (RFO)

A reconvergent fanout region is presented in Fig. 9.2.

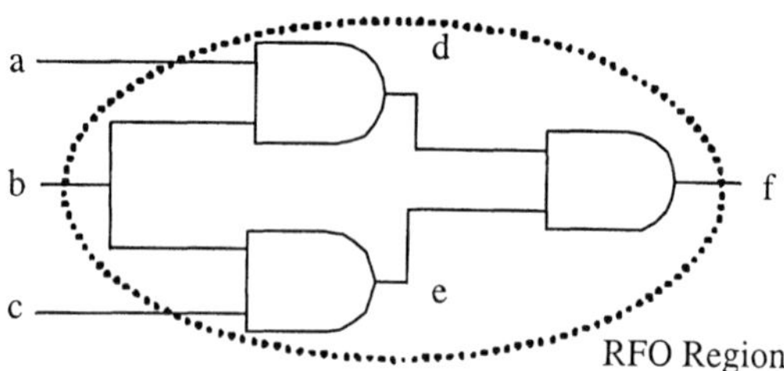

Figure 9.2.   Reconvergent Fanout

The RFO introduces additional dependencies among the circuit signals. Even though the primary inputs a, b, and c are independent, the signals d and e are correlated due to reconvergent fanout node b. This kind of correlation is reflected at the switching behavior of node f. Ignoring this dependency the calculation of signal and transition probability of node f is inaccurate.

The Boolean functions of nodes d, e, and f are $d = ab$, $e = bc$ and $f = de = abc$. Assuming that the primary inputs are independent we have:

$$p_f^1 = p(abc = 1) = p_a^1 p_b^1 p_c^1 \tag{9.8}$$

Ignoring the RFO effect the signals d and e are assumed uncorrelated; then we obtain:

$$p_f^1 = p(de = 1) = p(d = 1)p(e = 1) = p(ab = 1)\,p(bc = 1) = p_a^1 \,(p_b^1)^2\, p_c^1 \tag{9.9}$$

The signal probability calculation of (9.9) is inaccurate. The error occurs due to the assumption that the signals d and e are uncorrelated. However, the signals are structurally correlated that means $p(de = 1) \neq p(d = 1)\,p(e = 1)$. The problem of capturing the structural dependencies among the circuit signals is overcome by expressing node f in terms of structurally independent signals as in (9.8).

9.3.3 Sequential Correlations

The switching activity of a sequential circuit is the sum of the switching activity of the primary inputs, E_{in}, primary outputs, E_{out}, internal nodes of the combinational circuit, E_{int_nodes}, and the activity of the state lines, E_{state_lines}. However, the transition form one state to another one is uniquely determined by the *State Transition Graph* (STG) of the sequential circuit. This introduces an additional correlation among the state lines. This type of correlation is explained in Figure 9.3.

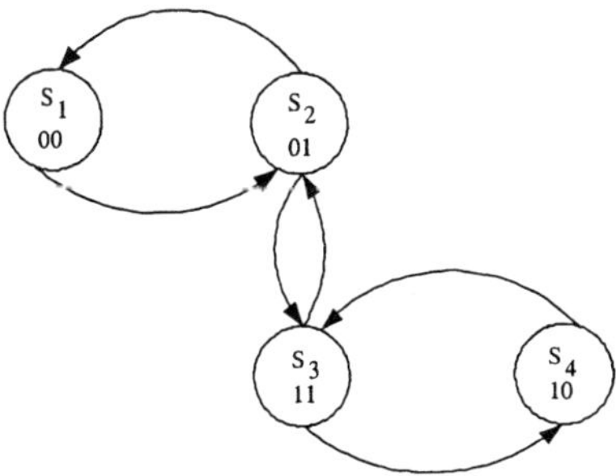

Figure 9.3. STG of a Sequential Circuit

Considering the structure of the above STG it is clear that there is not a direct transition form state S_1 to S_4, which means that the probability of appearing the combination $(00) \rightarrow (10)$ at the state lines is zero. In other words, the coding of the states and the structure of the STG introduce additional spatial and temporal dependencies at the state lines. Hence, this kind of dependency must be considered in the estimation of power consumption of sequential circuits.

9.3.4   Gate Delay Model

Another parameter that affects the switching activity of a circuit is the delay of the gates. When zero gate-delay model is assumed only the *functional transitions* at the gate output are considered. However, in real designs the delay of a gate is not zero resulting in the appearance of additional transitions called *spurious transitions* or *glitches*. Specifically, under a real gate-delay model the delays of the circuit paths are not equal. It means that the input signals of a gate may not arrive at the same time instance causing the gate output to perform spurious transitions. In the Fig. 9.4 the switching behavior of a circuit under zero and unit gate-delay model is illustrated, where a glitch at signal f is presented.

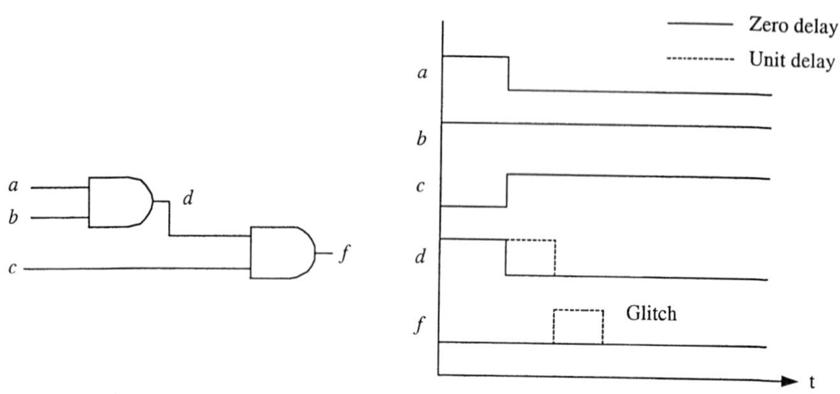

Figure 9.4.   Switching Activity Under Different Gate Delays

9.4   Classification of Power Estimation Methodologies

The power estimation methodologies at logic level are divided in two general classes, the simulation- and probability- based methodologies.

In the former class the circuit is simulated under the application input stream or a typical one and the switching activity and/or the average power dissipation are derived. The accuracy of the estimated power depends on the accuracy of the simulator. The major advantages of this technique are its accuracy and generality, since it can be applied to any circuit regardless of technology, design style, functionality, and architecture. Also, all the factors that affect the switching activity and power estimation are fully captured.

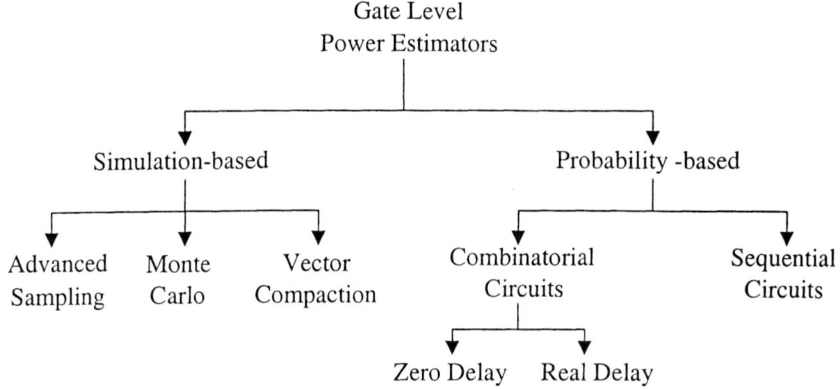

Figure 9.5. Classification of Gate Level Power Estimation Methodologies

However, this approach suffers by a major drawback that is the cost of the simulation time. The simulation results are strongly dependent on the input signals used to feed the simulator. In order to capture the real behavior of the circuit, a large number of input vectors must be fed to simulator resulting in long simulation times. This makes the simulation-based approaches impractical for large circuits and long input vector sequences. To overcome this limitation several methods aiming at reducing the number of the required input vectors have been proposed. These methods are classified in three categories, the *Monte-Carlo*, the *Advanced Sampling,* and the *Vector Compaction* methods.

In the second class specific probabilistic characteristics of the input stream (e.g. signal and transition probabilities) are used to compute the switching activity of the circuit without performing simulation. To achieve this, appropriate mathematical models are developed to capture the signals' correlations, structural dependencies, and gate delays. Also, low complexity algorithms are developed to estimate the switching activity of the circuit using probabilistic characteristics of the input stream and the developed mathematical models.

Compared to the simulation-based approaches, probabilistic-based methods have the advantage of performing the switching activity computation in reduced time. However, the accuracy of these methods is less than that of the simulation-based methods. It comes from the fact that the accurate modeling of the previous mentioned factors, which affect the switching behavior of a signal, is extremely difficult and in many cases impossible to be achieved with reduced complexity.

According to the type of the circuit the probability-based methods are categorized as probabilistic methods for combinational and sequential circuits. Also, according to the adopted gate-delay model they are classified as zero and real gate-delay power estimation methods. When zero gate-delay model is adopted only functional transitions are considered causing inaccuracies in power estimation. However, the estimation is performed in reduced time compared to the real gate-delay methods allowing the use of zero-delay power estimators during synthesis.

9.5 Simulation-based Power Estimation

9.5.1 Monte-Carlo Power Estimation

The basic idea of Monte-Carlo based power estimation is to simulate the circuit applying appropriate patterns at the primary inputs. During the simulation the switching activity of each node is monitored and a stopping criterion is used to determine when node activity has converged to its steady value. When the criterion is met for all circuit nodes the simulation stops and the average power dissipation is obtained [7]. The flow of the Monte-Carlo Power estimation is presented in Fig. 9.6.

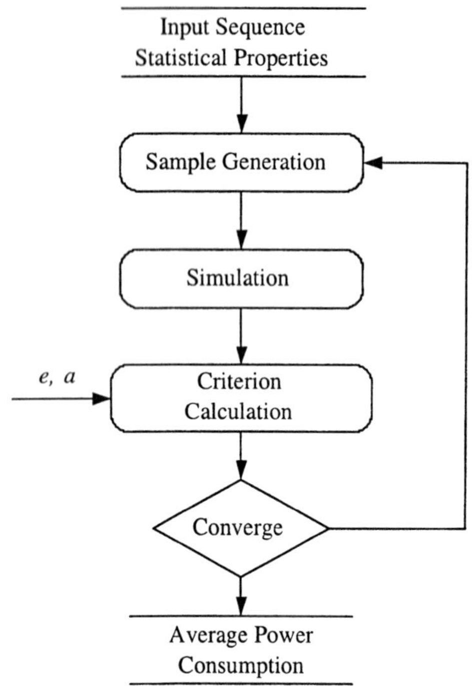

Figure 9.6.   Monte-Carlo Power Estimation

The input of the method is the statistical properties of the input sequence such as the average signal and transition probabilities. By adopting Markov models and using random number generators fixed length input vector sequences, which confirm the statistical properties of the initial input sequence, are generated. These fixed length input vector sequences are called *samples*. The simulation is performed in iterative fashion and in each iteration a sample, is simulated. The simulation results are monitored to calculate the sample mean and variance. The iteration terminates when the stopping criterion is met.

During a given period of duration T_D (clock cycles), the number of transitions at each node n is counted and the n/T_D value called *sample value* is obtained (T_D denotes the sample length). The process is repeated K times to obtain K independent samples. The sample mean is defined as $\bar{a} = (\sum a_j)/K$ with $a_j = n_j/T_D$ and $j = 1, 2, ..., K$. For large K, according to the Law of Large Numbers the expected value, α, and the standard deviation, σ, of the initial population are approached by the corresponding sample values $\bar{a}$ and s, respectively. Furthermore, according to the Central Limit Theorem [5] $\bar{a}$ is a random variable with mean α and has a distribution that approaches the normal distribution if K is large enough. Likewise $\sigma \approx s/\sqrt{K}$. It has been shown in [7] that for $(1 - \alpha)$ confidence level the following inequality holds:

$$\frac{|\alpha - \bar{\alpha}|}{\bar{\alpha}} \leqslant \frac{z_{\alpha/2}s}{\bar{\alpha}\sqrt{K}} \tag{9.10}$$

where, $z_{\alpha/2}$ is the area under the standard normal distribution from $z_{\alpha/2}$ to ∞ is $\alpha/2$. Therefore, if

$$K \geqslant \left(\frac{z_{\alpha/2}s}{\bar{a}\varepsilon'}\right)^2 \tag{9.11}$$

we have $\frac{|\alpha-\bar{\alpha}|}{\bar{\alpha}} \leqslant \frac{z_{\alpha/2}s}{\bar{\alpha}\sqrt{K}} \leqslant \varepsilon'$, hence $\frac{|\alpha-\bar{\alpha}|}{\bar{\alpha}} \leqslant \frac{\varepsilon'}{1-\varepsilon'} = \varepsilon$. Equation (9.11) is the stopping criterion for $(1 - \alpha) \times 100\%$ confidence and ε is an upper bound of the relative error.

If a node in the circuit has a very low activity ($\alpha << 1$) then the sample variance increase and the number of the required samples can be very large, which results in a slow convergence. However, since the low-activity nodes contribute little to power dissipation, a modified stopping criterion is proposed in [8]. The idea is to specify a particular threshold value, $\alpha_{\min}$, below which the switching activity is less important. Hence, it not required waiting these nodes to converge within a certain percentage of error. If

$$K \geqslant \left(\frac{z_{\alpha/2}s}{\alpha_{\min}\varepsilon'}\right)^2 \tag{9.12}$$

we have $\frac{|\alpha-\bar{\alpha}|}{\bar{\alpha}} \leqslant \frac{z_{\alpha/2}s}{\bar{\alpha}\sqrt{K}} \leqslant \frac{\alpha_{\min}\varepsilon'}{\bar{a}}$, hence $|\alpha - \bar{\alpha}| \leqslant \alpha_{\min}\varepsilon'$. Therefore, equation (9.12) becomes the stopping criterion with $\bar{\alpha} < a_{\min}$ for $(1 - \alpha) \times 100\%$ confidence level.

In case of sequential circuits the approach is slightly modified. Since the FSM may contain groups of *closed state sets*, the simulation may be locked to one of them if the simulation patterns are not enough. Thus, the stopping criterion is not correctly evaluated and the simulation terminates erroneously.

Consider the FSM of Fig. 9.7, where two closed state sets $G_1 : \{S_1, S_2\}$ and $G_1 : \{S_3, S_4\}$ exist. If the initial state of the machine is in $G1$ then it is unlikely to reach a state of $G2$ by simulating a small number of vectors. Using few vectors the evaluation of the stopping criterion results in an early stabilization of the FSM and in wrongly termination of the simulation.

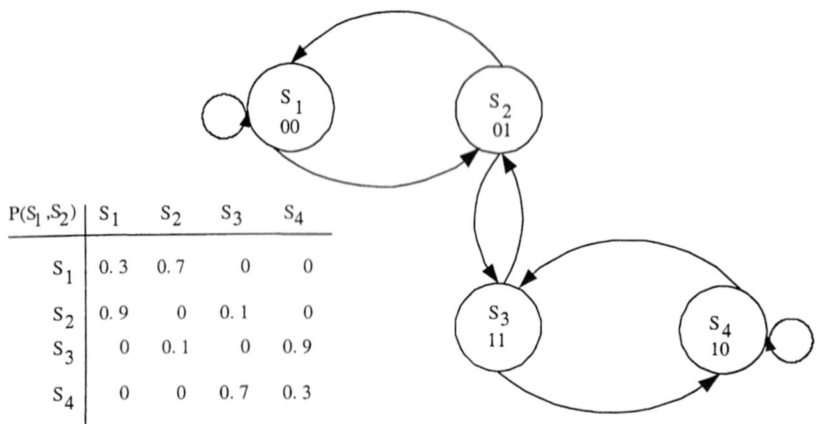

$P(S_1,S_2)$	S_1	S_2	S_3	S_4
S_1	0.3	0.7	0	0
S_2	0.9	0	0.1	0
S_3	0	0.1	0	0.9
S_4	0	0	0.7	0.3

Figure 9.7.   State Transition Diagram with Closed State Sets

This problem is overcome in [11], where a *warm-up* simulation is proposed to identify the closed sets and their probabilities. Assuming that the machine does not contain periodic states each state is visited during the warp-up period. It must be stressed that during the warm-up period no power estimation is performed. When the warp-up simulation finishes the probabilities of the closed sets are known. Then Monte-Carlo estimation is performed with start states from each closed set according to the probability that has been computed during warm-up period.

9.5.2   Advanced Sampling Techniques

Although Monte-Carlo power estimation addresses the problem of long simulation time, it suffers by the following drawbacks [9]. First, since the simulation vectors are generated based on statistics of the input stream, a large number of vectors must be examined to extract reliable statistics. Second, in each iteration the sample is generated ignoring the previous sample. Thus, the spatiotemporal correlations are not be adequately captured. Finally, there is a concern about the normality assumption on the sample distribution. Since the stopping criterion is derived based on the normality assumption, if the sample distribution significantly deviates from normal distribution, the simulation may terminate incorrectly. An example of such distributions is the bi-modal distribution.

To overcome the shortcomings of Monte-Carlo method, power estimation methodologies based on sampling theory have been proposed. In [9] *stratified random sampling* for power estimation is introduced. The basic idea is to partition the population (i.e. the initial input vector sequence) into disjoint subpopulations called *strata* so that the power consumption in each subpopulation is more homogeneous than in the initial population. The partitioning is based on a low-cost predictor (zero-delay power estimation) for each member (input vector pair) of the population. After that, *Simple Random Sampling* (SRS) is performed in each stratum and a sample is derived. The average power consumption is obtained by simulating the circuit with a number samples until a stop criterion is met.

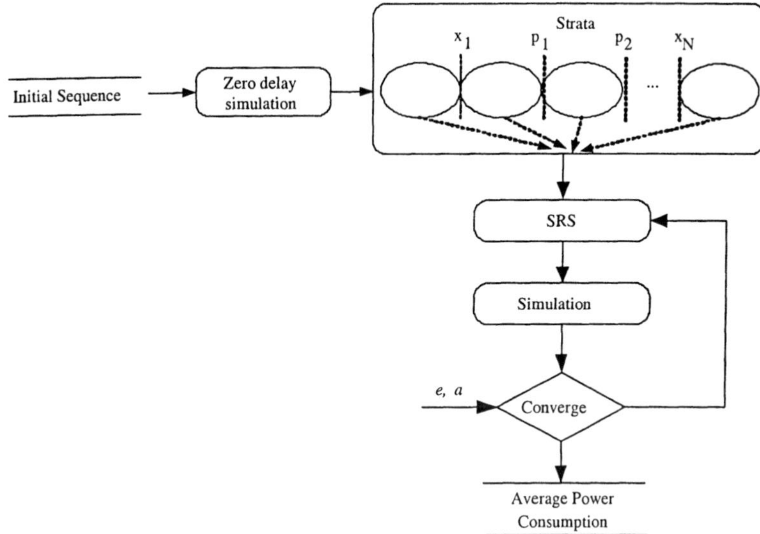

Figure 9.8. Stratified Random Sampling for Power Estimation

Regarding the construction of strata the authors propose the following procedure. Let the predictor value of unit, u_i, which corresponds to an applied input vector pair, be x_i. The population U (initial input sequence) is sorted according to the x_i values. Let the new order be $u_1, u_2, ..., u_n$. Then, K-1 separators, $p_1, p_2, ..., p_K$, are selected such that $x_1 < p_1 < p_2 < ... < p_{K-1} < x_N$. All the units whose x_i values are between two consecutive separators are put into the same stratum; thus the strata disjointly cover the population. Let the size of each stratum be N_i, then $N_1 + N_2 + ...N_K = N$, where N is the size of the original population. Sample units are drawn from each stratum independently and the sample size in the i−th stratum become n_i ($i = 1, 2, ..., K$) and

$n_1 + n_2 + ... + n_k = n$, where n is the total sample size and n_i is the sample size of the i-th stratum.

There are two more issues that must be discussed. The first is the selection of n_i values from each stratum, which is referred as *sample size allocation* problem, and the second is the finding of the optimal values of the separators p_i, which is referred as *stratum selection* problem. Regarding the first problem the authors propose a solution that is a variation of the *minimum variance allocation* problem [10]. Concerning the second problem, it has been suggested in [10] that an increase of K beyond 6 is not useful. Regarding, the values of the separators the following method is proposed. The stratum variances are approximated by the variance of the predictor in each stratum. Next, the units of the population are sorted according to their zero delay power estimates and put in a large number of bins. Adjacent bins are merged iteratively until K strata are formed and $W_i S_i$ are within 25% of each other, where S_i and W_i are the standard deviation and the weight of i-th stratum respectively.

When the size of the population is very large the cost of calculating the predictor becomes important. To reduce this overhead, two-stage sampling is proposed. In the first stage, a sub-population of size $M \ll N$ is randomly sampled from the original population U. In the second stage, stratified sampling is applied to this subpopulation to select a sample of size n.

Experimental results show that the efficiency of stratified random sampling and two-stage stratified sampling techniques are 3-10 times higher than that of simple random sampling and Monte-Carlo based power consumption estimation.

9.5.3   Vector Compaction

Vector compaction is another approach to overcome the simulation effort. The key idea is that given N input vectors, a new sequence of length M ($M \ll N$) is generated such that the new sequence is a good approximation of the initial one in terms of power consumption. Considering that power consumption depends on the correlations of the applied input vectors, the statistical properties (i.e. first order spatiotemporal correlations) of the initial sequence should be preserved.

In [12] a vector compaction methodology based on the adaptive modeling of binary input sequences as Markov sources of information was introduced. Also, a hierarchical structure of Markov chains to capture complex spatiotemporal correlations and dynamic changes of the input sequence is proposed. Additionally, variable-order dynamic Markov models to compact in an efficient way the input sequences of an FSM were introduced.

Let S_2 be a 5-bit sequence shown in Fig. 9.9; where the corresponding word-level transition graph is also presented. Each state corresponds to a distinct pattern of the sequence and each edge represents a valid transition between any two patterns. The label of each edge denotes the conditional probability of transition from the source to destination node. We also, suppose another sequence, S_1, which is deterministic and highly correlated. Note that the activity of S_1 is smaller than that of S_2. Specifically, 1.33 and 3 transitions on average per time step take place in S_1, and S_2, respectively.

Suppose that we duplicate 25 times the sequence S_1 and 100 times the sequence S_2, getting two new sequences S_1^* and S_2^*, respectively. Based on S_1^* and S_2^*, a new sequence S^* is constructed by concatenating S_1^* and S_2^* in the order of $S_1^* \rightarrow S_2^* \rightarrow S_1^*$.

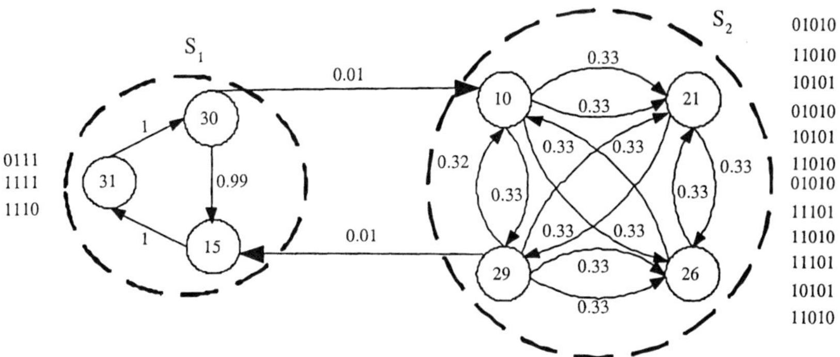

Figure 9.9. Transition Graph of the Composite Sequence S^*

The question that arises is: which will be the average power consumption when S^* is applied to a circuit? Obviously, the circuit has two different modes of operation: (i) where a low activity sequence (S_1) is applied at the primary inputs, and (ii) where high activity (S_2) sequence is applied. Adopting a flat compaction model, where the two different modes of operation are not explicitly taking into account, the generated compacted sequence does not efficiently represents the applied input sequence S^*. Thus, the power consumption, which is derived by simulating the compacted sequence, may be by far different than that when S^* is used.

To address this problem the authors of [12] proposed the use of a hierarchical Markov model. Under this model the input space is structured into a hierarchy of *macro-* and *micro-states*. At the first (high) level in the hierarchy a Markov chain of macro-states exists. At the second (low) level in the hierarchy, each macro-state is represented by a Markov chain for all its constituent micro-states. After constructing the hierarchy for an input sequence, starting with some macro-state, a compaction procedure with a specified compaction ratio is applied to compact the set of micro-states within that macro-state. Next, the control returns to the higher-level in the hierarchy and, based on the conditional probabilities that characterize the Markov chain at this level, a new macro-state is entered and the process repeats.

To derive the hierarchical Markov model the (ε, δ)-*property* is proposed. A weighted transition graph is said to have the (ε, δ)-property if there exists a grouping $\{S_1, S_2, ..., S_p\}$ on the set of states $\{s_1, s_2, ..., s_n\}$ of the transition graph satisfying:

$$(\varepsilon\text{criterion})\forall\, s_i \in S_k,\ s_j \in S_l \quad p(s_i|s_j) < \varepsilon \text{ and } p(s_j|s_i) < \varepsilon \qquad (9.13)$$

$$(\delta\text{criterion})\forall S_k,\ \exists\, W_k \ni |W_k - w_{ij}| < \delta\ \forall s_j, s_j \in Sk \qquad (9.14)$$

where, W_k is a *Weight Random Walk* in a weighed transition graph and w_{ij} is the associated weight of $p(s_i|s_j)$ transition. Each S_k called macro-state and each $s_j \in S_k$ called microstate within macro-state S_k.

The physical meaning of the above property is that conditional probabilities from any micro-state in S_k to another micro-state in $S_l(S_k \neq S_l)$ are negligible (ε criterion), and all transitions among micro-states belonging to the same macro-state have similar weights (δ criterion). Hence, in Fig. 9.9 micro-states '10', '21', '26', '29' from the macro-state S_2^*, while '15', '30', '31' from macro-state S_1^*. The corresponding hierarchical Markov model is shown in Fig. 9.10.

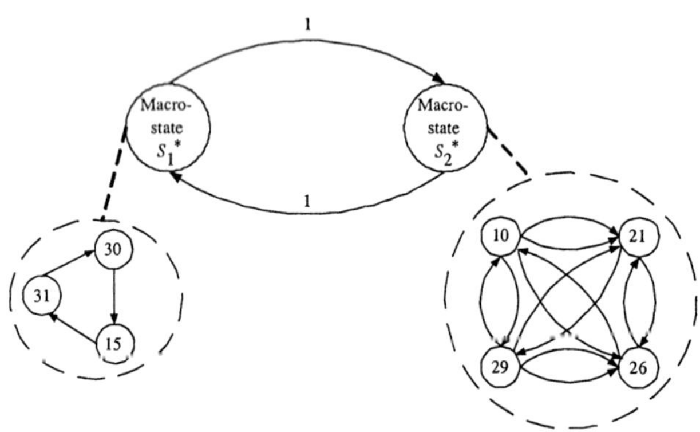

Figure 9.10.   A Hierarchical Markov Chain

An algorithm for binary sequence compaction within each macro-state is also presented. It relies on the modeling of a binary input sequence as a first-order Markov source of information and it is applicable both in combinational and sequential circuits. The algorithm captures spatial and first order temporal correlations.

This method has also extended to handle FSMs. More precisely, it was shown that: 1) Under the stationarity and ergodicity assumptions, complete capture of the characteristics of the external input sequence feeding into a target FSM is sufficient to correctly characterize the joint transition probabilities; 2) If the external input sequence has order k, then a lag-k Markov chain model of this sequence will suffice to exactly model the joint transition probabilities in the target FSM; The key problem is thus to determine the order of the Markov source, which models the external input sequence. To address this problem the notion of *block entropy*, a technique for identifying the order of a composite source of information has been proposed.

In [13] a different approach based on the Fourier transform is proposed. Its feature is that it uses the spectral analysis to synthesize a reduced vector sequence. The flow of the methodology is presented in Fig. 9.11.

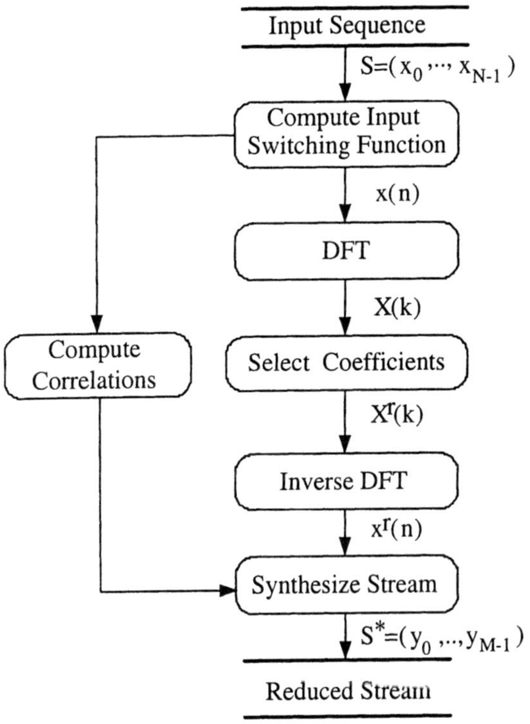

Figure 9.11. Stream Synthesis based on Fourier Transform

The input is the initial input vector set, $S(x_0, \ldots, x_N)$. From S, a function, $x(n)$, called *input switching function* is derived. This is an integer-valued function that represents the number of bit transitions between two successive vectors as a function of time. After that, the Discrete Fourier Transform (DFT) of $x(n)$ is calculated and a function, $X(k)$, in the frequency domain is obtained (i.e. the spectrum of $x(n)$). Next, a subset of coefficients is selected and a new function, $X^r(k)$, called *partial spectrum* of $x(n)$ is derived. The next step is the computation of the Inverse DFT of $X^r(k)$. Finally, considering the pairwise spatiotemporal correlations of the input stream the reduced stream $S*(x_0,...,x_M)$ is obtained with $M \ll N$.

The key point of the method is the choice of the subset of coefficients. If the selection of the coefficients is done properly, then the function $x^r(n)$ becomes an accurate approximation of $x(n)$ with limited number of samples. Selecting a frequency value, k_c, as the basic frequency and suppressing all the remaining frequencies that are not multiplies of k_c, a sampled function, $X^r(k)$, is obtained (obviously the dc-coefficient $X(0)$ is kept). The counterpart of $X^r(k)$ in time domain (i.e. function $x^r(n)$) is a periodic function with period $P = N/k_c$.

Thus, the frequency k_c determines the compression ratio N/M. Selecting a small value of k_c a large number of coefficients is included in the partial spectrum. Thus, the initial function is approximated with high accuracy with the cost of the reduced compression ratio. On the other hand, a large value of k_c provides large compression ratio but limited accuracy. The authors address this trade off as follows. Since the exact periodicity of $x^r(n)$ is required then the largest frequency N must be included in the partial spectrum of $x(n)$. It is achieved by selecting the frequency k_c such that the largest frequency N to be a multiple of k_c.

9.6   Probabilistic methods

The flow of the probabilistic methods for computing the switching activity and the power consumption consequently is presented in Fig. 9.12.

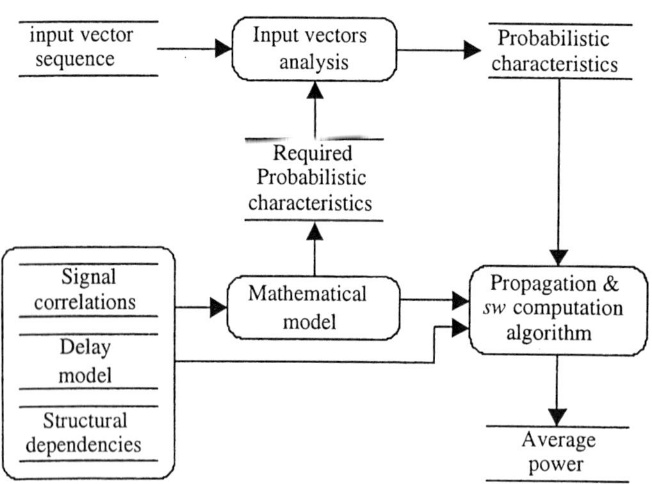

Figure 9.12.   Probability-Based Power Estimation

Taking into consideration the factors that affect the switching behavior of a digital signal (signal correlations, gate delay model and structural dependencies), an appropriate mathematical model is established. In almost all published methods, a Markov based stochastic model is adopted. Based on this model the required probabilistic characteristics such as the average signal and transition probabilities and the pairwise correlations are extracted from the input stream.

Next, a low complexity algorithm that propagates the extracted probabilistic characteristics into the circuit and computes the switching activity of each circuit node is developed. The algorithm also considers the mathematical model and gate delays. The output is the average switching activity and the average power dissipation.

9.6.1   Combinational Circuits

Zero delay power estimation.   Assuming spatiotemporal independence among the circuit signals, the switching activity, $E^x(sw)$, of a circuit node, x, is given by $E_x(sw) = 2p_x^1 p_x^0 = 2p_x^1(1 - p_x^1)$. Thus, the problem is reduced to the computation of the signal probability, p_x^1, of signal x.

The first approach presented in [14], where an algebraic variable, which corresponds to the signal probability, is assigned to each primary input. For each internal node f, an algebraic expression is derived, considering the functionality of gate f and the algebraic expressions of all fan-in nodes of f.

Adopting the fundamental concepts of Boolean algebra, a gate output, f, is described by the sum of disjoint cubes, c_i. Since the cubes are disjoint and the primary inputs x_i are assumed independent, the signal probability of a f is computed as follows:

$$p_f^1 = \sum_{\forall c_i} p_{c_i}^1 \text{ ,where } c_i = g\left(\prod x_j\right) \text{ and } p_{c_i}^1 = \prod_{\forall x_j \in c_i} p_{x_j}^1 \qquad (9.15)$$

However, due to reconvergent fanout nodes, the variable x_j may be appeared more than once in each c_i (i.e. in exponent form) causing inaccuracies in the signal probability computation. To overcome this problem, first the exponents are suppressed from each c_i and after that the signal probability calculation takes place. Let's assume the circuit of Fig. 9.2, where the algebraic expression of node f is $f = de = ab^2c$. Applying *SupExp* (Suppress Exponent), we obtain $SupExp(f) = SupExp(ab^2c) = abc$. After that, the probability $p_f^1 = p_a^1 p_b^1 p_c^1$ is computed correctly.

Although this method is simple, its computational complexity is high because each node is expressed in terms of primary inputs, which require high computational effort for their manipulation.

In [15] the high computational complexity issue is addressed by the *Tree Algorithm*, where each node is expressed only in terms of its fan-in inputs.

Thus, the corresponding expressions are smaller and the complexity is reduced. For each basic gate, a formula to compute the signal probability of its output given the signal probabilities of its inputs was proposed. However, this approach suffers by the fact the structural correlation is not captured, since every gate is expressed in terms of its inputs. The problem of capturing the structural dependencies is addressed in [16] where the concept of *supergate* is proposed. In particular, a supergate for a node f is defined as the minimal sub circuit that feds f and whose inputs are structurally independent. The problem with this approach is that in some cases the supergate reaches the primary inputs resulting in large expressions.

All the previous methods compute the transition probability of a signal assuming spatiotemporal independency resulting in inaccurate estimations. In the following, a set of methodologies that address the problem of capturing the temporal, spatial, and structural correlations under zero gate-delay are presented.

In [17] the concept of Transition Density was introduced. The transition density, $D(x)$, of a node x is defined as the average number of transitions of x per time and is expressed as follows:

$$D(x) = \lim_{T \to \infty} \frac{n_x(T)}{T} \tag{9.16}$$

where, $n_x(T)$ is the number of transitions in the time of period. Thus transition density can directly substitute the term $f\, E_i(sw)$ of eq. (9.1). Furthermore, the behavior of a binary signal is modeled as a stochastic process and is proved that the result of a differentiating a Boolean function f with respect to a variable, x_i, has a constant mean. Taking into consideration the Shannon expansion ($f = f|_{x_i=1} \oplus f|_{x_i=0}$) of a Boolean function, $f(x_1, x_2, ..., x_n)$, the transition density is computed as follows:

$$D(f) = \sum_{i=1}^{n} p(\frac{\partial f}{\partial x_i}) D(x_i) = \sum_{i=1}^{n} p(f|_{x_i=1} \oplus f|_{x_i=0}) D(x_i) \tag{9.17}$$

Thus, given the transition densities and signal probabilities of the variables x_i of function f, the transition density of f can be computed using (9.17). A transition density simulator called DENISM [17], [21], which propagates the transition densities and signal probabilities from input to outputs and computes the transition density for each internal circuit node was developed. In [17], the use of the Ordered Binary Decision Diagrams (OBDDs) [18], [19], [20] was proposed for the computation of (9.17) and it is included in DENISM.

OBDDs offer great advantages in representing and manipulating Boolean functions. First, OBDDs can be transformed in canonical forms that uniquely determine a Boolean function. Second, operation on OBDDs can be done in polynomial time in terms of their size, i.e. vertex set cardinality. OBDDs have been used in the past in many design tasks such as logic synthesis, verification, testing as well as to evaluate the probability a Boolean function f. One of the main properties of OBDDs is that a function f is represented by a set of cubes that disjointly cover the on-set of f. Exploiting this property the probability of a logic function is computed in an easy and accurate manner.

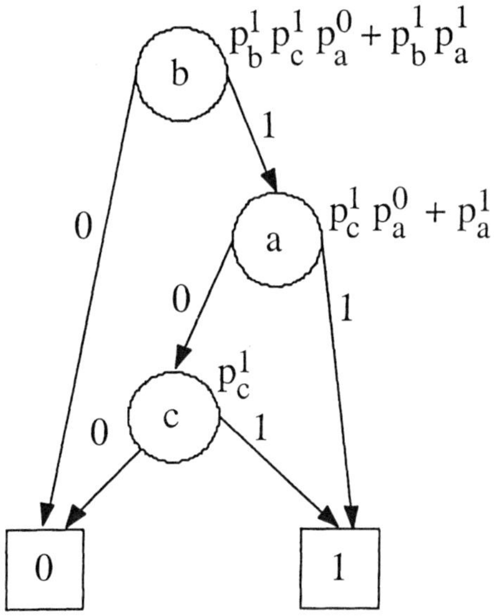

Figure 9.13.   OBDD of $f = ab + bc$

In Fig. 9.13 the OBDD of function $f=ab+bc$ is illustrated. Traversing the OBDD from the leaf node 1 to root node we obtain $f = \bar{a}bc + ab$. Note that the cubes $\bar{a}bc$ and ab are disjoint. Hence, the probability of f assuming independent input is evaluated as follows:

$$p_f^1 = p\left((\bar{a}bc + ab)\right) = 1 = p(\bar{a}bc = 1) + p(ab = 1) - p(abcab = 1)$$
$$= p(\bar{a}bc = 1) + p(ab = 1) = p_a^0 p_b^1 p_c^1 + p_a^1 p_b^1$$

$$(9.18)$$

Thus, the problem of structural dependencies in the function probability evaluation is handled efficiently. In practice, the function probability evaluation is performed in recursive manner walking from the leaf node 1 to the root node. In each internal node the probability is evaluated considered the probability evaluation of the lower levels as shown in Fig. 9.13.

The main drawbacks of [17] are: i) the primary inputs are assumed spatially independent, ii) the structural dependencies are not efficiently captured, and iii) for each function f only one input is allowed to perform transition. The reported results show a speed up of 2-3 times order of magnitude compared to a logic simulator while the error in power estimation is about 30% for random inputs.

A modified approach that considers simultaneous switching at the gate inputs was proposed in [22]. It was achieved by extending the Boolean difference to higher order difference terms. The proposed formula is:

$$\frac{\partial f^k \, | b_{i_1}, ..., b_{i_k}}{\partial x_{i_1}, ..., \partial x_{i_1}} = f \, | \, x_{i_1} = b_{i_1}, ..., x_{i_k} = b_{i_k} \oplus f \, | \, x_{i_1} = \overline{b_{i_1}}, ..., x_{i_k} = \overline{b_{i_k}}$$

$$(9.19)$$

with $b_{ij} \in \{0, 1\}$. Eq. (9.19) determines a set of signals, $(x_{i_1}, ..., x_{i_k})$, under which the function f performs a transition if the inputs toggle simultaneously from $(b_{i_1}, ..., b_{i_k})$ to $(\overline{b_{i_1}}, ..., \overline{b_{i_k}})$, while the other inputs remain stable.

Since OBDDs are used for Boolean function representation and transition densities computation, a method to partition the circuit into structural independent sub-circuits to obtain small OBDDs was introduced. The experimental results of this method show an error of 8% when random input vectors are applied at the primary inputs. It should be stressed that this error corresponds to global error analysis and not to node-by-node analysis. It means that the real error should be higher.

In [23] the switching activity is computed considering the first temporal correlations. Simulating symbolically the circuit each internal node f is expressed in terms of primary inputs. The transition probability is computed by calculating the signal probability of the function $p\,(f(t) \oplus f(t+T))$ using OBDDs. In other words, since we are dealing with static CMOS circuits it means that two input vectors $V(t)$ and $V(t+T)$ should be considered to estimate the transition probability of node f. Furthermore, the node f performs a transition when its value in two consecutive time instances t and $t+T$ is different. Thus, by building a function $f(t) \oplus f(t+T)$ for every circuit node f and evaluating the signal probability of this function the transition probability f is computed:

$$E_{sw}(f) = p\,(f(t) \oplus f(t+T)) \tag{9.20}$$

Since the one-step transition probabilities are used, the method captures the first order temporal correlation. Also, the structural dependencies are considered since each circuit node is expressed in terms of the primary inputs. However, as each circuit node is expressed in terms of primary inputs, the corresponding the Boolean functions become large and complex resulting in huge and intractable OBDDs (global OBDDs), which prevents the use of the method for large circuits. Moreover, the spatial correlation of the primary inputs is ignored.

The method of [24] models the behavior of the signal as a strict sense stationary Markov process and computes the transition activity by evaluating the signal probability of (9.20) using OBDDs. To overcome the problem of building global OBDDs the authors propose a technique to partition the circuit into structurally independent sub-circuits adopting concepts of the theory of circuit testing. Thus, local instead of global OBDDs are built allowing the switching activity computation with reduced complexity.

Finally, the most accurate method to estimate the switching activity of combinational circuits under zero-delay model was proposed in [25]. The method captures the first-order spatiotemporal dependencies at the primary inputs, the structural dependencies, while the multiple simultaneous transitions at the primary inputs are also taken into account.

To handle the first order temporal correlation the signal is modeled as a lag-one strict sense stationary Markov process having to states, s, with $s \in S = \{0, 1\}$. Under this consideration the conditional probability, π_x^{ij}, for a signal x to perform a transition from state i to state j $(i, j \in S)$ in two successive time instances t and $t + T$, given that the signal was in state i in t, exists and it is provided by the probability matrix of the Markov process. Also, the transition probability, p_x^{ij}, is related with π_x^{ij} as shown below:

$$\pi_x^{ij} = p\left(x(t+T) = j \,|\, x(t) = i\right) = \frac{p(x(t+T) \wedge x(t) = i)}{p(x(t) = i)} = \frac{p_x^{ij}}{p_x^i} \quad (9.21)$$

To capture the first order pairwise spatiotemporal dependencies the *Transition Correlation Coefficient*, (TC) $TC_{x_1,x_2}^{ij,kl}$, between two signals x_1 and x_2, which perform the transitions $i \to k$ and $j \to l$ in two successive time instances, was introduced. Since two signals are considered and each signal may be in two different states, a four state Markov model is proposed. The mathematical definition of $TC_{x_1,x_2}^{ij,kl}$ is:

$$TC_{x_1,x_2}^{ij,kl} = \frac{p(x_1^{ik} \wedge x_2^{jl})}{p(x_1^{ik})\, p(x_2^{jl})} \, with \, i, j, k, l \in S \quad (9.22)$$

Hence, the probability for an 2-input AND gate, $y = x_1 x_2$, to perform the $1 \to 1$ transition is computed as follows:

$$p_y^{11} = p(x_1^{11} \wedge x_2^{11}) = p(x_1^{11}) p(x_2^{11})\, TC_{x_1,x_2}^{11,11} \quad (9.23)$$

Thus, the problem of computing the joint probability $p(x_1^{11} \wedge x_2^{11})$ has been reduced to the multiplication of single transition probabilities and *TC*s, which are known for the primary inputs. Since the higher order spatiotemporal correlations are neglected, the *TC* of more than two signals is defined as the product of the pairwise *TC*s. Although the method can be generalized to capture higher order spatiotemporal dependencies with increased complexity, the experimental results show that the first order spatiotemporal model computes the switching activity with adequate accuracy.

Two algorithms, namely the global and incremental, were proposed to propagate the transitions probabilities and coefficients from input to output and compute the switching activities of circuit nodes. In the first case, each internal node is expressed in terms of the primary inputs offering high accuracy in power estimation with the cost of high complexity. In the second case the internal nodes are expressed in terms of immediate fan-in signals resulting in reduced complexity and accuracy. Also, a dynamic programming approach is proposed to compute the transition probabilities using the OBDD of function $f(t)$ instead the OBDD of $f(t) \oplus f(t + T)$.However, the backtracking of the incremental approach often results in the global approach and the accuracy was reduced due to the multiplications with small numbers. Thus, the new concepts of *conditional independency* and *signal isotropy* were introduced to speed up the method and improve its accuracy. Two signals a, b are conditionally independent with respect to c when the following condition holds:

$$p\,(ab|c) = p(a|c)p(b|c) \tag{9.24}$$

Under this condition the problem of handling higher order correlations is reduced to the problem of handling the first-order correlations (e.g. $p(abc) = p(ab|c)p(c) = p(ac)p(bc)/p(c)$, which problem is easily handled using the *TC*s. However, since the problem of finding a variable x such that the rest of the support set of a function is conditionally independent is an NP-complete problem, the concepts of *almost conditional independency* and *almost isotropy* were proposed [25]. A subset $\{x_i\}$ $1 \leqslant, j \leqslant n$, $i \neq j$, of signals x_1, ..., x_n is almost conditionally independent with respect to x_i, if there is exists ε $(0 \leqslant \varepsilon \leqslant 1)$ such that:

$$\left| \frac{\prod p(x_j|x_i)}{p\left(\prod x_j|x_i\right)} - 1 \right| \leqslant \varepsilon \tag{9.25}$$

If the above condition holds for every x_i, the set of signals x_1, ...,x_n is almost isotropy. The usefulness of the above result is twofold. First the small number ε is an upper bound of the relative error of the transiting probability and second, it was proved that it is not profitable to express a node with signals beyond some l predecessor levels (in practice l is smaller than 8), based on signal isotropy. The reported results show an average error less than 7% for highly correlated input sequences with small running times.

In [26] the computational complexity of the above method in terms of the number of multiplications has been reduced. By developing specific formulas to calculate the switching activity of the basic gates and developing appropriate data structures a reduction up to 70% is achieved in terms of the multiplications without affecting the accuracy of the method.

9.6.2 Real-Delay Gate Power Estimation

In [27] the notion of *probability waveform* is introduced to estimate the power consumption under real delay model. Each probability waveform consists of the signal and transition probability values over time. Thus, the signal probability can be computed at any time instance t from the initial signal probabilities taking into account the transition probabilities before t. Given the probability waveforms of the primary inputs an algorithm to propagate them and compute the probability waveforms of the internal nodes is presented. The propagation starts from inputs to outputs in an event-driven fashion. For each transition at time t of an input of a gate f a transition is scheduled at time $t+d$, where d is the delay of the gate. It must be stressed that the propagation mechanism requires statistical independency at the gate inputs. Thus, the structural dependencies are ignored, while the primary inputs are also assumed independent.

In [28] the *tagged probabilistic simulation* is proposed. This method is an extension of the probability waveforms and handles the correlation due to reconvergent fanout nodes. For each node the associative probability waveform is broken into four tagged waveforms, which correspond in the cases the node to stay at high level, to perform a low transition, to perform a high transition and to stay at low level. For each gate a *Forcing Set Table* introduced and the four probability waveforms are computed considering the waveforms of the gate inputs. To handle the structural correlation OBDDs are used at the tagged waveforms construction.

The method of [23], which has been presented for zero delay power estimation, can be also used under real gate delay model. Let a assume unit delay model. For each node only the t, $t+1$, ..., $t+l$ time instances are considered, where t is time where an primary input switches and l is the circuit level of the node. For each node i symbolic simulation is performed to construct the $f_i(t+j)$, $j \in \{0, 1, ..., l\}$ Boolean functions. The function $f_i(t+j)$ is evaluated to one if node i is one at time $t+j$. Finally, the transition occurred at the time interval $[t+j, t+j+1]$ is computed by exclusive-OR'ing the functions $f_i(t+j)$ and $f_i(t+j+1)$.

To capture the inertial real gate delay model a filtering scheme based on which short pulses are discarded is proposed. The idea is to include a pre-processing step where possible transition times are calculated symbolically by propagating events form inputs to outputs. However, pulses whose the duration is smaller than the inertial delay of the gate are ignored. The reported error of the experimental results is smaller than 5% for random primary inputs (i.e. the input spatial correlation is ignored).

In [29] the method of [14] was extended to handle real gate delay models. For each node f four algebraic polynomials P_f^{00}, P_f^{01}, P_f^{10}, P_f^{11} are assigned, which correspond to the probability that node f stays low, makes a rising transition, makes a falling transition and stays high, respectively. The set of these four polynomials is referred as polynomial group, P_f, for gate f. At time t the polynomial group is denoted as $P_f(t)$. Also, a set of formulas to evaluate the polynomial group of basic gates given the polynomial groups of its inputs is proposed.

The switching activity of f is evaluated by calculating the $P_f(t)$ polynomials in any transition time t and summing up the values of the corresponding probabilities. To achieve this, the circuit is traversed from input to output. Considering the gate delays, the time instances where node f may perform a transition are derived. After that, $P_f(t)$ polynomials are computed using the polynomials of the Boolean variables of f. To reduce the size of the polynomials and the complexity of the method, the substitution of the polynomial variables with their arithmetic values is proposed. Since it introduces inaccuracies in the power estimation the concept of *active nodes* was proposed. The active nodes are those nodes that must be used as variables in the polynomial of a gate output in order to capture the structural dependencies. They are derived after a reconvergent analysis of the circuit.

In [30] the method of [25] is extended to estimate the switching activity of a combinational circuit under real-delay gate model considering temporal, structural and input pattern dependencies. It is proved that the switching activity evaluation problem is reduced to a zero-delay switching activity estimation problem at specific time instances. A mathematical model based on Markov stochastic processes, which describes the first order temporal and spatial correlation in terms of the associated zero-delay parameters, was presented. To handle the influence of time on glitch generation, the theory of the *Timed Boolean Function* (TBF) is adopted [34]. Also, an algorithm to evaluate the switching activity at specific time instances using TBF-Ordered Binary Decision Diagrams (TBF-OBDDs) was presented.

To illustrate the method, we consider the simple circuit of Fig. 9.14, where the delay of each gate equals to one time unit, d, and the inputs are applied synchronously at times $t = 0, T, 2T, \dots$.

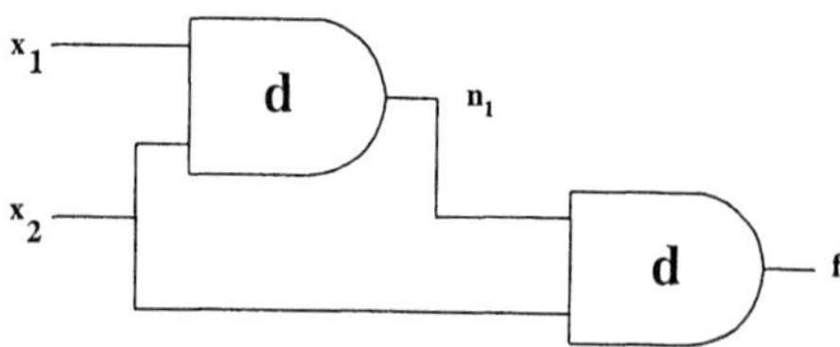

Figure 9.14.   Real Delay Power Estimation

Then the behavior of node f can be described in time domain by the following modified logic function:

$$f = F(x_1, x_2, t) = x_1(t - 2d)\, x_2(t - 2d)\, x_2(t - d) \qquad (9.26)$$

Substituting into eq. (9.26), the values of the time instances, $t_1^f = d$ and $t_2^f = 2d$, where the node f may perform a transition, we obtain the logic functions:

$$f_1 = F(x_1, x_2, d) = x_1(-d)\, x_2(-d)\, x_2(0)\, and \qquad (9.27a)$$
$$f_2 = F(x_1, x_2, 2d) = x_1(0)\, x_2(0)\, x_2(d) \qquad (9.27b)$$

Eventually, computing and summing the transition probabilities of functions f_1 and f_2, the switching activity of node f can be calculated.

Therefore, the logic behaviour of signal f is described by the modified logic function of eq. (9.26) at every time instance, which can be reduced to an ordinary Boolean function, where the Boolean variables are the corresponding values of the input signals at specific time points. Manipulating each Boolean function and considering the probabilistic properties of its variables, at specific time instances the switching activity of each time point is evaluated.

To compute the switching activity of functions f_1 and f_2 the transitions probabilities and pairwise transition correlation of the primary input signals x_1 and x_2 at specific time instances, $t = \pm kd$ in time interval $[-T, T]$ are required. Taking into account that each input signal is in steady state within the time intervals $(-T, 0)$ and $(0, -T)$, a set of mathematical formulas was proposed, using the transition probabilities and correlation coefficients at any time instance $t = \pm kd$ ($\pm kd \in [-T, T]$) of the primary inputs are computed as a function of the transition probabilities and correlation coefficients at time $t=0$. To describe formally the behaviour of each circuit node in time the theory of TBFs is adopted. Finally, a similar algorithm with that of [25] is proposed to compute the switching activity using the TBF-OBDDs.

The reported errors are less 5% for medium correlated input sequences, while in the case of highly correlated input sequences the error is less than 7% under real inertial gate delay model. The main drawback of the method is that each circuit node expressed in term of the primary inputs resulting in large Timed Boolean Function equations that increase the complexity of the method.

In [31] the accuracy of the above method has been relaxed to be used in large combinational circuits. A heuristic method to express each circuit node with structurally independent fanin signals is proposed. Also, the first order pairwise spatial correlation is ignored (i.e. the TCs are not used).

9.6.3   Sequential Circuits

A sequential circuit consists of a combinational logic and latches, whose states can be described by a STG as shown in Fig. 9.15.

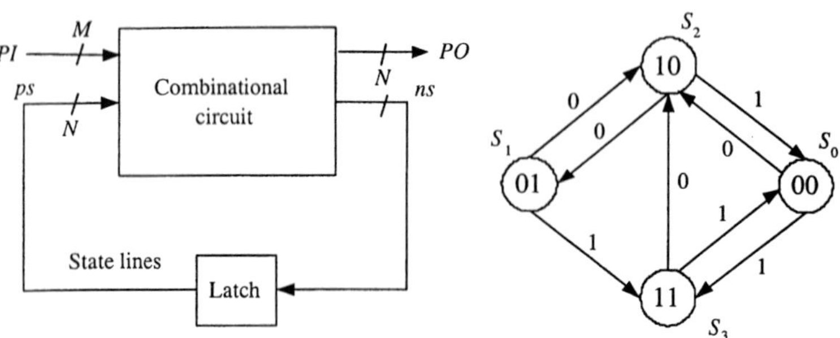

Figure 9.15.   Sequential Circuit and State Transition Graph

As it has been mentioned the coding and ordering of the states of the STG introduce additional correlations in the state lines, which should be considered in the switching activity computation.

In [33] the symbolic simulation method of [23] was used for the calculation of the switching activity. The method of [23] computes the switching activity of a circuit by performing symbolic simulation considering two successive applied input vectors and calculating the probability of the derived symbolic function for each circuit node.

Since we are dealing with static CMOS circuit, two successive input vectors must be considered. The applied input vector pair, $(v(0), v(T))$, to the combinational part consists of the external primary inputs, $i(0)$ and $i(T)$, and the current and next state lines, ps, and ns. However, given $i(0)$ and ps, the value of each ns is uniquely determined by the functionality of the combinational circuit. In other words, the next state, NS, is a function of the current primary input, $i(0)$, and current state, PS, i.e. $NS = F(i(0), PS)$. This is modeled by prepending the next state logic to the symbolic simulation equations as shown in Fig. 9.16. Specifically, the gate output switching activity could be determined given the vector pair $(i(0), i(T))$ for the primary inputs but only PS for state lines. Therefore, to compute the switching activity of a gate output, the transition probabilities of the primary inputs and the steady state probabilities of the current state lines are required.

To compute the steady state probabilities, an FSM is modeled as discrete-time discrete-state, Markov chain and Chapman-Kolomogorov equations are used [4], [5]. For instance, the associated equations of the STG of Fig. 9.15 are:

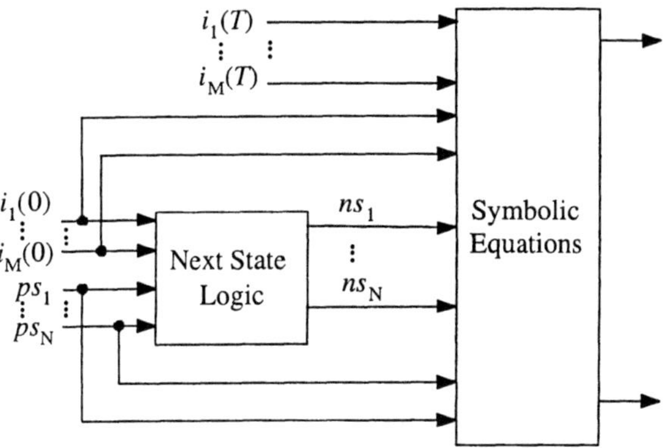

Figure 9.16. Symbolic Simulation for Consumption Estimation for Sequential Circuits

$$
\begin{aligned}
p(S_0) &= & p_i^1 p(S_2) + p_i^1 p(S_3) \\
p(S_1) &= & p_i^0 p(S_2) \\
p(S_2) &= p_i^0 p(S_0) + p_i^0 p(S_1) + & p_i^0 p(S_3) \\
p(S_3) &= p_i^1 p(S_0) + p_i^1 p(S_1) \\
p(S_0) &+ p(S_1) + p(S_2) + p(S_3) = 1 &
\end{aligned}
\tag{9.28}
$$

Assuming that $p_i^0 = p_i^1 = 0.5$ and solving the above linear system, it is obtained $p(S_0) = 0.28$, $p(S_1) = 0.17$, $p(S_2) = 0.33$, and $p(S_3) = 0.22$

The method of Chapman-Kolmogorov requires the solution of a linear system of equations of size 2^N, where N is the number of the flip-flops of the circuit. Thus, the method becomes impractical for circuit with large number of flip-flops. To overcome the above limitation an approximate method was proposed [33]. Instead of solving the system of Chapman- Kolmogorov equations the authors propose to solve a system of n non-linear equations to derive the steady state probabilities for the state lines. However, this approach poses an error in the evaluation of the state probabilities. Considering the above FSM, the probabilities of the state lines are $p_{s_1}^1 = 0.45$ and $p_{s_2}^1 = 0.55$, but since the state lines are correlated this means that $p(S_3) = p(s_1 = 1 \wedge s_2 = 1) \neq p_{s_1}^1 p_{s_2}^1$. An error less than 10% of the estimated power consumption, was reported.

The basic computation step is the solution of a non-linear system of equations as follows:

$$
\begin{aligned}
p_{ns_1}^1 &= prob(f_1(i_1, ..., i_M, ps_1, ..., ps_N)) \\
p_{ns_2}^1 &= prob(f_1(i_1, ..., i_M, ps_1, ..., ps_N)) \\
&\quad .... \\
p_{ns_N}^1 &= prob(f_N(i_1, ..., i_M, ps_1, ..., ps_N))
\end{aligned}
\tag{9.29}
$$

where $prob(f_1(i_1, ..., i_M, ps_1, ..., ps_N))$ is the probability of $f_1(i_1, ..., i_M, ps_1, ..., ps_N)$ being **1**.

Using Picard-Peano or Newton-Raphson methods, the set of equations of (9.29) can be solved. It was proposed the unrolling of the next state k-times to capture higher orders of correlation of the state signals. The reported errors of the estimated power consumption are less than 10% for temporally uncorrelated inputs.

Nevertheless, the above method exhibits a major drawback that is that the Chapman-Kolmogorov approach assumes that the present state and present input vector are independent. This does not hold for temporally correlated inputs, where the present state and present primary inputs are dependent on the previous primary input. Considering the STG of 9.15 and the applied input sequence is "...0101010101...", that means that $p_i^1 = 0.5$ and $E_i=1$. The STG of 9.15 shows that after at most 3 clock cycles the FSM is locked in states S_0 and S_2. Thus, the state probability values are $p(S_0) = p(S_2) = 0.5$ and $p(S_1) = p(S_3) = 0.5$, which is different than the values provided by the Chapman-Kolmogorov equations.

To overcome this limitation a novel technique has been proposed in [32], taking into account the temporal correlation of the primary inputs and structural dependencies among the circuit signals. The basic idea is to unroll the circuit k times and to compute the switching activity of the unrolled circuit handling it as a flat combinational circuit. In this circuit a buffer substitutes each flip-flop. Then, adopting a method of computing the switching activity of combinational circuits the power dissipation is evaluated.

However, the method of [24] requires mutually spatial independent variables. For this reason the function $f(t) \oplus f(t + T)$, based on which the switching activity of node f is evaluated, is decomposed to mutually uncorrelated sub-functions. To achieve this, the variables of f are divided into packets. Each packet contains all variables that correspond to one signal at different time instances. Thus, each packet contains temporally correlated variables but each variable of a packet is uncorrelated to any variable of a packet. Recursively, each packet is decomposed and the transition probabilities of the final functions are computed.

A system of $2N$ non-linear equations is obtained, which is solved by using the Picard-Peano method. The equations of this system are smaller than those of method [32] resulting in reduced time for its solution. A set of experimental result was performed were the accuracy of the method proved as well as the inaccuracy of ignoring the input temporal correlation is presented.

9.7   Conclusions

Power estimation is a challenging problem with great importance in modern electronic design. Methodologies for estimating the power consumption at gate level were presented. Starting for the problem formulation the main factors that affect the accuracy of the estimation, the existing methods were classified in two main classes, the simulation- and probability-based methods. For each class the most important and representative methodologies were discussed.

Although, we limit our study in static CMOS circuits it is clear the power estimation is a multidimensional and complex problem. In spite of a lot of research has been performed in the past there are some open issues such as the development of robust, accurate and low complexity real gate-delay power estimator that have to be addressed. Finally, the problem becomes more challenging and complex when deep submicron technologies are considered.

References

[1] J. Rabaey and M. Pedram, "Low Power Design Methodologies," *Kluwer Academic Publishers*, 1996

[2] A. Chadrakasan, and R. W. Brodersen, "Low Power Digital CMOS Design", *Kluwer Academic Publishers*, 1995.

[3] W.Nebel, J. Mermet, "Low Power Design in Deep Submicron Electronics", *NATO ASI Series, Kluwer Academic Publishers*, 1997.

[4] K. Triverdi, "Probability and Statistics with Reliability, Queuing, and Computer Science Applications", *Prentice-Hall*, 1982.

[5] A. Papoulis. "Probability, Random Variables and Stochastic Processes", *McGraw-Hill*, 1984.

[6] P. Schneider, and R. Krishnamurthy, "Effects of Correlations on Accuracy Of Power Analysis- An Experimental Study", in *Proc.* of *ISLPD*, 1996.

[7] R. Burch, F. Najm, P. Yang, and T. Trick, "A Monte Carlo Approach for Power Estimation", in *IEEE Trans.* on *VLSI*, Vol. 1, No.1, pp. 63-71, March 1993

[8] M. Xakelis, And F. Najm, "Statistical Estimation of the Switching Activity in Digital Circuits", in *Proc.* of *DAC*, pp 728-733, 1994.

[9] C-S. Ding, Q. Wu, C-T. Hsiesh and M. Pedram, "Stratified Random Sampling for Power Estimation", in *IEEE Trans.* on *VLSI*, Vol. 17, No.6, pp. 465-471, June 1998.

[10] B.W. Lindgren, "Statistical Theory", London, U.K.: *Chapman & Hall*, 1993.

[11] T. Chou, and K. Roy, "Accurate Estimation of CMOS Sequential Circuits", in *Trans.* on *VLSI*, Vol.4, No.3, pp. 369-380, Sept. 1996.

[12] R. Marculescu, D. Marculescu and M. Pedram, "Sequence Compaction for Power Estimation: Theory and Practice", in *IEEE Tans.* on *CAD*, Vol 8., No 6., pp. 749-812.

[13] A. Macii, E. Macii, M .Poncino and R. Scarsi, "Stream Synthesis for Efficient Power Simulation Based on Spectral Transforms", in *IEEE. Tans of VLSI*, Vol.9, No. 3, pp. 417-426, June 2001.

[14] K. Parker and J .Mc.Clsuskey, "Probabilistic Treatment of General Combinational Circuits", in *IEEE Trans.* on *Computers*, C-24, pp. 668-670, 1975.

[15] L.H. Goldstein. "Controllability /Observability of digital circuits", in *IEEE Trans.* on *Circuits and Systems,* pp.685-693, 1979.

[16] S. Setch, L. Pan and V. Agrawal, " PREDICT-Probabilistic estimation of digital circuit testability", in *Proc.* of *Fault Tolerant Computing Symposium*, pp. 220-225, 1985.

[17] F. Najm, "Transition Density: a stochastic measure of activity in digital circuits", in *IEEE Trans.* on *VLSI Systems*, Vol. 12, No. 2, pp. 446-455, February 1994.

[18] R. Brayant "Graph based algorithms for Boolean function manipulation", in *IEEE Trans.* on *Computers*, volume C-35, pp.677-691, 1986.

[19] R. Drechsler, "Binary Decision Diagrams: Theory And Applications", *Kluwer Academic Publishers*, April 1998.

[20] S. Chakravarty, "On The Complexity of Using BDDs For The Synthesis And Analysis of Boolean Circuits", in *Proc.* of 27th *Annual Alletron Conf. on Communication, Control and Computing*, pp. 730-739, 1989.

[21] H. Kriplani, F. Najm, P. Yang and I. Hajj, "Resolving Signal Correlations for Estimating Maximum Currents in CMOS Combinational Circuits", in *Proc.* of *DAC*, pp. 384-388, 1993.

[22] T. Chou, K. Roy and S. Prasad, "Estimation of Circuit Activity Considering Signal Correlations and Simultaneous Switching", in *Proc.* of *ICCAD*, pp. 300-303, 1994.

[23] J. Monteiro, S. Devadas, K. Keutzer and I. White, "Estimation of average switching activity in combinational circuits using symbolic simulation", in *IEEE* Trans. on *CAD*, Vol. 16, No. 1, pp. 121-127, Jan. 1997.

[24] P. Schneider, U. Schlichtmann, and B. Wurth, "Fast Power Estimation of Large Circuits", in *IEEE Design and Test of Computers,* pp.70-77, Spring 1996.

[25] R. Marculescu, D. Marculescu and M. Pedram, "Probabilistic Modelling of Dependencies During Switching Activity Analysis", in *IEEE Trans.* on *CAD,* Vol.12, No.2, pp. 73-83, Feb. 1998.

[26] G. Theodoridis, S. Theoharis, D. Soudris and C.E. Goutis, "A Fast and accurate method of power estimation for logic level networks", in *Journal of VLSI Design*, Vol.12, no2. pp.205-219, 2001.

[27] F. Najm, R. Burch, P.Yang and I. Hajj "Probabilistic Simulation for Reliability Analysis of CMOS VLSI circuits," in *IEEE Trans. on CAD*, 9 (4) pp. 439-450, Apr. 1990.

[28] C. Tsui, M. Pedram and A. Despain, " Efficient Estimation of Dynamic Power Dissipation Under Real Delay Model", in *Proc.* of *ICCAD*, 1993, pp.224-228, 1993.

[29] J. Costa, J. Monteiro and S. Devadas, "Switching Activity Estimation using Limited Depth Reconvergent Path Analysis", in *Proc.* of *ISLPD*, pp. 184-189, 1997.

[30] G. Theodoridis, S. Theoharis, D. Soudris, and C. Goutis, " Switching Activity Estimation Under Real-Gate Delay Using Timed Boolean Functions", in *IEE Proc. on Computers & Digital Techniques*, Vol. 147, No. 6, pp. 444-450, Nov. 2000.

[31] S. Theoharis , G. Theodoridis , Zervas N. and Goutis C., "Accurate And Fast Power Estimation of Large Combinational Circuits",in *Proc.* of *PATMOS'99*.

[32] P. Scheider, M. Shenn, and b. Wurth, "Power Analysis for Sequential Circuits", in Proc. of EURO-DAC 96, pp. 22-27, 1996.

[33] C-Y.Tsui, J.Monteiro, m. Pedram, S. Devadas, and A. Despain, " Power Estimation for Sequential Logic Circuits", in *Trans.* of *VLSI*, Vol. 3, No.3., pp. 404-416, Sept. 1995

[34] W. Lam and R.K. Brayton "TIMED BOOLEAN FUNCTIONS-A Unified Formalism for Exact Timing Analysis", *Kluwer Academic Publishers*, 1994.

II

LOW POWER DESIGN STORIES

Chapter 10

LOW-POWER DESIGN FOR SAFETY-CRITICAL APPLICATIONS

Athanassios Kakarountas

Kyriakos Papadomanolakis

Costas Goutis

University of Patras, Patras, Greece
{ kakaruda,papk,goutis } @ee.upatras.gr

Abstract   Low-power design has been crucial to nowadays applications for both extended autonomy of the product and green issues. Due to continuous transistor size reduction and hence defect susceptibility, a major design factor, reliability has to be additionally considered. An approach to achieve high levels of safe operation while system's power dissipation is maintained low is presented in this chapter.

Keywords: low-power design, on-line testing, self-checking circuits, reliability

10.1   Introduction

Design for Low-Power results in the increase of system's autonomy and thus extended portability. Portable systems that mainly target the multimedia and telecommunication markets are quite common nowadays. However, the complexity of such systems has increased significantly over the last decade. Outstanding reduction of transistor dimensions, from 5 microns in 1985 to 0.13 micron in 2001, enabled the integration of great transistor densities in a single chip. Unfortunately the reduction of transistor dimensions has turned out the whole system to be susceptible to new types of errors. Thus, systems present a higher rate of susceptibility to transient errors, due to cross-talk noise and the like, and temporal errors due to normal radiation on earth surface.

D. Soudris et al. (eds.), Designing CMOS Circuits for Low Power, 205–234.
© 2002 *Kluwer Academic Publishers. Printed in the Netherlands.*

The evolution of nowadays technology characteristics render reliability of a system to be a significant design factor. Especially for low-power systems and safety-critical applications, traditional off-line testing cannot guarantee the detection of transient errors. Various approaches to increase system's reliability propose the utilization of appropriate materials to tolerate radiation or the use of high-quality components. However, such approaches have a major disadvantage; high cost. On the other hand, on-line testing strategy, which focuses on the incorporation of hardware redundancy, is advisable under several conditions. The required conditions to apply on-line testing in conjunction to low-power design rules are exhibit in this chapter. The controversial nature of low-power design and design for on-line testability is explored and an approach to implement both design strategies to a targeted system is presented.

The rest of this chapter is organized as follows. The terms related to the reliability of a system and design for testability are presented in detail. The controversial nature of low-power design and design for on-line testability is exhibited presenting unsuitable low-power techniques. Design of self-checking circuits, combinational and sequential, is presented and how they can be combined to form a safe operating system. A portable system designed to infuse medicine to patients, which is a safety demanding application, has served as a reference system. Finally, conclusions are offered.

10.2　　Basic background on testing

The properties of testing met throughout the text are initially presented and defined. Primarily, the terms *defect, error* and *fault* of a *Circuit-Under-Test* (CUT) have to be defined in order to ease streamlined reading.

Definition 1. *A* defect *in an electronic system or circuit is the unintended difference between the implemented hardware and its intended design.*

Defects may occur either during manufacturing process or during operation. Typical defects in VLSI chips are [3]:

- Process defects, e.g. parasitic transistors, oxide break-down etc.

- Age defects, e.g. dielectric breakdown, electromigration etc.

- Package defects, e.g. contact degradation, seal leaks etc.

The effect of the defects is the production of erroneous output.

Definition 2. *An* error *is defined as the incorrect behavior of the output of the CUT, due to the occurrence of a defect.*

The term error corresponds to the highest abstraction level of testing from where a designer/tester can approximate only the result and not the cause. It is therefore necessary to model the cause in the bounds of circuit in order to address erroneous behaviour. The vehicle to achieve this scope is the *fault*.

Definition 3. *A fault is defined as the representation of a physical defect, which may cause erroneous behaviour in the functional level.*

10.2.1 Safety properties

In order to represent all the defects needed to be faced, a fault model is formed. The most common fault models follow, while basic definitions of the circuit properties concerning safe operation are offered. Extensive information for the terms and the fault models presented here can be found in [2, 4, 6, 5]. Also, detailed requirements for safe operation of a system can be found in [7].

10.2.1.1 Fault models. Faults can be categorized in conjunction to other parameters e.g. time [2, 4]. The parameters of time occurrence and locality of the fault are considered. Thus, according to their duration, faults can be categorized as:

- *permanent*, i.e. exist even after their first occurrence;

- *intermittent*, i.e. exist only during some intervals;

- *transient*, i.e. exist only when there is a temporal change in some environmental factors;

The faults can be additionally discriminated by their locality of occurrence. The most common fault models are:

- stuck-at fault model, i.e. assigning a fixed value to a signal;

- bridging fault model; i.e. a short between a group of signals;

- stuck-open fault; i.e. one transistor is permanently open;

The most common model used for logical faults is the *single stuck-at fault*. This model is adopted throughout this chapter; due to the design abstraction level it is applicable (gate level). Both other models are mainly applied to the transistor level model but their behavior can be successfully modeled as a stuck-at fault. The latter presumes that a circuit is constructed by interconnecting Boolean gates (or other primitive cells) of a CMOS technology. Thus, any fault occurrence is possible only to the inputs or the outputs of these gates. In a n-line circuit there are $3^n - 1$ possible stuck combinations, due to the three states a line can be: s-a-1, s-a-0 and fault-free. As long as the factor n remains low, it is possible to test occurrence of stuck-at faults. When n increases further, a prohibitive large number of multiple stuck-at faults is resulted.

A special hypothesis is required to address the prohibitive increase of the possible combinations and render the fault model applicable.

Hypothesis 1. *Two conditions have to be met so that the single stuck-at fault model to be valid;*

> *1 A fault may occur only at an input or an output of a gate and its value is either set to 0 or 1.*

> *2 Only one fault may occur at any time instance and there is enough time to be detected before a second one appears.*

The above hypothesis results in a maximum of $2n$ stuck-at faults in a n-line circuit. The set that includes the $2n$ possible stuck-at faults is referred from hereinafter as $\mathcal{F}$.

10.2.1.2   Error detection through data encoding.

In order to detect fault occurrence, encoding schemes can be used. In the literature, the more frequently reported encoding schemes are the *Single-Error-Detecting* (SED) parity code, which is used in a variety of applications, as well as other *Single-Error-Correcting\Double-Error-Detecting* (SEC\DED) codes (e.g. Hamming). Detection of an error using an encoding scheme is based on the addition of extra bits (information redundancy).

Definition 4. *The word that is constructed from the data and the extra code bits is called* encoded word *while the application of the code is called* encoding.

The transmission of encoded words (or blocks) is widely met in communication applications. In order to transmit words (or blocks) through unreliable channels, where corruption of the information bits can occur, the words (or blocks) are encoded. The received word (or block) must belong to a field of values that corresponds to the encoding scheme. The field of values is formed by appending additional bits to the original information bits. The code word is then transformed using a known function (by the receiver) to the original information block. In case of error occurrence, the information block is corrupted.

Definition 5. *The opposite process that extracts the original information bits from the encoded word is called* decoding.

Supposing that it is required to detect error occurrence on a word of k-bits long. The values that have to be encoded are 2^k. Thus, the field of values that has to be formed must contain at least this amount of elements and also extra elements to allow error-detection. Selecting to encode the information bits using an n-bit code, where $n > k$, there are 2^k valid encoded words or *code words* and $(2^n - 2^k)$ invalid or *non-code words*.

Definition 6. *The ratio of the information bits k to the encoded word bits, n is called* code rate.

A code is called *separable* when the information bits can be easily extracted from the code word. In contrary, a code is called *non-separable* when the information bits can be extracted only by using a decoder. A typical separable code is the parity encoding, while an *m-out-of-n* code is a typical non-separable code. The latter mentioned examples are also typical error-detecting codes, which are the only class of codes considered in this chapter. In the following, the error-detecting parity, cyclic, Berger and *m-out-of-n* codes are presented and their characteristics are discussed.

Parity code   The parity code is obtained by maintaining the number of ones ('1') odd or even by appending a bit with the appropriate value to the original data (odd or even parity). Although odd and even parity are mathematically equivalent, odd parity is preferred to ensure that there is at least one '1' in every code word. Encoding is done easily using an EX-OR gates tree. The main advantage of this encoding scheme is that it is a separable code and the information bits can be acquired easily. However, it can't detect even number of errors, turning it obsolete for multiple error detection. Due to its low area requirements, parity is used for on-line testing of arithmetic units.

Cyclic Codes   Let us express a binary vector $R = r_m r_{m-1} ... r_0$ as the polynomial $r_m x^m + r_{m-1} x^{m-1} + ... + r_0$. Any cyclic code is a multiple of a generator polynomial $\mathbf{G}(\mathbf{x})$ of degree $(n - k)$. Thus, a cyclic code $\mathbf{c}(\mathbf{x})$ is given by:

$$\mathbf{c}(\mathbf{x}) = x^{n-k} \mathbf{D}(\mathbf{x}) + \mathbf{R}(\mathbf{x}) = \mathbf{G}(\mathbf{x})\mathbf{Q}(\mathbf{x}) \tag{10.1}$$

where $\mathbf{D}(\mathbf{x})$ corresponds to a data polynomial, $\mathbf{Q}(\mathbf{x})$ is the quotient and $\mathbf{R}(\mathbf{x})$ is the remainder. The advantage of these codes is the ability to detect multiple errors, while they are also separable codes. The main drawback is the complexity of encoding, which is done by dividing and shifting, introducing significant area overhead. This code is frequently used in memory blocks (e.g. register file) where multiple error-detection is of great importance.

Berger code   The Berger code is formed from the k information bits and c check bits, where $c = [\log_2(k + 1)]$. The check bits represents the number of 1s of the information bits. If the number of information bits in a Berger code is $I = 2^c - 1$, $c \geq 1$, then it is called *maximal length Berger code*. This code is separable and it is the least redundant code for detecting single and unidirectional multiple errors [5]. This code is ideal for successful propagation of the control signals throughout a safe operating system.

m-out-of-n codes   The *m-out-of-n* code has exactly m 1s and $(n\text{-}m)$ 0s in a n-bits code word. Thus, the number of possible code words is equal to $n!/(n\text{-}m)!m!$. This type of codes is efficient for input/output encoding and state assignment in sequential circuits (e.g. system's control unit). Contrary to the latter mentioned codes, *m-out-of-n* is a non separable code.

10.2.1.3   Properties of a safe operating system.

Based on the DVE801 [7] standard, applicable for portable devices, the circuits that construct the system must have the ability to check their operation during normal operation to achieve the maximum safety level. The procedure of testing a system during its normal operation is called *on-line testing*. In contrast, if testing procedure is performed when the system is not operating or it is halted and imposed in a test mode, then it is called *off-line testing*. In order for on-line testing to be successful, the circuits of the system have to be *self-testing*.

Definition 7. *A circuit is called self-testing with respect to fault set $\mathcal{F}$, if and only if each fault in $\mathcal{F}$ can be detected by at least one input vector.*

The detection of an error occurrence is achieved using the appropriate encoding scheme. Because error detection latency is crucial and due to the requirements of the targeted fault model, a circuit must also be *fault-secure*.

Definition 8. *A circuit is called fault-secure with respect to fault set $\mathcal{F}$, if and only if each fault in $\mathcal{F}$ produces either the correct output code word or a non-code word.*

The fault-secure property ensures that any fault occurrence is either detectable or does not corrupt the output of the circuit. Thus, it enables *concurrent error detection*. Usually, an extra circuit that checks the output of the circuit is used to detect an error. The type of circuits that detect error occurrence by observing their output is called *self-checking*. Self-checking circuits are the most common type of circuits when designing for on-line testing. Such circuits can be constructed by using either information redundancy (encoding schemes) or hardware redundancy [1], as shown in figure 10.1.

An important subclass of self-checking circuits is that of the *totally self-checking* (TSC). This class of circuits is targeted throughout this chapter. The more characteristic TSC circuits are the checkers of the self-checking circuits (see figures 10.1a and 10.1b).

Definition 9. *Totally self-checking circuit. A circuit is called totally self-checking with respect to fault set $\mathcal{F}$, when it is both self-testing and fault-secure.*

The main characteristic of TSC circuits is the ability to detect errors occurring even in themselves. The advantages they offer are:

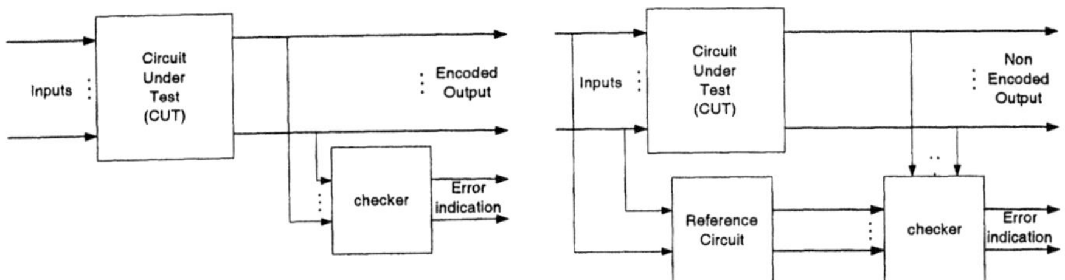

Figure 10.1. (a) Self-checking circuits using information redundancy; (b) Self-checking circuits using hardware redundancy.

- permanent and temporary faults are detectable;

- faults are concurrently detected preventing corruption of information bits

The TSC checkers are typical circuits in systems embedding on-line testing. Their ability to detect error occurrence in their own structure turns them to be significant for successful testing of the system. A widely used TSC checker is the *Two-Rail Checker* (TRC) or also known as the double-rail checker. The TRC checker has two types of inputs X and X^* and two outputs f and f^*. Under error-free operation, the X and X^* are supposed to be complementary as well as the pair f and f^*. In figure 10.2 the implementations of the TRC with two input pairs, using CMOS logic gates, is illustrated.

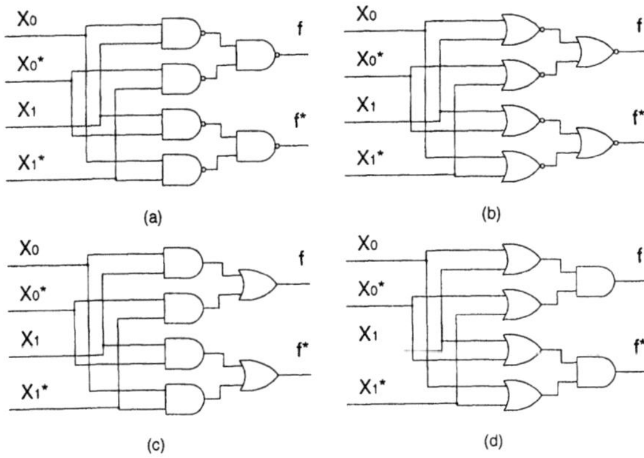

Figure 10.2. Implementations of the TRC with two input pairs using various CMOS logic gates.

The TRC checker is totally self-checking with respect to the following set of faults:

- stuck-at faults at input and output lines;

- bridging between input lines;

- breaks in input lines;

10.2.2 Safe operation constraints vs. low-power techniques

The application of low-power techniques results in power-effective designs. This may effect some TSC properties of a circuit. Usually, such applications or modifications on a circuit have effect on the required safety levels. When designing fault-tolerant systems, a measure to evaluate the achieved safety levels is *dependability*. Under dependability, terms like *reliability, availability* etc. are included. In the case of TSC circuits there is the need for a measure analogous to the reliability of fault-tolerant systems. In [8] a probabilistic measure is proposed for the TSC class of circuits. This measure was named the probability to achieve TSC Goal at the tth cycle: $TSCG(t)$. The probability to achieve TSC Goal in a circuit is defined as follows:

$$\mathbf{TSCG(t) = R(t) + S(t)} \tag{10.2}$$

where $R(t)$ represents the conditional probability that the circuit guarantees the TSC goal when the circuit is fault-free at cycle t, and $S(t)$ is the conditional probability that a circuit guarantees the TSC goal when faults occur before or on cycle t. The term $S(t)$ is calculated by adding all the probabilities that the circuit is fault-secure and/or self-testing with respect to the first fault and the probabilities that a fault is detected before a second one occurs. The term $R(t)$ is the qualitative representation of the reliability of a circuit and it is relative to the constant failure rates. A simplified approach to define the TSC goal, which is similar to maintain the TSC property, is:

Definition 10. *Given the fault assumption, a self-checking circuit always produces a non-code word as the first erroneous output due to a fault.*

The effect on TSC properties of a circuit due to the reduction of its hardware redundancy levels was mainly examined in [8]. Based on this work, [9] has showed that there is an impact on the TSC properties of a circuit when design for low-power dissipation is applied.

Typical Low-Power Techniques   The most common and broadly met technique used to dynamically manage power is the reduction of the activity on the circuits' inputs. A plethora of similar or related techniques have been proposed and more information can be found in [10, 11]. By applying such a low-power technique the vector set that normally appears at the input of a TSC circuit is reduced to a uni-valued set. Unfolding the equation of the $TSCG(t)$ to its terms $R(t)$ and $S(t)$, the effect of the input activity can be better presented. The $S(t)$ term can be analyzed into two terms $S1(t)$ and $S2(t)$, $S(t){=}S1(t) + S2(t)$. The term $S1(t)$ represents the possibility to detect the error using one vector, while $S2(t)$ represents the possibility to detect it using two vectors. Term $S2(t)$ is not contributing to the $TSCG(t)$ when the input of the circuit is 'locked' to a certain state, while $S1(t)$ is slightly contributing to the $TSCG(t)$ due to the small possibility to detect errors with one input vector. The term $R(t)$ remains unaffected, due to the fact that it is referring to the fault-free state of the circuit. Thus, to keep the $TSCG(t)$ of a circuit in high levels its input bits must present a certain activity. In addition, a system that includes TSC checkers must maintain the $TSCG(t)$ in high levels.

In [9], the degradation of the $TSCG(t)$ due to the application of low-power techniques was examined. As a qualitative example the effect of the degradation of the $TSCG(t)$ due to the application of low-power techniques is offered for the TRC. In figure 10.3 the $TSCG(t)$ is illustrated before and after the application of the low-power technique. The TRC presents the $R(t)$ and $S(t)$ (curves 2 and 3 respectively) and their sum results $TSCG(t)$ (curve 1). The application of low-power design technique degrades the $S(t)$ (curve 4) resulting a new degraded $TSCG'(t)$ (curve 5). In the following a design technique to maintain the level of $TSCG(t)$ to its original value is presented.

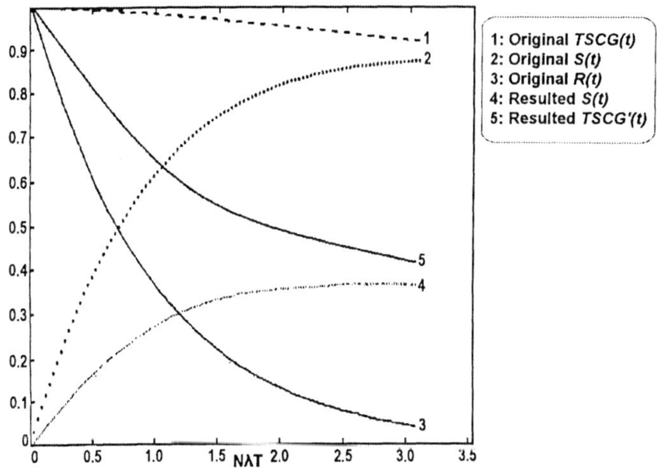

Figure 10.3. The degrading effect of the activity reduction on the input of a TRC.

10.3 Unsuitable design techniques for safety-critical applications

The design of systems that require both low-power dissipation as well as high levels of safe operation, has showed off a number of unsuitable design techniques. In this section, the design strategies that are not suitable due to the violation of the safety constraints are presented. The section is divided in two parts; in the first part general design strategies unsuitable for safety-critical systems are presented while the second part focus on the unsuitable low-power techniques.

10.3.1 Unsuitable design strategies

The design of systems targeting safety-critical applications turns out formal design techniques to be unsuitable. The topologies of global signals are mainly considered, while special reference is given to locally generated control signals and interconnection throughout the design.

Clock Distribution   Typical distribution of the system's clock (global clock) throughout the design is based on two methods: *geometric buffering* and *tree buffering*. Both methods are offering solution to the distribution of a heavily loaded signal, such as system's clock. The combination of the two methods is common in designing most of the systems met on consumer electronics market. Although, the methods are highly efficient for successful integration of the systems, when a system requires safe operation, the ability to detect possible corruption of the clock signal is obligatory. Several techniques to detect errors on the clock signal have been proposed but most of them cannot be applied to digital CMOS design.

A simple technique to detect errors on the clock signal is local encoding in pairs and additionally propagation of the clock to the output of the chip. In figure 10.4 the typical distribution of the clock signal throughout a chip is illustrated, while in figure 10.5 the use of local encoding units (e.g. counters) to detect error occurrence is showed.

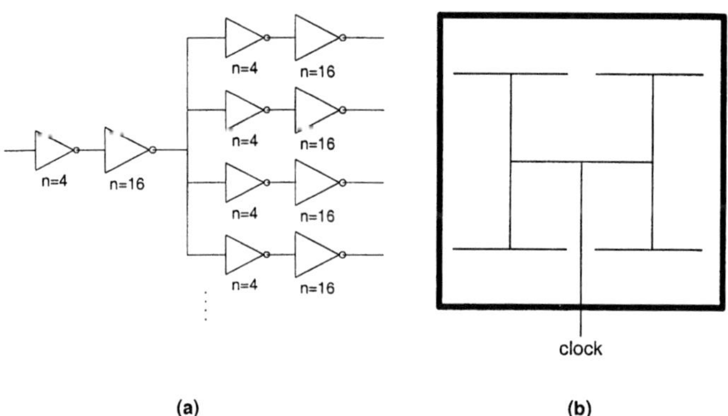

Figure 10.4.   a) Combined geometric and tree buffering, b) Typical distribution of the global clock throughout a system.

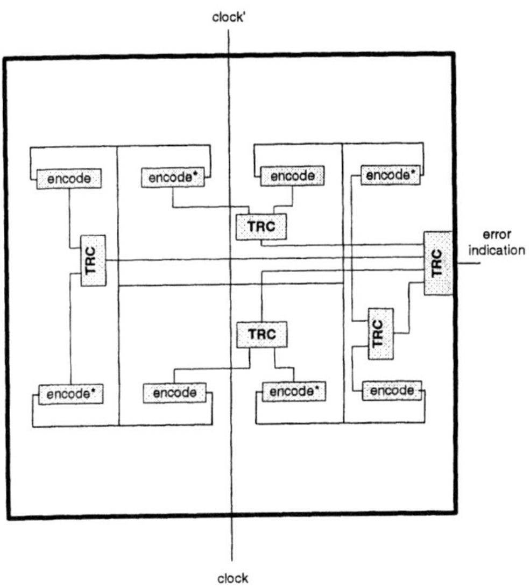

Figure 10.5. Local encoding of the clock and propagation to chip's outputs.

Global reset signal distribution An equally significant signal as the clock signal, is the system's reset. The topologies are similar to those of the clock signal. However, when there is a small number of memory elements the geometric buffering is preferred. The solution of propagating the signal to the output and local encoding is also applicable for reset.

In-chip generated control signals The signals, which are generated inside the chip, to control the operation are known as *control signals*. These signals are generated mainly in the Control Unit of the system or in some cases, locally to complex parts of the system. Due to their critical role for correct operation they cannot be implemented without special concern for error detection. Formal implementations were only considering fan-out problems, applying the latter mentioned geometric buffering method. Test procedures were only applying encoding schemes to the input and any error occurrence was supposed to be detected by the monitoring mechanisms of the output. However, this increases error aliasing, thus turning the system less reliable. An approach to address control signal corruption is continuous check of correct propagation. In figure 10.6 a topology to achieve continuous checking of control signals' propagation, is shown.

Flip-Flop's clocking The clock signal is a crucial one as it was previously mentioned. Thus, it is not advisable to implement local clock signals to control flip-flops (e.g. output of one flip-flop feeding clock of another). This design

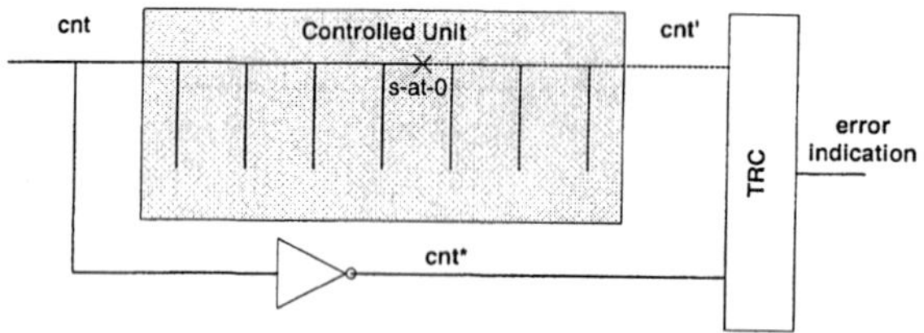

Figure 10.6. Checking error occurrence on a control signal (e.g. cnt).

technique reduces system's observability, while it complicates testing procedure. In a previous paragraph, a solution to address corruption of the clock signal was presented. Applying this technique, tricks concerning flip-flop controlling via the clock input must be avoided. An approach to control writing to a flip-flop should be done by controlling the data inputs of the flip-flop. In figure 10.7a the interconnection of two D-type flip-flops is illustrated, which should be avoided when designing safe operating systems. In contrast, the correct interconnection, in terms of safety, is showed in figure 10.7b.

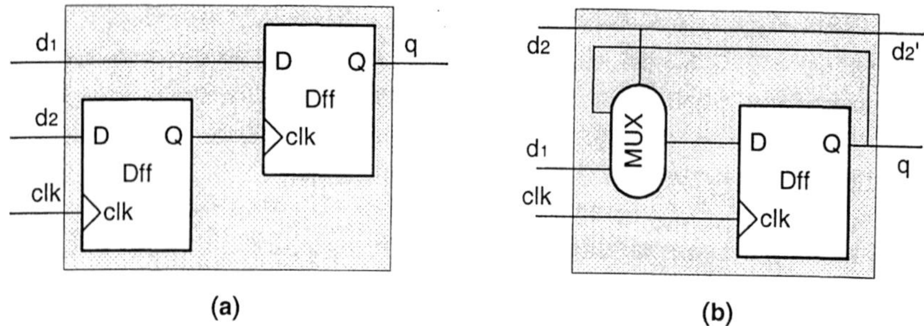

Figure 10.7. Control writing to a flip-flop (a) False interconnection, low observability (b) Testable structure of (a).

10.3.2   Unsuitable low-power techniques

There is a number of design techniques utilized to reduce power dissipation, or to be precise to manage power dissipation of a system. Usually, they are based on the principle of avoiding operation of power consuming units by halting their activity. This is achieved by increasing the complexity of a system in terms of control logic and system's controllability. Another approach is not to stop operation but to decrease bit activity on the buses (decrease Hamming distance) through appropriate data encoding. Both approaches utilize several methods that violate the constraint of safe operation in safety-critical systems.

Gated clock A common design technique to isolate parts of a system is the gated clock. Management of several circuits of the system is achieved by locally generating clock signals. This technique results in poor design quality of the system, due to the generation of short spikes and the reduced observability. It is similar to the unsuitable techniques, which were latter described. In addition, the *TSCG(t)* of the circuit is reduced due to the possibility of corruption of the controlling signal. The result of such corruption is a codeword at the output but not the correct one. In figure 10.8 the unwanted spikes and the effect of a s-at-0 fault to the generated clk2 signal are illustrated.

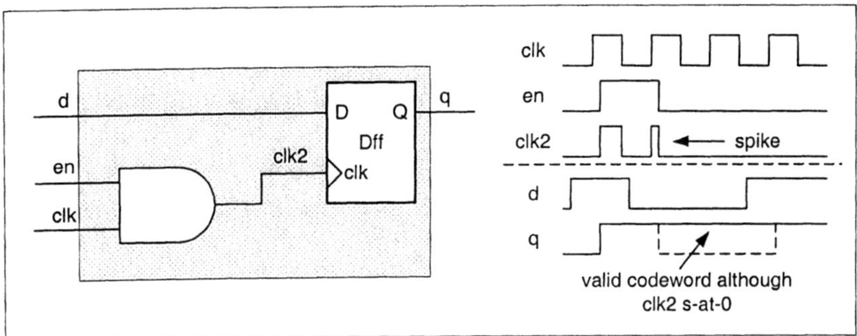

Figure 10.8. Gated clock technique and its effects.

This fault allows store of a valid codeword (information + check bits) and does not allow further writing to the flip-flops. This renders the circuit unreliable. The approach that is advised to be followed is the propagation of a global clock and the generation from the central control unit of the enable control signals that have effect on the data buses.

Locally distributed control This technique follows the same design approach as that of the gated clock. It is showed to a following section that the input and the output of the central control unit are encoded using multiple error detecting codes. This safety measure introduces extra area and power overhead. Dividing the control of the system to local control units to manage system operation and power dissipation may result quite the opposite. The encoders (decoders) are usually large compared to the control unit and present dissipate considerably. Their power dissipation is analogous to the control bits they encode (decode) and the encoding scheme. Special consideration has to be given for the selection of the encoding scheme.

Low Hamming distance A technique that leads to low-power dissipation on large data buses is the application of a special code with low Hamming distance (e.g. Gray code). This technique is highly dependent from the nature of the targeted application and it is preferable to multimedia devices or generally for data-dependent applications. Although power dissipation is reduced, the same effect is observed to the $TSCG(t)$ of the system. This happens due to the reduction of the possible input sequence, reducing the terms $S1(t)$ and $S2(t)$. Thus, when selecting a low Hamming distance code to achieve low-power dissipation, the exploration of its effects on the $TSCG(t)$ of the system must be performed.

Revolutionary approaches   Novel low-power design techniques that come to contrast with the previously mentioned techniques, are called *revolutionary*. Two revolutionary techniques are best known; the adiabatic circuit and the asynchronous circuit design. Unfortunately, although these techniques are quite promising to reduce power dissipation of a system, there is a lack of computer-aided tools to develop such systems. Furthermore, the asynchronous circuit design adopts techniques similar to clock gating which were showed to be unsuitable for safety-critical applications. On the other hand, adiabatic circuits are not CMOS based, although they can be implemented within standard CMOS systems [12]. Thus, although adiabatic circuits can save significant amount of power, it is unlikely to become an alternative to mainstream CMOS.

10.4   Low-power and safely-operating circuits

In this section, a library of TSC circuits, which can be used for the design of systems targeting safety-critical applications. These circuits have also the characteristic of low-power dissipation. First, the low-power techniques that have been adopted are presented. Then, the TSC circuits are illustrated covering all the possible units found in a system. Additionally, the structures that resulted safe operation and low-power dissipation are explicitly described. Finally, a system that was designed using these circuits is presented and the results of its integration are reported.

10.4.1   Appropriate low-power techniques for safe operation

Two techniques are mainly used to reduce power-dissipation of the TSC circuits; (i) the *glitch free design* and (ii) the *dynamic power management*. Both techniques are based on power optimization by switching activity reduction. Dynamic power dissipation of a circuit is given by:

$$P_{dynamic} = K \frac{C_{out}V_{dd}^2}{T} = KC_{out}V_{dd}^2 f \qquad (10.3)$$

where V_{dd} is the power supply, which depends on the technology, C_{out} is the output capacitance, T is the clock period of the circuit and $f = 1/T$ is the clock frequency and K is the average number of transitions of the output (divided by two). Due to the fact that V_{dd} is technology dependent and f is constrained the more important term is the switching capacitance. Thus, controlling the *effective capacitance* ($C_{eff} = KC_{out}$) the power dissipation of the circuit can be reduced. The control of effective capacitance is focused in this section mainly on the elimination of useless activity. The main sources of useless activity are:

- Spurious transitions due to unequal propagation delays (glitches);

- Transitions of redundant or not active circuits.

Dynamic power management technique is based on the elimination of useless switching activity by introducing new operation modes to the system. This technique, when applied with respect to the required time to detect the occurrence of a fault, results in significant power savings in large sequential circuits or in circuits that are not always active in a system. On the other hand, glitch free circuits minimize the appearance of spikes, reducing the power dissipation.

10.4.2   Safe-operating circuits

When designing a safety-critical system it is mandatory to select the appropriate encoding schemes to ensure safe operation. In section 10.2 it was mentioned that there is a variety of encoding schemes that allow single and/or multiple error detection. However, application of an encoding scheme results in area increase and extra power dissipation. In this case, when there is additionally the low-power dissipation constraint, it is preferable to select the appropriate encoding scheme based on the required fault coverage.

Thus, there is no need to use cyclic codes to detect single error occurrence in arithmetic operators when the parity code is sufficient. The selection of the encoding scheme is done based on two parameters; the required safety level and the achieved power saving. In [7] there is a discreet flow for any kind of application to select the appropriate encoding based on these parameters. A variety of encoding schemes is used to achieve a reasonable increase of area cost and power dissipation. The topology used for a functional unit, to achieve the self-checking property, is shown in figure 10.9, which is applicable for separable codes. In this topology, a *code symbol* is generated, based on the selected encoding scheme, from the inputs and results in the check bits. The circuit also includes the *Information Symbol Generator* (ISG), where the output is generated. This output is then used to generate in the complement of the check bits in *Check Symbol Complement Generator* (CSCG). The resulted check bits and their complements are then compared in a n-variable TRC checker.

Hereinafter, several admissions about encoding are done for successful designing of a low-power safety-critical system. The memories are encoded using the CRC code including the address bits with the information bits to avoid aliasing of the errors. In [2, 4, 13] various topologies to ensure successful memory

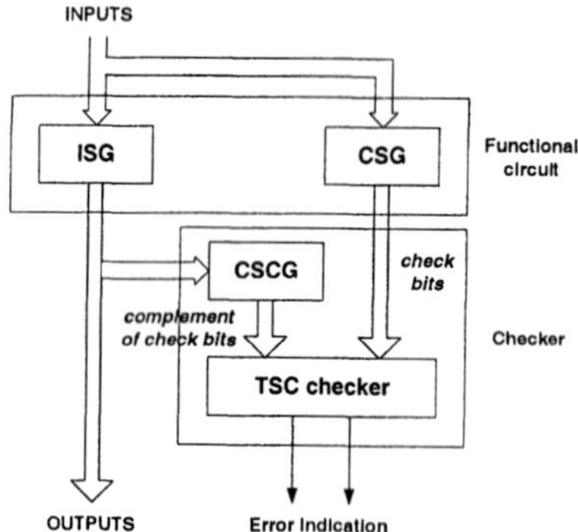

Figure 10.9. Self-checking circuit based on separable codes.

testing have been presented. The cyclic codes are preferred from the SED codes, while a Hamming code is suitable if the requirements were those of a fault-tolerant system. In figure 10.10 a topology to detect errors occurring to memory blocks is illustrated. The bits of data and address are encoded as one block and the check bits are stored in memory. If an error occurs in either reading or writing process, there is a disagreement of the stored check bits to those resulted from decoding phase.

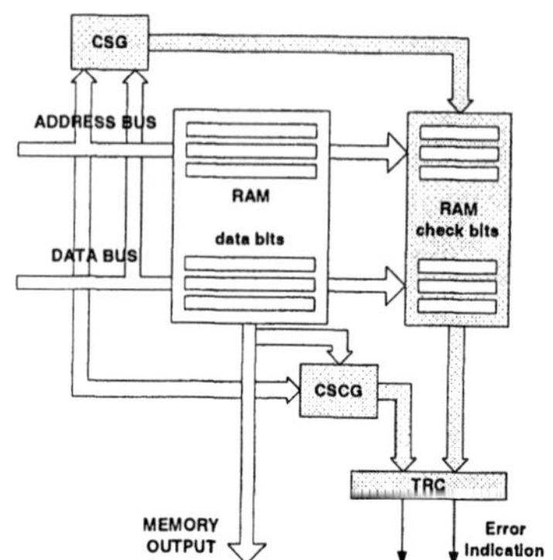

Figure 10.10. A topology to check memory operation.

The arithmetic operators and generally the combinational circuits are encoded using parity code. The check of these circuits is based on *parity prediction* technique which exploits several properties of the combinational circuits and is implemented introducing low area overhead. This technique is based on the existence of correlation between the inputs and the output of an arithmetic operator. For example, the parity of an adder's output is equal to $PS = PA \oplus PB \oplus C_{in} \oplus PC_p$, where PA and PB are the inputs parities, C_{in} is input carry and PC_p is the parity of all partial carries.

Finally, sequential circuits, such as the control unit are either based on encoding using multiple-error detecting codes, as the *m-out-of-n* or the Berger code, or parity prediction technique only in circuits with special characteristics, such as the counter. Multiple-error detecting encoding schemes are required, due to the susceptibility of *Finite State Machines* (FSMs) to unidirectional errors. In [5] it is advised to use *m-out-of-n* coding for the inputs of the FSMs. The *m-out-of-n* encoding presents minimum Hamming distance of two (2), which is required to achieve the fault-secure property for the FSMs. For more details and the proof consult the [5]. Furthermore, to optimize power dissipation, minimization of the redundant bits is required. An hypothesis is made for every sequential circuit. A unidirectional error can occur only to the inputs and the outputs or in the combinational part of the FSM. Even error occurrence in the flip-flops can be modeled as an error in the combinational part of the FSM.

10.4.3   Transforming circuits into TSC circuits

The circuits that can be found in a system are categorized in combinational and sequential. Thus, the combinational TSC circuits are presented first including circuits such as adders, multipliers etc. Finally, the sequential TSC circuits are presented emphasizing in the design of FSMs and counter units.

10.4.3.1   Combinational circuits.   Combinational circuits do not include clock or reset signals and their operation is not dependent by previous values of the inputs. The most well known combinational circuits are the multiplexer, the adder, the multiplier and the shifter. Larger combinational circuits are usually composed by these units. This type of circuits is categorized into bit-sliced and random logic, based on their structure. A bit-sliced structure is easily distinguished by its granularity.

Bit-sliced structures   As mentioned earlier, bit-sliced structures are easily distinguished because of the frequent use of fundamental blocks (e.g. fulladder cell). Thus, they are generated by repetitively placing the same block side-by-side using a common rule for interconnection. In figure 10.11 a typical 2-to-1 multiplexer of N-bit input length is shown. This bit-sliced structure is the simplest, where inputs are associated to specific inputs of the fundamental blocks and the output is directly connected to the blocks output. Notice that the control signal `sel` is propagated throughout the multiplexers to the output. This allows checking of the control signal through a TRC checker, similar to figure 10.6. Similar interconnection can be used to design fault-secure adders. In figure 10.12 a fault-secure carry-ripple adder is illustrated.

In [14] an extended survey of designing parity prediction arithmetic operators was presented. This work was concerned in [15] where the circuits' require-

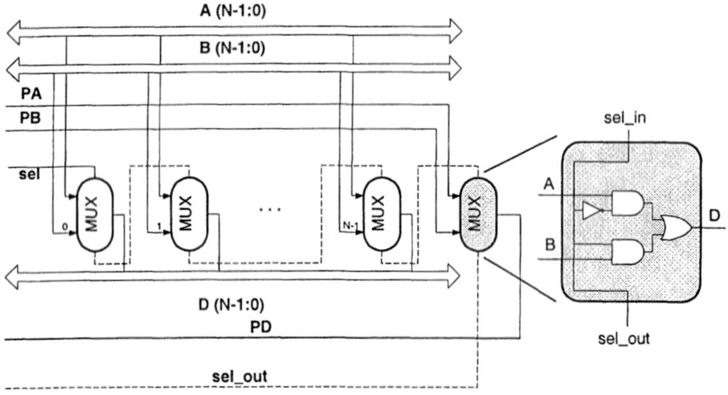

Figure 10.11. A TSC 2-to-1 multiplexer of N-bit input length.

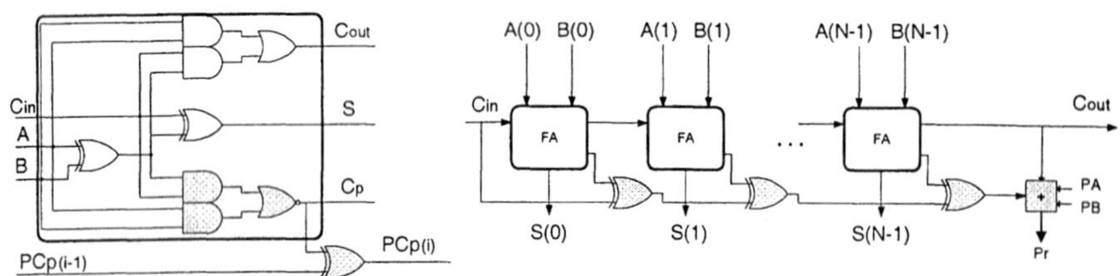

Figure 10.12. A fault-secure carry-ripple adder.

ments in terms of area integration and power dissipation were further explored. Furthermore, in [16, 17, 18] several modifications on the fault-secure circuits were performed to reduce their power dissipation. Thus, the fault-secure adder of figure 10.12 can be implemented using either a balanced EX-OR tree to produce the PCp signal or selecting the carry-complete structure, see figure 10.13, that eliminates glitches inside the adder. In general, the use of balanced structures is ideal to reduce power dissipation of a circuit. Even for circuits with low power savings, as in the case of some adders, their use in complex system design results significant system power savings.

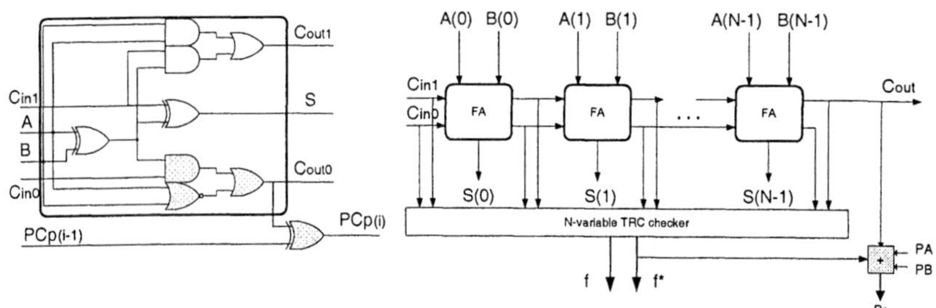

Figure 10.13. A low-power fault-secure carry-complete adder.

A significant class of arithmetic operators is that of multipliers. The simplest multiplication structure is the Braun multiplier, which is formed by an array of full-adder cells. In figure 10.14 the structure of the Braun multiplier is illustrated as it is reported in [14]. The use of the latter mentioned carry-propagate adders as well as balanced EX-OR trees to produce the PCp signal decreases significantly power dissipation. Comparison of the implementation of the fault-secure Braun multiplier, as shown in figure 10.15, to the simple non-safe implementation is given in table 10.1. Recall that common fault-secure implementations require duplication of the simple non-safe circuit. Thus, a typical double-channeled fault-secure implementation dissipates twice or more the power than that of the simple non-safe implementation. In all the implementations presented hereinafter the crucial factor is x2.

Table 10.1.   Characteristics of the fault-secure Braun multiplier (compared to the simple Braun-array multiplier).

Comparison Factor	4x4	8x8	16x16	32x32
Chip Area	1.654	1.643	1.637	1.637
Time delay (1/Performance)	1.500	1.314	1.233	1.189
Power dissipation	1.577	1.642	1.656	1.656

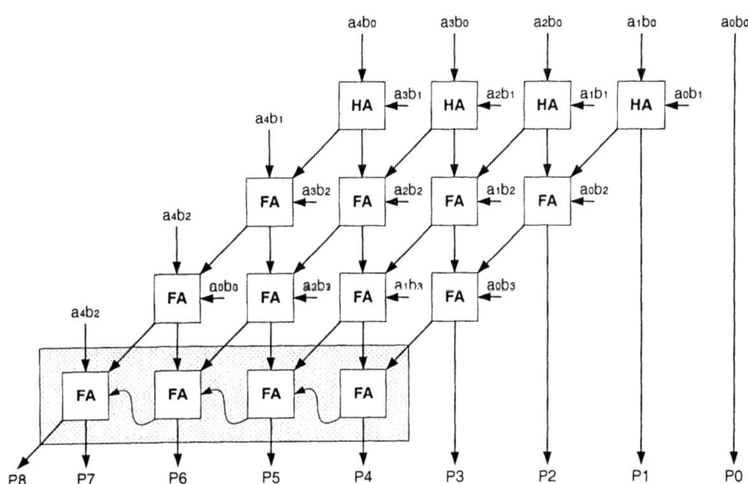

Figure 10.14.   A Braun multiplier network.

Random logic circuits   This kind of circuits does not present a discreet interconnection rule. However, there is the possibility to reduce their power

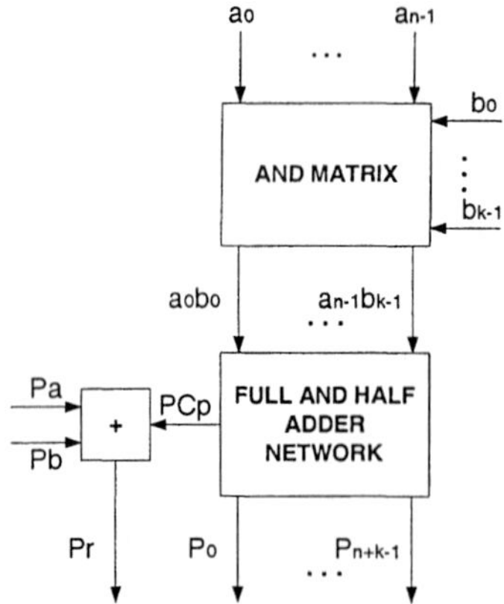

Figure 10.15.   A fault-secure Braun multiplier.

dissipation by decreasing their useless activity. An excellent example to illustrate power dissipation on a random logic circuit is the Wallace-tree multiplier. Adding extra hardware in order to balance the Wallace-tree, as proposed in [19], a glitch free structure is achieved. A fault-secure Wallace-tree multiplier is implemented as illustrated in figure 10.16 and its characteristics compared to the simple non-safe are shown in table 10.2.

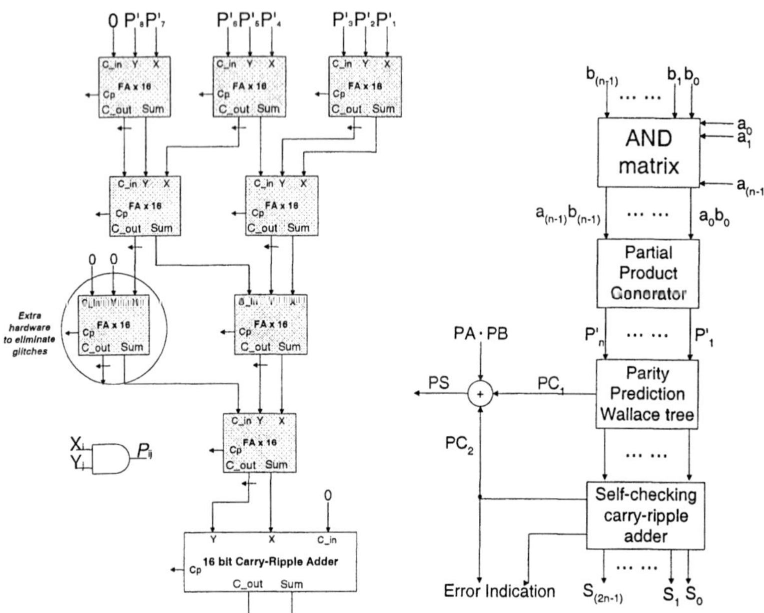

Figure 10.16.   (a) A balanced Wallace-tree; (b) A fault-secure Wallace-tree multiplier.

Table 10.2. Characteristics of the fault-secure Wallace-tree multiplier (compared to the simple Wallace-tree multiplier).

Comparison Factor	4x4	8x8	16x16	32x32
Chip Area	1.336	1.400	1.719	2.909
Time delay (1/Performance)	1.277	1.262	1.205	1.160
Power dissipation	1.050	1.108	1.164	1.197

Similar techniques can be applied to more complex circuits where control signals are required and encoding of the information bits is performed. In this case, duplication of small parts is preferable than encoding. Such actions are justified when there is the need to check small parts of the circuit due to the resulted C_{eff}. Sometimes, although switching activity is reduced the final area overhead is such that the resulted power dissipation is not reduced. Thus, when trying to reduce power dissipation of a circuit it is compulsory to perform an exploration of the trade off between increasing C_{out} and decreasing K.

The most complex arithmetic operator is the Booth multiplier, shown in figures 10.17, 10.18 and 10.19. The structure of this multiplier includes encoding and decoding parts, control signals propagating throughout the circuit and complex interconnections. Duplicating the small encoders/decoders and using the balanced structures, the Booth multiplier presents a 20% power saving compared to the double-channeled safe implementation (see table 10.3).

Table 10.3. Characteristics of the fault-secure Booth multiplier (compared to the simple Booth multiplier).

Comparison Factor	4x4	8x8	16x16	32x32
Chip Area	1.751	1.711	1.559	1.455
Time delay (1/Performance)	1.712	1.226	1.174	1.141
Power dissipation	2.423	1.981	1.728	1.616

From the latter exploration several conclusions are derived concerning low-power and safe multiplication. The balanced Wallace-tree fault-secure design was proven to be the most suitable, if the power consumption is the critical issue of the safety critical application we want to develop, but this design is of poor performance and it has an extremely large area overhead in comparison to the other designs. On the other hand the Booth fault-secure design is advisable, if our application demands low area overhead and high performance.

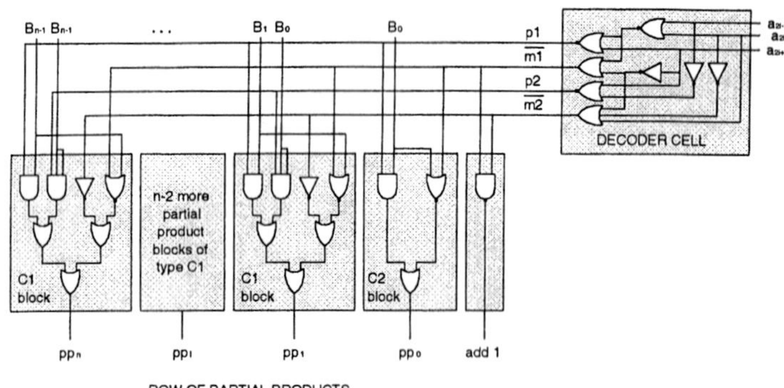

Figure 10.17.   Compact version of Booth decoder and selector circuits.

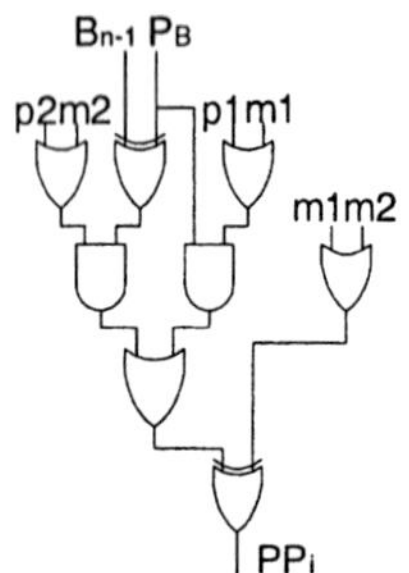

Figure 10.18.   Row Parity Prediction unit.

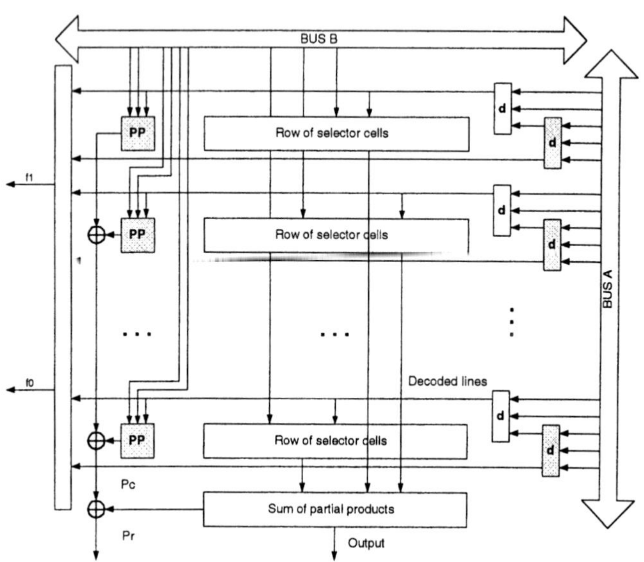

Figure 10.19.   Decoder cell duplication scheme and Double Rail checking scheme.

The last combinational circuit that interest a designer is the shifter. The barrel shifter structure can further serve as an example to show the application of the dynamic power management technique. The structure of a fault-secure barrel shifter was presented in [20] and is illustrated in figure 10.20. Adding flip-flops to shifter's inputs, the designer gains the ability to control the unit. This enables eliminating unnecessary operation of the shifter. This technique is not in general advisable but it can be applied to rarely operating units. In order to ensure safe-operation even for these units, the structures proposed in [9] to face up violations of the $TSCG(t)$ can be used.

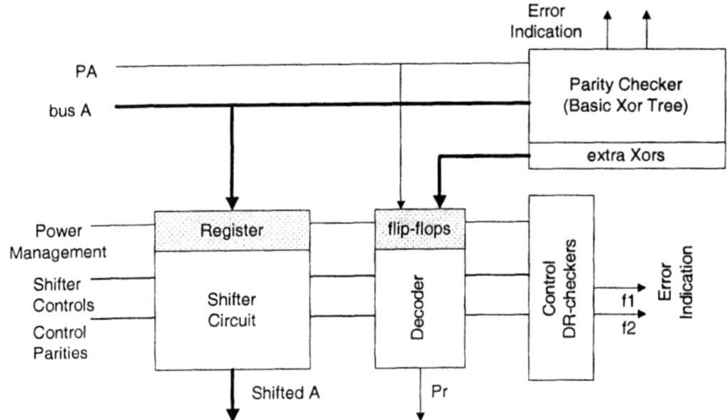

Figure 10.20. A fault-secure shifter.

Using the latter mentioned circuits it is easy to design an *Arithmetic & Logic Unit* (ALU). Exploiting the bit-sliced structures and the re-usability of several parts a low-power and small sized fault-secure ALU can be designed. The re-usability of non-active parts to calculate or generate the predicted parity introduces low area overhead for the checking mechanisms. In figure 10.21 the bit-slice of the ALU is illustrated. Notice the re-usability of the OR gates to predict the parity of the AND operation of the ALU and vice versa.

10.4.3.2 Sequential circuits. Sequential circuits have the propriety of time dependency meaning that the circuit's output depends on previous output values. Due to time dependency of the circuits, which may affect multiple lines, multiple-error detection codes have to be used to face error masking. The general principles to design self-checking sequential circuits are first presented. Finally, special focus is given to the most common sequential circuit in a system, the counter, which can be designed to be fault-secure using the parity prediction technique.

Designing FSMs The design of an FSM consists of three discreet specification phases: specification of the state, of the state transition and of the output. The specification of the state is achieved using the appropriate number of flip-

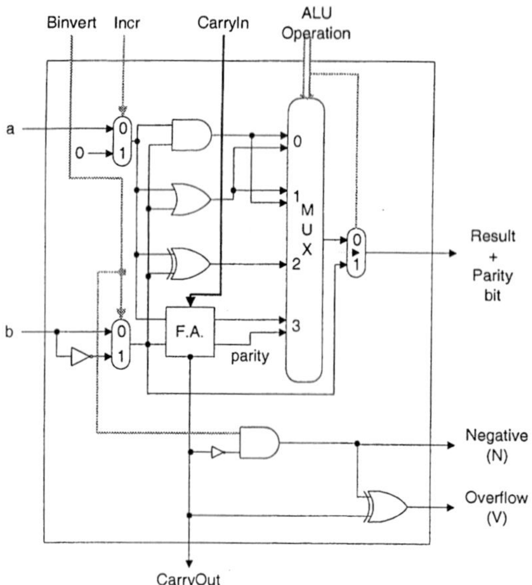

Figure 10.21.   Fault-secure bit-slice ALU.

flops [21]. Both state transition and output specification are small combinational circuits that depend on the current state of the FSM and the output. Such FSMs are called *deterministic* and only this type is concerned. The topology of a self-checking FSM Mealy machine is shown in figure 10.22, where checking is performed to both the output and the state of the FSM.

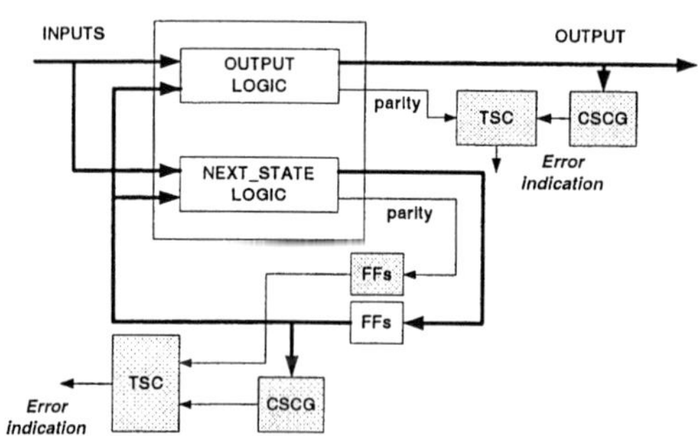

Figure 10.22.   A topology for self-checking Mealy FSMs.

Due to the dependency of the FSMs from the current state these circuits are susceptible to unidirectional errors. In [5] it is stated that in order to design a self-checking FSM it is advisable to use *m-out-of-n* encoding. This encoding scheme has a minimum of two (2) bits as Hamming distance, which is a basic requirement for input encoding targeting bidirectional error elimination. Thus, encoding the inputs of the FSM using the *m-out-of-n* code, a unidirectional multiple fault at the inputs cause either a single-bit error or a unidirectional multiple error at the outputs. Assigning the output value space appropriately, the unidirectional multiple error at the outputs can be mapped to non-code words. Thus, the output has either the expected correct value or it presents an detectable erroneous value (non-code word). Due to the non explicit topology of an FSM and due to the random character of the sequential units, there is no explicit technique to turn these kind of self-checking circuits into low-power. Thus they are only referred and not considered further for the completeness of the text. However, there is a critical circuit for every system which is sequential and there are straightforward techniques to lower its power dissipation; the counter.

Counting with safety The counter is a critical circuit found in almost any system and in most of the cases it affects significantly the critical path. It can be characterized as the simplest FSM, due to its nature. The most common topology to design a counter is a register that stores in every clock cycle its current value incremented (or decreased) by a fixed step. As far as the length of the counter increases, the critical path of the required adder follows the increase. Regarding the rate of the increase of the critical path, always in comparison to the increase of the counters length, it depends on the adder selection. Although, the self-checking sequential units are encoded using the *m-out-of-n* code, the counter can be designed based on the parity prediction technique as it was presented in [22, 23].

In [22], a fault-secure binary counter is proposed, which is based on the parity prediction technique. This counter achieves fault-secureness taking into consideration both current state Q^t and the next expected output value T^t, predicting the output's parity (figure 10.23). The main control signals `clock`, `reset` and `enable` follow the design guidelines for signal propagation throughout the system. This counter can efficiently detect single-fault errors, however this structure targets mainly binary counters with a fixed increase step of one (1). This means that the complexity of the counter increases significantly when the step of counting changes. In addition, the selected approach decreases significantly the critical path due to the introduced safety mechanism.

In [23], an approach to design fault-secure counters based on segmented structures was presented. A basic block was formed that was based on Johnson-Mobius encoding. This type of encoding is a Gray-like and it has a significant property; it has a fixed parity transition, which eliminates complexity in calculating the expected parity, so it does not affect significantly the critical path. In figure 10.24 the basic block of the segmented counter is illustrated, while in figure 10.25 a 8-bit fault-secure counter is shown built from the proposed segment.

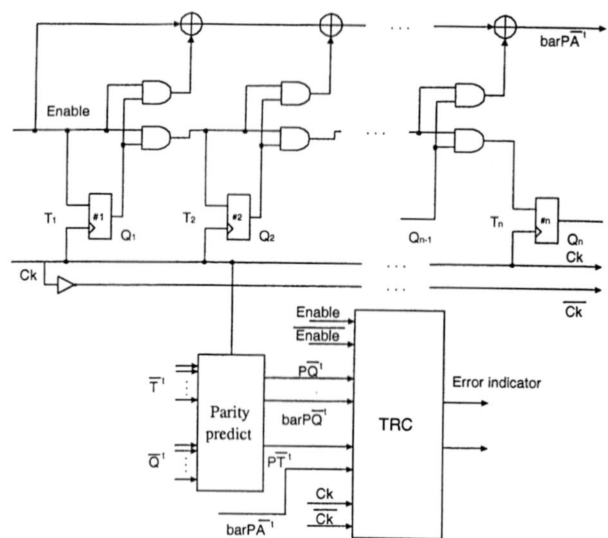

Figure 10.23. A binary fault-secure counter.

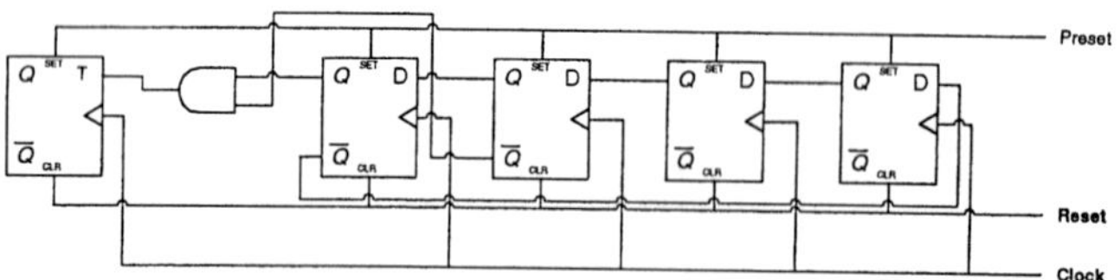

Figure 10.24. The basic block of the proposed segmented counter.

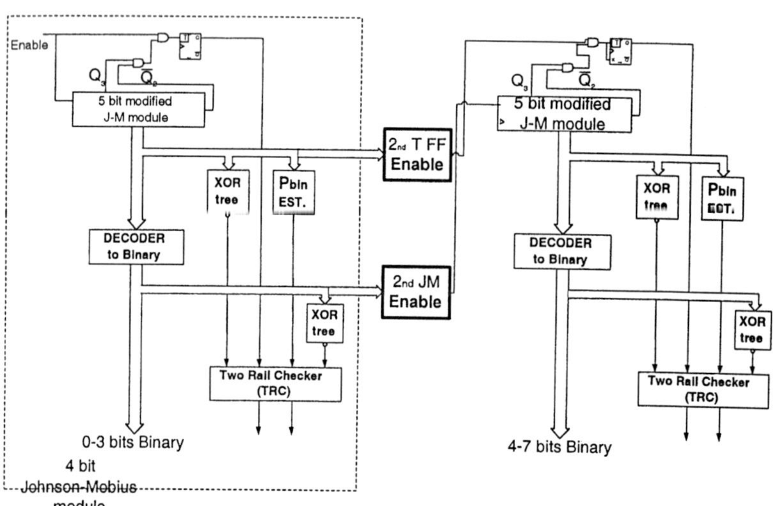

Figure 10.25. An 8-bit segmented fault-secure counter.

In tables 10.4,10.5,10.6 the two latter implementations are compared to the commonly met double-channeled counter. If integration area is the main design issue in the implementation of a safety critical application, then the simple FS binary counter proposed in [22] is the most suitable. On the other hand as it is derived from table 10.5, the binary FS counter is better suited for low-power safety critical applications since it is by far the least power dissipating counter, in comparison to the other two FS designs. In table 10.6, the main asset of the proposed Johnson-Mobius FS binary counter is highlighted. This counter is by far the fastest counter when compared to the other two fault-secure designs. Although, it seems convenient to compare the Johnson-Mobius counter in terms of performance (delay time) to the other two counters, when it is designed as non-safe and then duplicated it exceeds power dissipation and integration by a factor of x6. Thus, only the fault-secure implementation of this counter is considered.

Table 10.4. Comparison of the FS counters in terms of integration area (compared to the simple binary non-safe counter).

Implementation	4-bit	8-bit	16-bit	32-bit	64-bit
Double-channeled FS counter	2.75	2.91	2.93	2.97	2.98
FS Binary counter	2.41	2.29	2.18	2.14	2.11
Segmented FS counter	2.54	2.54	2.54	2.57	2.57

Table 10.5. Comparison of the FS counters in terms of power dissipation (compared to the simple binary non-safe counter).

Implementation	4-bit	8-bit	16-bit	32-bit	64-bit
Double-channeled FS counter	3.49	3.72	3.86	3.94	3.98
FS Binary counter	3.04	2.39	1.94	1.67	1.52
Segmented FS counter	3.89	3.58	3.28	3.11	3.03

10.4.4 Evaluating the presented design techniques

Following the guidelines that have been presented, a real-life application, which required both safe-operation and portability, was developed and integrated. In the bounds of CoSafe, an ESPRIT IV funded project, an ambulatory pump to infuse medicine to patients had served as an evaluation vehicle of the methodology. An existing commercial ambulatory pump, distributed and pro-

Table 10.6. Comparison of the FS counters in terms of performance (compared to the simple binary non-safe counter).

Implementation	4-bit	8-bit	16-bit	32-bit	64-bit
Double-channeled FS counter	1.08	1.05	1.03	1.02	1.01
FS Binary counter	1.65	1.69	1.67	1.62	1.52
Segmented FS counter	1.71	1.16	0.86	0.69	0.59

duced by Micrel S.A., has served as the referenced system. A prototype chip was integrated in Fraunhofer-Institut(FhG-IIS) using the AMS 0.6 um technology. A significant benefit that is derived from the nature of the design technique was the reduction of the required time to complete testing procedure.

The final prototype [24] integrates all the presented fault-secure circuits. In table 10.7 the main characteristics of the prototype are showed. The striking goal is the reduction of the power dissipation by 45% of the whole system maintaining the safety levels of operation as the initial product, which was based on the double-channeled technique.

Table 10.7. Characteristics of the safety-critical portable medicine infusion pump.

Feature	Measure
Core	8-bit Data Bus, 16-bit Address Bus, 256 Direct Address Peripherals, Watchdog Timer, Built-In Signature Analysis Unit, Wallace Multiplier, 8-bit Barrel Shifter, RS-232, PIC.
Area	73,000 eq. gates($49.48\ mm^2$, AMS 0.6um)
Power consumption	12.5 mW/MHz (upper bound)
Clock frequency	Up to 100 MHz
Soft-Error coverage	~98%

10.5   Conclusions

It was showed that mixing design techniques targeting both low-power dissipation and testability, especially on-line, is quite tricky. Safety critical applications are demanding redundancy to increase observability and achieve the required safety levels. On the other hand low-power design targets on minimization of the switching activity and the effective capacitance. As it was showed, the two design approaches often are proved to be controversial.

Design rules that can guarantee successful development of portable safety critical systems were presented. First, the controversial nature of the two design approaches were showed. Then, the design strategies concerning the propagation of critical control signals, such as `clock` and `reset` were presented. Finally, the units that are commonly met in every system were analyzed in terms of fault-secureness. In every design approach, the power dissipation term was considered.

A real-life application has served as an evaluation vehicle and it was showed that the presented circuits and guidelines have resulted in a reduction of 45% of the existing product's power dissipation, while safe operation levels have been maintained. Thus, the benefits of the application of the presented circuits and guidelines are turned to be significant, when targeting safety critical applications.

References

[1] J. von Neumann. *Probabilistic Logics and Synthesis of Reliable Organisms from Unreliable Components*. Automata studies, in Annals of Mathematical Studies, no.34. Princeton University Press, 1956.

[2] M.L. Bushnell and V.D. Agrawal. *Essentials of Electronic Testing for Digital, Memory & Mixed-Signal VLSI Circuits*. Kluwer Academic Publishers, 2001.

[3] M.J. Howes and D.V. Morgan, editors. *Reliability and Degradation - Semiconductor Devices and Circuits*. UK: Wiley-Interscience, 1981.

[4] M. Abramovici, M.A. Breuer and A.D. Friedman. *Digital Systems Testing and Testable Design*. IEEE Press, 1990.

[5] P.K. Lala. *Self-Checking and Fault-Tolerant Digital Design*. Morgan Kaufmann Publishers, 2001.

[6] P.K. Lala. *Digital Circuit Testing and Testability*. Academic Press, 1997.

[7] DVE 801. *Principles for computers in safety related systems*. DIN V DVE 0801 Standard, 1990.

[8] J.-C. Lo and E.B. Fujiwara. Probability to Achieve TSC Goal. *IEEE Trans. on Computers*, 45:450–460, April 1996.

[9] A.P. Kakarountas, K.S. Papadomanolakis, S. Nikolaidis, D. Soudris, and C.E. Goutis. Confronting Violations of the TSCG(t) in Low-Power Design. In *IEEE ISCAS 2002 Proc.*, Scottsdale, USA, 2002.

[10] A. Raghunathan, N.K. Jha and S. Dey. *High-Level Power Analysis and Optimization*. Kluwer Academic Publishers, 1998.

[11] A. Chandrakasan and R.W. Brodersen. *Low Power Digital CMOS Design.* Kluwer Academic Publishers, 1995.

[12] J.M. Rabaey and M. Pedram. *Low Power Design Methodologies.* Kluwer Academic Publishers, 1996.

[13] A. Benso et. al.. Online and Offline BIST in IP-Core Design. *IEEE Design & Test of Computers*, vol.18, pp.92–99, Sept. 2001.

[14] M. Nicolaidis, R.O. Duarte, S. Manich and J. Figueras. Fault-Secure Parity Prediction Arithmetic Operators. *IEEE Design & Test of Computers*, vol.14, pp.60–71, April 1997.

[15] A.P. Kakaroudas, K.S. Papadomanolakis, E. Karaolis, S. Nikolaidis, N. Alachiotis and C.E. Goutis. Hardware and Power Requirements of Self-Checking Circuits. In *ICECS 1999 Proc.*, Pafos, Cyprus, pp.1655–1658, 1999.

[16] A.P. Kakaroudas, K.S. Papadomanolakis, V. Kokkinos and C.E. Goutis. Comparative Study on Self-Checking Carry-Propagate Adders in Terms of Area, Power and Performance. In *PATMOS 2000 Proc.*, Göttingen, Germany, pp.187–194, 2000.

[17] K.S. Papadomanolakis, A.P. Kakarountas, V. Kokkinos, N. Sklavos and C.E. Goutis. A Comparative Study on Fault Secure Signed Multiplication Designs. In *11th IFIP Intl. Conf. on VLSI Proc.*, Montpellier, France, pp.183–188, 2001.

[18] K.S. Papadomanolakis, A.P. Kakarountas, V. Kokkinos, N. Sklavos and C.E. Goutis. The Effect of Fault Secureness in Low Power Multiplier Designs. In *PATMOS 2001 Proc.*, Yverdon-Les-Bains, Switzerland, 2001.

[19] T. Sakuta, W. Leeand P.T. Balsara. Delay Balanced Multipliers for Low Power/Low Voltage DSP Core. In *IEEE Symposium of Low Power Electronics: Digest of Technical Papers*, pp 36–37, 1995.

[20] R.O. Duarte, M. Nicolaidis, H. Bedder and Y. Zorian. Fault-Secure Shifter Design: Result and Implementations. In *European Design & Test Conference Proc*, Paris, France, 1997.

[21] Z. Kohavi. *Switching and Finite Automata Theory.* McGraw-Hill, 1978.

[22] E. Karaolis, S. Nikolaidis, and C.E. Goutis. Fault Secure Binary Counter Design. In *ICECS 1999 Proc.*, Pafos, Cyprus, pp.1659–1661, 1999.

[23] K.S. Papadomanolakis, A.P. Kakarountas, N. Sklavos and C.E. Goutis. A Fast Johnson-Mobius Encoding Scheme for Fault Secure Binary Counters. In *DATE 2002 Posters*, Paris, France, 2002.

[24] A.P. Kakarountas, K.S. Papadomanolakis, V. Spiliotopoulos, S. Nikolaidis and C.E. Goutis. Designing a Low Power Fault-Tolerant Microcontroller for Medicine Infusion Devices. In *DATE 2002 Designer's Forum Proc*, Paris, France, pp.205–211, 2002.

Chapter 11

DESIGN OF A LOW POWER ULTRASOUND BEAMFORMER ASIC

Robert Schwann

Thomas Heselhaus

Oliver Weiss

Tobias G. Noll

University of Technology RWTH Aachen
Chair of Electrical Engineering and Computer Systems
tgn@eecs.rwth-aachen.de

Abstract   This chapter describes a design experiment on the implementation of the digital part of a low power 16-channel ultrasound beamformer ASIC. The design is based on a physically oriented design methodology applying a datapath generator for automatic layout generation and a quantitative optimization of all building blocks for lowest possible power dissipation. State-of-the art power reduction strategies were applied on all levels of the design process from the algorithmic system level down to the physical layout level. In addition to an optimized degree of pipelining, the use of the datapath generator preserves locality, the key to disruptive technology power features. The application of appropriate number representation in arithmetic building blocks, clustered supply voltage techniques, and customized memory blocks, among other power reduction strategies, allowed the attainment of unequaled power figures.

Keywords: Digital interpolating beamformer, datapath generator, on-chip A/D converter, on-chip delay calculation.

D. Soudris et al. (eds.), Designing CMOS Circuits for Low Power, 235–269.

11.1   Introduction

Ultrasound systems are built using transducer arrays (to transmit and receive ultrasound waves), frontends (to perform the beamforming and signal processing on the received signals), and backends (to perform image processing, video display and interfacing). Special options like Doppler and color flow processing can be added to give real-time information on blood flow properties.

Conventional receive beamforming is performed with a "delay-and-sum" of the weighted outputs of the transducer array (figure 11.1). In combination with the transducer, the frontend of an ultrasound system mainly determines the image quality which is the most important metric of such a system.

Electronic beamforming has become standard in today's medical diagnostic ultrasound scanner systems. Because of increased reliability and especially image quality in comparison to mechanical sweeped-transducer based systems, this technology has recently migrated from high-end systems to high-volume low-end systems. Due to the relatively large computational power required to perform electronic beamforming, reasonable power dissipation, especially for battery-powered hand-held systems, becomes very challenging.

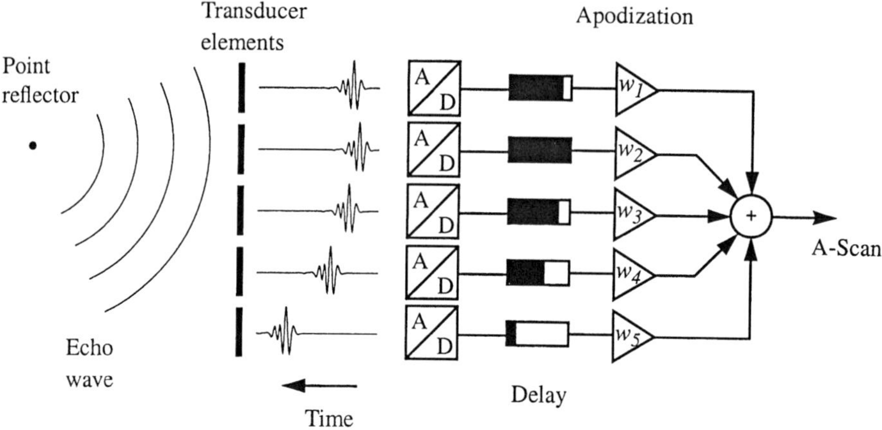

Figure 11.1.   Principle of delay-and-sum beamforming

Besides beamforming and further signal processing, a digital frontend consists of an array of Analog-to-Digital (A/D) converters. After a preamplification, the output signal of each active transducer element is fed into an A/D converter, resulting in a typical setup with 64, 128 or today even 256 discrete A/D converters used simultaneously. Current beamformers usually consist of several ASICs. Due to the large amount of interfacing (by connecting several

A/D converters with typically 10 bit wordlength) and the necessary processing complexity, the number of channels per ASIC is limited. A complete beamformer subsystem requires additional ICs for the digital control logic.

Thus, digital frontends are expensive and feature very high power dissipation. Using 128 discrete A/D converters in combination with a large number of complex ASICs for the signal processing results in a power consumption and PCB area being not acceptable at least for hand-held systems, and in a design effort not acceptable for high-performance products. Therefore, the applicability regarding high-performance and low-cost systems would not be given without further integration, combining the A/D converters, the beamformer and the additional digital signal processing on a single chip.

On-chip A/D converters with specifications typical for medical ultrasound are now realizable. The number of A/D converters to be implemented on one chip depends on the maximum allowable power dissipation and the available chip area. The goal of the LUCS project (Low Power Ultrasound Chip Set) was to integrate at least 8 channels and preferably 16 channels on one chip that allows the realization of a scalable ultrasound beamformer architecture. Scalability in channel number can be attained by cascading several identical beamformer chips.

The primary industrial exploitation of the beamformer chip described here is the use in a hand-held scanner, where low power consumption is a cardinal constraint with respect to acceptance and applicability. Therefore, the chip was highly optimized for the application in portable, battery-powered hand-held ultrasound scanner systems (single chip solution), but should be suitable for high-performance systems (cascaded multi-chip solution) as well.

Key technologies for this development are on-chip A/D converters and dedicated VLSI circuits featuring lowest possible power dissipation, and leading to a key component with the A/D converters, a beamformer and the required control logic on one chip.

The applied technology is a three metal layer, 0.5-μm CMOS 3.3 V process suitable for low power mixed-signal design which is provided by one of our project partners who designed the on-chip A/D converters.

The high throughput rates and complex tasks of digital signal processing as required in the beamformer chip would be associated with high power consumption if conventional design methods would be applied. For battery-powered hand-held systems with very limited power budget the resulting power dissipation would be not acceptable. Therefore, beside the application of state-of-the-art power reduction strategies, application-specific solutions have been elaborated to reach the goal of a very low power consumption:

On architectural level, an efficient strategy for low power circuits is based on the following two-step strategy: In the first step, power consumption is taken into account as a major part of the efficiency criterion while optimizing the circuit components. This is done by selecting an optimal degree of pipelining, a well-known and very power efficient parallelization method. In the second step, the throughput rate is matched to the requirements by applying parallelization or by voltage scaling in a clustered supply voltage approach.

For power optimization on logic level, a power optimal circuit technique was chosen and design strategies to reduce the local switching activities were applied. For example, additional switching activity due to glitches was reduced by using localization and avoiding structures with high logical depths. Once again, pipelining is the method of choice to limit the logical depth and enforce localization. Due to locality, interconnect lines can be kept short allowing minimum device sizing. Hence, decreased wiring capacitances as well as reduced gate capacitances lead to low power consumption. The choice of the number representation also has a strong impact on the switching activity. To take full advantage of all these optimization strategies, a full-custom design style was utilized. By using a datapath generator described in section 11.2, the design effort was kept comparable to that of a synthesis-based standard-cell design and a fast iterative quantitative optimization became possible.

On circuit level, the most attractive way to reduce the power consumption of digital CMOS circuits is a reduction of the supply voltage. However, a supply voltage reduction leads to degradations in the maximum clock frequency and therefore in throughput rate. In order to design circuits which are optimal with respect to an efficiency criterion including power consumption as well as throughput rate and silicon area, several supply voltages ("clustered supply voltage") have to be used. For the hand-held application, the different supply voltage levels were chosen to minimize power consumption. Since high-performance applications require a higher clock frequency, the design ensures faster operation by an increase in the supply voltage levels, payed for by a more-than-linear increase of power consumption.

Additionally, the choice of optimized register circuits and the concept of the clocking system is essential with respect to the overall power consumption in highly pipelined systems, as required for the proposed chip set. The register circuits have to be adapted carefully to the given technology. Once more, this optimization is only applicable in a full-custom design style. For the clock system, distributed clock generation and buffering has to be applied.

Finally, on layout level, the device sizing has to be optimized quantitatively to allow minimal power consumption. Generally, increased transistor-widths lead to higher power consumption due to the increased gate capacitances. By applying pipelining to enforce localization, long interconnections are mostly avoided and transistor dimensions can be kept minimal. For the transistors influencing the time-critical path, however, a careful enlargement of device sizes leads to higher throughput rates and efficiency and therefore, by supply voltage scaling, to decreased power consumption. Of course, only a full-custom design style offers the possibility to apply these strategies.

11.2 The Key to Physically Oriented Design: Automated Datapath Generation

The design of the beamformer chip requires a low power design methodology applied on all levels of CMOS design from the algorithmic system level down to the physical realization level.

Applying a full-custom design strategy guarantees lowest possible power consumption but suffers from an exhaustive design effort. On the other hand the synthesis of standard cell based designs suffers from suboptimal throughput rate, large area and considerably high power consumption. Additionally, back-annotation of physical design properties spans a rather long design cycle. With a physical design oriented implementation, macro blocks providing better area utilization are created with precisely controlled timing in critical paths, even at low power consumption. Therefore, automated datapath generation is the key to physically oriented design at a design effort comparable to that of today's standard cell based synthesis strategies.

The EECS Datapath Generator (DPG) mainly exploits the following properties of typical algorithms for digital signal processing:

DSP algorithms are iterative in time and functionality which leads to commonly regular datapaths. Typical DSP datapaths can be split into non-iterative control and glue logic, and iterative datapaths. The control logic may be synthesized with state-of-the-art tools while the time-, area-, and power-critical datapaths are well-suited for an efficient physically oriented design. Such implementations of highly regular datapaths show two main properties: The core region which is composed of only a few different basic cells typically occupies more than 90% of the whole datapath area. The border regions include many individual basic cells which are slightly modified core cells at most. Yet these border cells typically require up to 90% of the design effort. This requires the following DPG key feature:

(a) The DPG includes powerful routines for layout modification to exploit regularity very efficiently. This ensures a minimized set of basic layout elements (leaf cells) required for the physically oriented design of datapath macros at very low design effort.

DSP algorithms are based on only a few types of simple basic functions. This means the DSP datapath basic functions can be optimized at low design effort and cover a broad range of DSP applications. Consequently, this leads to the next DPG key feature:

(b) The DPG supports the highly optimized design of DSP basic functions as parameterizable macros based on only a few leaf cells.

DSP algorithms inherently contain a high degree of locality. Locality at the physical layout level leads to short interconnections with small wiring delays and small wiring capacitances which is even more significant in the deep submicron domain. Hence, locality is a benefit of itself and ensures high throughput, small silicon area, and low power dissipation. Therefore a third DPG key feature is derived:

(c) The DPG preserves locality down to the physical layout level by directly deriving the placement of the basic cells from the signal flow graph of the DSP algorithm.

The key features (a), (b), and (c) lead to the basic DPG flow depicted in figure 11.2. The DPG processes the structure of the datapath which is specified by a parameterizable signal flow graph at word level. A dedicated datapath description language (DDL) is used as language for the SFG description. In contrast to well-known HDLs, such as VHDL and Verilog, the DDL syntax provides intuitive commands for node definition, node instantiation, and branch definition. During the first DPG run phase, a structure at bit level containing the relative placement of the leaf cells is derived while preserving the locality of the original signal flow graph.

According to the structure at bit level, leaf-cell layouts are taken from the leaf-cell library, adapted to the individual environment, and stored in the basic-cell library. A key function of the adaption is a dedicated routing engine optimized to

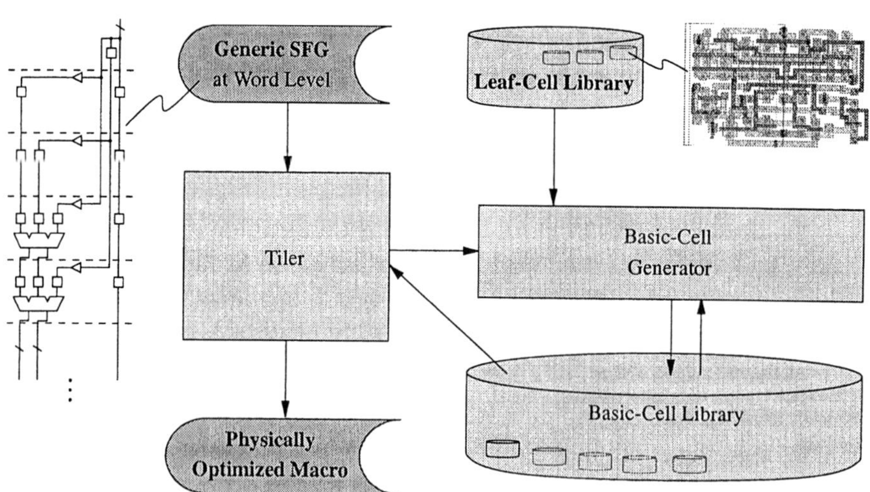

Figure 11.2.　　Main DPG Flow

route a small number of interconnections in a very small area, which allows the construction of optimum, shape-based net-to-net routing solutions in a compact layout.

Finally, the datapath macro is assembled by tiling the generated abutment basic cells according to the topology given in the bit level structure. A final routing pass is not necessary. In this way the required abutment basic cells are automatically derived from a very small set of leaf cells ensuring a low design effort. The DPG output consists of a layout, a schematic, and an abstract view. Timing and energy views are generated employing industrial standard tools. Thus the DPG smoothly interoperates with state-of-the-art design frameworks, and the generated macros can be integrated into any existing standard cell based design flow as hard-macros. An industrial standard flow is applied for logic and layout verification.

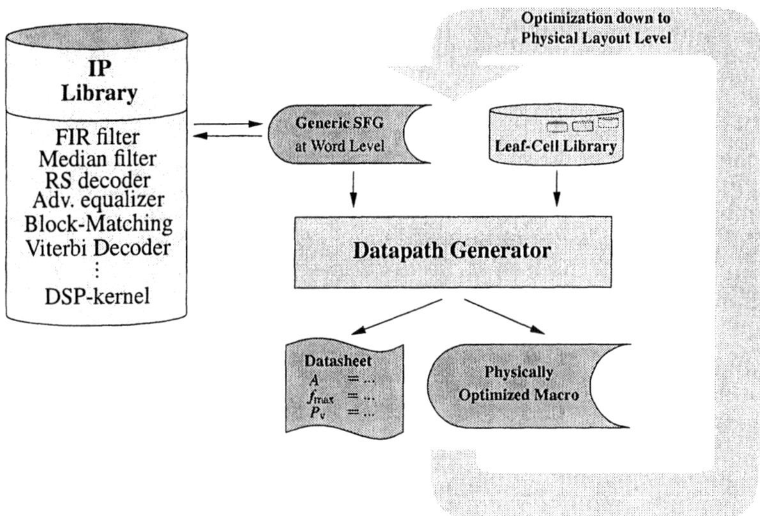

Figure 11.3. DPG-based IP

As shown in figure 11.3 the DPG enables the quantitative optimization of generic macros down to the physical layout level in short iteration cycles by modifying the SFG and/or modifying the leaf-cell library according to the features derived from the previous DPG run. Archiving the parameterizable SFGs lead to a growing IP library of generic architectures such as multipliers, dividers, memories, FFT modules, Viterbi decoders, Reed-Solomon decoders, etc. Numerous industry related benchmarks prove the performance of the DPG-based design methodology [8, 7, 30].

11.3   Beamformer's System Overview and Requirements

Figure 11.4 shows a functional block diagram of the beamformer chip. Only one channel of the digital beamformer is displayed in detail. In the following section, the components of the block diagram are outlined.

After amplification and lowpass filtering, the receive data are sampled by on-chip A/D converters at uniform time steps given by the sampling period. N channels (8 or 16) are processed by one beamformer chip. The parallel channels are shown here in a stack.

The sampled data are stored in the delay unit, allowing a programmable latency. Thus, the temporal resolution of the delay unit is given by the sampling period of the A/D converter. The length of the delay unit is programmable.

The sampling clock signal Φ_S for the A/D converters and the synchronization signal "Synch" for the delay unit are provided by a control unit. The control units of the parallel channels can be synchronized with minimum global control. Start-up information and weighting factors are broadcasted to all channels by a register chain.

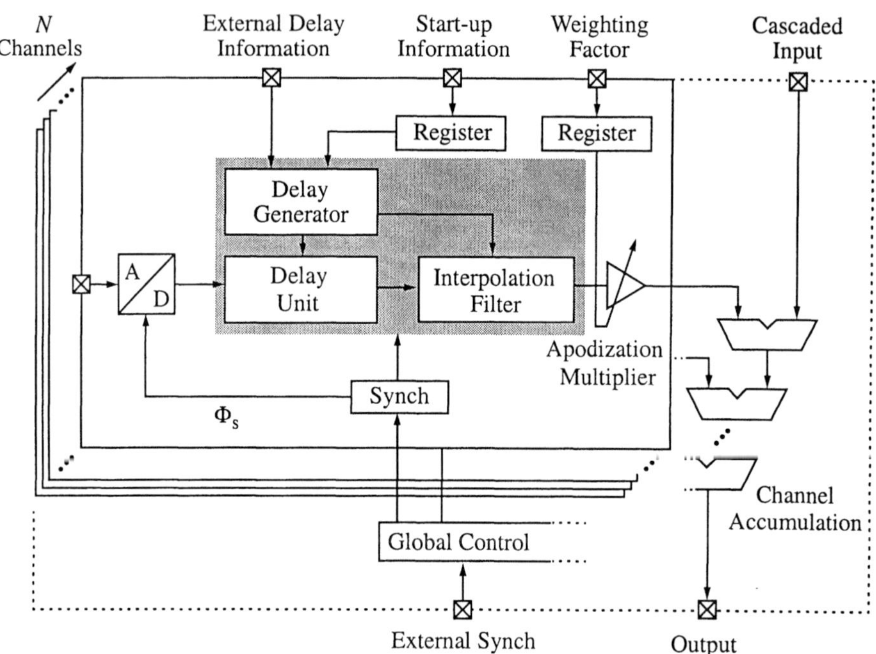

Figure 11.4.   Block diagram of the LUCS chip

For low power dissipation, the A/D converters should operate at the lowest possible sampling rate allowed by the sampling theorem. However, for high image quality, the temporal resolution must be much higher than that. Therefore, an interpolation filter is used to allow delay increments smaller than the sampling period. With respect to the sampling period, the delay information consists of an integer part and a fractional part. The integer part is used to control the delay unit, and the fractional part is used to control the interpolation filter.

The appropriate delay information is required for each channel and each sampling point individually. For maximum accuracy, the delay information could be loaded from an external memory. However, an external storage causes high power dissipation due to the resulting chip-to-chip data transfer at sampling rate which is not acceptable at least for hand-held applications. For lowest power dissipation, the delay information has to be provided by an on-chip delay generator. The delay generator is capable of computing the delay information iteratively and in real time (section 11.4.2).

The apodization multiplier allows to improve the directivity of the transducer aperture and thereby the lateral resolution by multiplying the delayed data with a channel-dependent weighting factor. To account for dynamic changes of the receive focus, this factor can be updated during reception.

The output of all channels is summed up on-chip using the channel accumulation. Scalability in the number of active system channels is attained by cascading several chips. A power-optimized multi-chip solution is attained with an appropriate partitioning and interface definition.

11.4 Optimization and Benchmarks of Basic Building Blocks

In the following, the low power design is presented for those functional blocks of the digital signal path that feature the largest impact on power dissipation.

11.4.1 Coarse Delay Unit: Optimization of Customized Memories

To delay the sampled data words according to the depth-dependent receive focus, the coarse delay is performed using a programmable delay unit for each channel which is controlled by its own delay generator. The sample word capacity of the delay unit is derived from the requirements of beam steering, channel count, and cascadability.

One method would be to provide a conventional random access memory for storing the sample words. According to the required delay, the read address could be generated by subtracting an offset from an address counter value. Figure 11.5 illustrates this implementation. At the start of a new scan line, the read address counter and the write address pointer are initialized. During operation, this solution allows arbitrary changes of the read address by changing the offset.

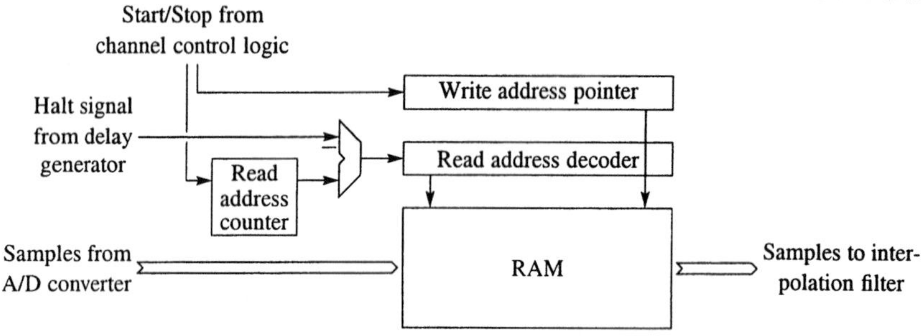

Figure 11.5.   RAM-based implementation of the delay unit

However, the proposed beamformer chip does not need the full functionality of this addressing scheme since the necessary delay is only increased monotonically by at most one sampling period per clock cycle. Because of this, the overhead for generating the read address can be reduced by applying an address pointer scheme for read operations. By the use of this simple addressing scheme, the power dissipation can be significantly reduced.

The usage as a shift register of dynamic length operating at sampling rate requires simultaneous read and write operations. A power efficient implementation with separate access ports can only be realized with a dedicated memory structure. Three different types of memory core cells were considered with respect to their power dissipation. A comparison with possible single (6T) and dual ported (8T) SRAM implementations shows that the 3-transistor (3T) DRAM implementation not only features the smallest area but, here more important, the lowest power dissipation, even with a partial reduction of the supply voltage (table 11.1).

To achieve a power-optimized solution compared to standard generator-based memories, concepts for the reduction of power dissipation applicable to this dedicated memory were investigated. Verified by simulations, a power dissipation model was developed to minimize the power dissipation while ensuring reliable operation under worst case conditions. The concepts as well as the obtained results are described in the following subsections.

Table 11.1.   Power optimized RAM implementation

Cell Type	f_{access}	P @ 3.3 V	P @ 1.8 V	A
6T-SRAM	80 MHz	395 µW	105 µW	100 µm^2
8T-SRAM	40 MHz	390 µW	110 µW	150 µm^2
3T-DRAM	40 MHz	210 µW	85 µW[1]	64 µm^2

[1] Read Circuitry @ 1.8 V, Write Circuitry @ 3.3 V (1.8 V / 3.3 V)

11.4.1.1 Power reduction concepts.

One major parameter for optimizing the power dissipation of the memory in the delay unit is the parasitic word and bit line capacitance. This is considered in the following sections applying concepts for partitioning the memory into blocks and by segmentation of the memory cell core. Furthermore, techniques for supply voltage reduction are considered in order to achieve a power-optimized solution.

Block partitioning. The power dissipation of the parasitic access line capacitances of the memory can be reduced by partitioning it into multiple blocks. All blocks are connected to a common data bus. For access operations, only one block is active while the other blocks are in standby mode, consuming less power. Each block consists of a memory core and three peripheral units as shown in figure 11.6. Row and column pointers are separated for the read and write port, and can be addressed independently. During simultaneous read/write operations the addressed data words of the memory are passed through I/O circuits to independent data lines of the common data bus. The I/O circuit is enabled by the corresponding column pointer. It contains a simple sensing/driving circuit, and latches to hold the data of a complete row. Row and column pointers as well as the required clock generation are controlled in each block independently.

The power dissipation caused by driving the common data bus depends on the number of blocks and I/O circuits inside a block which are connected to the bus as well as the bus length itself. To keep the bus length as short as possible, mirrored blocks are used to access the bus from both sides. Partitioning the memory into a large amount of blocks reduces the power dissipation in the memory cores but increases the power dissipation for driving the long common data bus. On the other hand, considering only one block, the power dissipation of the short "common" data bus is reduced, while the memory core dissipates more power, since there are no inactive blocks. Obviously, there will be an opti-

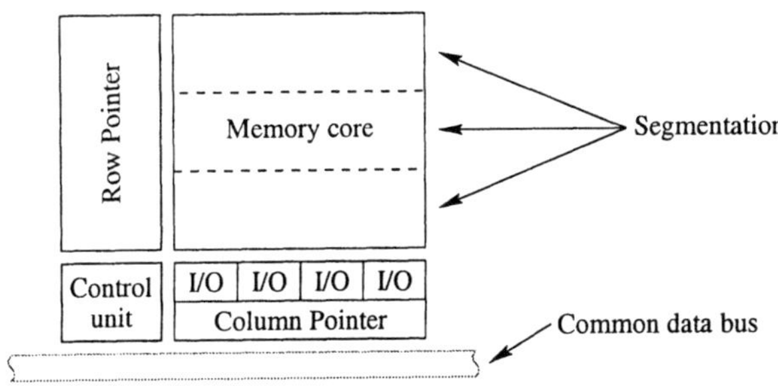

Figure 11.6. Delay unit: one block of the partitioning is shown

mal block partitioning for a minimum power dissipation which was determined by applying the power dissipation model of the delay unit.

Core segmentation.   Since the power dissipation is dominated by the parasitic capacitances of the access lines, the memory core could be segmented as shown in figure 11.6 to reduce the power dissipation caused by the parasitic access lines relative to an unsegmented memory core [13]. Because the memory is accessed sequentially, it is necessary to activate the global word line several times in order to read a complete row. Thus a hierarchical word line structure for the memory core of the delay is not profitable in this case. Besides that, a multi-divided data line archtecture only offers advantages above a specific number of rows. In conclusion, for small memories of a few Kbit as required for this application and taking into account the sequential access scheme, an unsegmented structure has been shown to be favorable.

Bit line voltage reduction.   The read operation in the memory core area is performed sequentially by precharging the read bit lines and activating the read word line. An additional supply voltage allows to reduce the precharge voltage and thereby the power dissipation of the sensing circuit (subnote in table 11.1).

Row pointer selection.   Since the memory of the delay unit is accessed in sequential order, each block (figure 11.6) implies two separate ring pointers for read and write operations. In order to obtain a low power solution for the ring address pointers of the delay unit, a new pointer structure with low clock capacitance was developed and compared to different address pointer concepts commonly applied to large FIFO memories (figure 11.7). They consist of

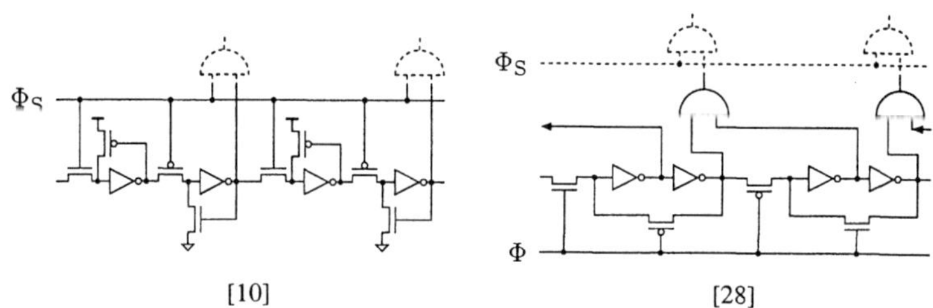

Figure 11.7.   Conventional low power pointer types

transfer gates and weak pull-up or pull-down devices to transport the pointer information which is stored dynamically on the input capacitances of subsequent gates. The shift operation is sensitive to both edges of a single phase clock.

However, if applied to the memory of the delay unit, it would be necessary to provide an additional evaluation phase clock for gating the word lines. With circuit simulations it was found that short-circuit currents at the active pointer position lead to an increased power dissipation which is relevant especially for small pointer chains such as those required for the memory of the delay unit.

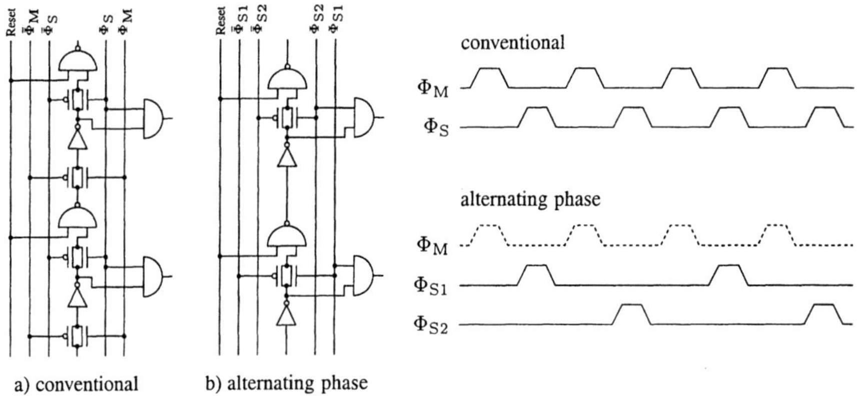

a) conventional   b) alternating phase

Figure 11.8. Row pointer comparison

An alternative way to implement the row pointer is based on a complementary two phase clock system. In this case, register cells consisting of subsequent master and slave latches are used (figure 11.8a). However, the master clock Φ_M is not required for the addressing scheme applied here since the word lines are activated in the slave phase only. Therefore, the power dissipation due to the capacitances of the master clock line could be saved if the transport of the pointer signal can be guaranteed by applying slave clock signals only. Hence, an alternating slave clock phase system based on two different pointer cells was developed (figure 11.8b). The slave latches are clocked with different slave phases which are derived from an appropriate complementary two phase clock system. Concerning only the shift operation, the power dissipation can be reduced by a factor of about two.

Table 11.2. Pointer comparison at 40 MHz and 3.3 V

Pointer type	Evaluation phase	Evaluation logic	P
Ref. [10]	missing	missing	151 μW
Ref. [28]	missing	missing	236 μW
Conventional	provided	omitted	119 μW
Alternating phase	provided	provided	76.3 μW

Four different pointer chains with a length of 16 were compared for power dissipation (table 11.2). The first two pointer types would have to be extended with an evaluation logic for gating the word lines, as required in our case, which would slightly increase the power figures for those pointers. The proposed alternating phase pointer type already includes the evaluation logic. It was simulated for this dynamic memory solution and achieves the lowest power dissipation (highlighted in table 11.2).

11.4.1.2   Memory cell characteristics.   To ensure reliable operation of the memory cell, the minimum data retention time and leakage current require a minimal charge on the storage node, and therefore require a specific capacitance for a given storage node voltage. Because the storage node voltage influences the MOS capacitance of the storage node, its degradation causes a reduction of the stored charge. The results of the related investigations allow a reduction of the bit line voltage in order to reduce the power dissipation.

11.4.1.3   Power dissipation breakdown.   Power dissipation figures were simulated for each part of the memory block and the common data bus. The approach considered parameters like operating voltages, the row count of the row pointer, and characteristic details of each part of the delay unit as well as the operating conditions. The total power dissipation of the memory core and the peripheral units of a single active block (figure 11.6) can be written as

$$P = P_{\text{Core}} + P_{\text{Rowptr}} + P_{\text{Interface}} + P_{\text{Colptr}} + P_{\text{Control}} + P_{\text{Bus}}. \qquad (11.1)$$

The power dissipation of the inactive blocks can be neglected since they are in standby mode with practically no power dissipation. The power dissipation of the common bus is determined by its length and therefore is a function of the block count and core size. Each part of the equation 11.1 depends on the operating frequency. However, for a core containing multiple sample words in one row, the power dissipation is proportional to the sampling frequency divided by the number of its columns, f_s/N_C. This also holds for the row pointer, since the word lines of the core are only activated every N_C-th clock cycle. Core segmentation is not considered here since the specified delay length is too small to expect a significant power reduction by this method.

Table 11.3 shows the simulated worst-case power dissipation of core memories with different partitioning. The parameter N_B specifies the required number of blocks, while N_R specifies the number of rows. The product $N_B \cdot N_R \cdot N_C$ determines the delay length and must contain at least 540 sample words to build a complete delay unit for the given beamformer specifications.

Optimal partitioning yields a power dissipation of approximately 3.2 mW with $N_B = N_C = 4$ as highlighted in table 11.3. A comparable dedicated memory, optimized for low power as reported in [15], was scaled to the re-

Table 11.3. Power dissipation in μW for different forms of partitioning at 3.3 V

N_B	N_R	N_C	P_{Core}	P_{Rowptr}	P_{Colptr}	$P_{\text{Interface}}$	P_{Bus}	P_{Sum}
2	68	4	1040	145	630	1115	385	3315
2	54	5	880	95	675	1115	465	3230
2	46	6	780	70	730	1115	540	3235
2	40	7	700	55	785	1115	595	3250
2	34	8	615	45	835	1115	695	3305
4	34	4	615	85	630	1115	770	3215
4	28	5	525	60	675	1115	925	3300
4	24	6	475	45	730	1115	1080	3445
4	20	7	420	35	785	1115	1235	3590
4	18	8	385	30	835	1115	1390	3755
6	24	4	475	65	630	1115	1155	3440
6	18	5	385	45	675	1115	1390	3610
6	16	6	355	35	730	1115	1620	3855
6	14	7	320	30	785	1115	1850	4100
6	12	8	285	25	835	1115	2080	4340

quired memory size, technology, and supply voltage of 3.3 V. It then consumes 15.4 mW which is almost a factor of five above our solution.

Since the memory has been designed using a parameterizable DDL as described in section 11.2, and as programmable delay units are of general interest in many applications, the applied power reduction strategies can be re-used in future memory designs of different sizes and partitioning.

11.4.2 Delay Generator

During receive beamforming, the necessary delay to be applied to the channel input signals varies during the acquisition of a scan line. For high image quality, the receive focus has to be updated continuously. In a discrete-time system, this means ultimately to update the receive focus at sampling rate. Different scan lines are acquired by either activating different transducer elements or by applying electronic beam steering. The second method is used for phased arrays and requires different sets of delay information for each scan line. A sidelobe reduction in the directivity pattern can be attained by increasing the resolution of the applied delay. Assuming a sampling rate of four times the transducer center frequency, an evaluation was performed with an ultrasound simulation tool. A typical directivity pattern is shown for a transmit steering angle of zero degrees in figure 11.9 with the delay resolution as a parameter given in fractions of the sampling period T_s. Delay shifts smaller than $T_s/16$ did not significantly improve the image quality but increase the hardware effort for the subsample delay adjustment (section 11.4.3.4).

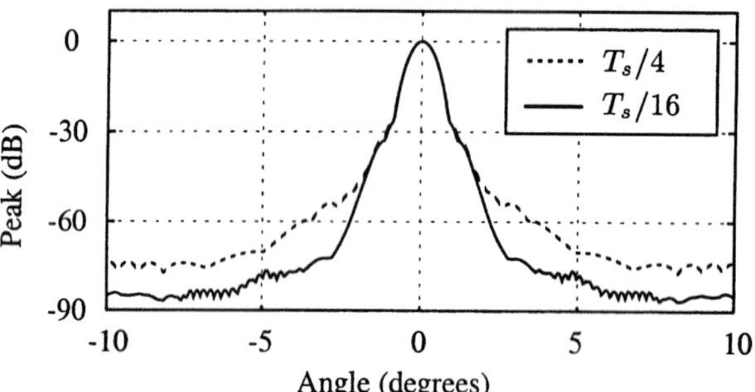

Figure 11.9.   Directivity pattern with respect to the delay resolution

Look-up table.   The delay information for a complete scan line can be pre-calculated and stored in a look-up table [20, 4]. A delta-encoding scheme can be used to reduce the delay information to 1 bit per update cycle [20]. In a typical phased array system with 128 scan angles and 5200 sampling points, the memory capacity required for a system with dynamic receive focus would be approximately 650 Kbit per channel. Obviously, for 8 or 16 channels this would require a large external memory resulting in excessive power dissipation due to the operation of the external memory and chip-to-chip communication. Thus, this solution is not feasible for a battery-powered hand-held scanner.

On-chip real time calculation.   For the on-chip calculation of delay information, iterative concepts have been described in the past which allow to increase the subsample delay resolution by additional hardware [2, 21]. The concepts rely on the midpoint algorithm discussed in [1]. In comparison with a new algorithm used here [6], our solution operates with a very small number of adder stages in the critical path and thus relaxes the timing constraints. Finally, an increase in delay resolution can be achieved with a cascadable structure comprising only one adder.

11.4.2.1   New algorithm for on-chip calculation of delay information.

Figure 11.10 shows the geometry which is used to determine the channel- and depth-dependent delay of a focused transducer array. After a wavefront is transmitted into the medium, an echo wave propagates back from the focal point p to the transducer. With c_0 denoting the speed of sound in the medium, the distance from p to the origin is given by $r(t) = t \cdot c_0/2$ and differs from the distance to a transducer element located at the position x. The difference

$$l(t) = \sqrt{r^2(t) + 2x \cdot r(t) \cdot \sin\theta + x^2} - r(t) \qquad (11.2)$$

leads to a difference in propagation time of $l(t)/c_0$. The receive beamformer has to compensate for these time differences for each active element individually.

Of course, a straightforward implementation of equation 11.2 is not feasible for low power applications.

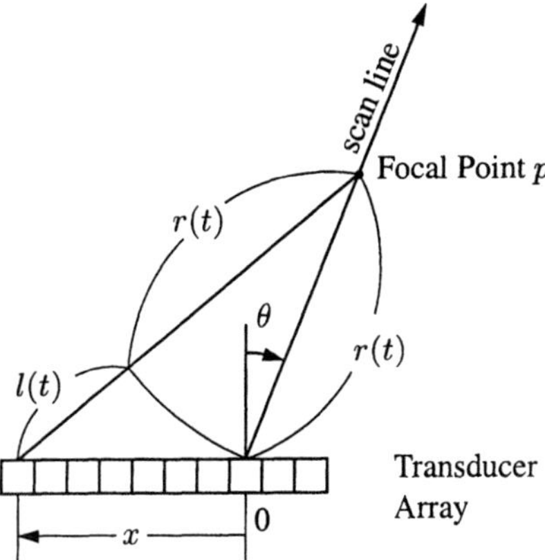

Figure 11.10. Geometry of a focused transducer array

Iterative calculation of delay information. A discrete-time beamformer evaluates equation 11.2 every update period T_u. Between two evaluations, the focal point p has moved by the distance $d = c_0 \cdot T_u/2$ along the scan line. To simplify arithmetic operations equation 11.2 is squared and normalized by the distance d.

$$\underbrace{n^2 + 2k(n) \cdot n + k^2(n)}_{A(n)} = \underbrace{n^2 + \alpha \cdot n + \beta}_{B(n)} \tag{11.3}$$

While $r(n \cdot T_u)/d$ becomes a sequence ($n = 1, 2, 3 \ldots$) of integers, $k(n) = l(n \cdot T_u)/d$ is quantized to integers by the algorithm, thus determining the delay accuracy and allowing a simple low power hardware implementation. The initial delay is given by $k(0) = \lfloor x/d \rfloor$. Constant terms like $\alpha = 2(x/d) \cdot \sin\theta$ and $\beta = x^2/d^2$ are pre-calculated for each scan line. The terms A and B are calculated iteratively,

$$A(n+1) = A(n) + \underbrace{2n + 2k(n) + 1}_{a(n)} \quad ; \quad A(0) = k^2(0)$$

$$B(n+1) = B(n) + \underbrace{2n + 1 + \alpha}_{b(n)} \quad ; \quad B(0) = \beta \tag{11.4}$$

The update terms are calculated as

$$\begin{aligned} a(n+1) &= a(n) + 2 \quad ; \quad a(0) = 2k(0) + 1 \\ b(n+1) &= b(n) + 2 \quad ; \quad b(0) = \alpha + 1 \end{aligned} \tag{11.5}$$

Changes in the difference $k(n)$ are detected by a comparison of terms A and B in equation 11.3. Whenever A becomes greater than B, the algorithm indicates that $k(n)$ has to be decreased by one. Then $A(n+1)$ remains at $A(n)$ and $a(n+1)$ remains at $a(n)$ for one iteration step. A signal flow graph of the algorithm is shown in figure 11.11. The initialization values are loaded into the four registers once at $n = 0$ for each scan line.

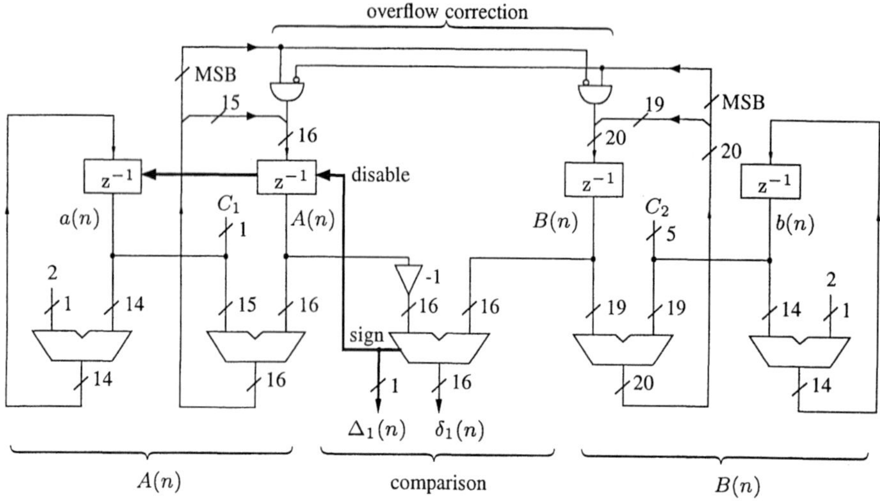

Figure 11.11.   Iterative calculation of delay information

Increased delay resolution.   With the ultimate operation at sampling rate $(T_u = T_s)$, the algorithm calculates 1 bit of subsample delay information, according to a delay resolution of $T_s/2$. To increase the delay resolution, various concepts were examined with respect to their power dissipation.

Each doubling of the update rate $1/T_u$ decreases d and increases the delay resolution by 1 bit. A calculation of the delay information in parallel is possible with focus shifts in fractions of d. To avoid the fact that the power dissipation of both methods increases at least quadratically with the delay resolution, a new concept has been elaborated. It is based on a comparison of the term $B(n)$ to one of the intermediate thresholds $A(n \pm 1/2) = (n + k(n) \pm 1/2)^2$ with the positive sign for $A \leq B$. To simplify the calculation, equivalent thresholds have been derived for the difference $\delta_1(n) = B(n) - A(n)$:

$$\frac{a(n)}{2} - \frac{1}{4} \quad \text{for} \quad A \leq B \text{ or } \Delta_1(n) = 0$$

$$-\frac{a(n)}{2} + \frac{3}{4} \quad \text{for} \quad A > B \text{ or } \Delta_1(n) = 1$$

$$(11.6)$$

The evaluation of $\delta_1(n)$ provides one additional bit of subsample delay information. For practical transducer array geometries, the term $a(n)/2$ is dominating and the additional terms (equation 11.6) can be omitted without a significant error. A signal flow graph of the resulting structure is shown in figure 11.12. For the next bit of subsample delay information, four cases have to be examined, since the threshold depends on $\Delta_1(n)$ and $\Delta_2(n)$. Neglecting the small terms again, the threshold just depends on $\Delta_2(n)$. Thereby, the structure can be cascaded to meet the required delay resolution. This has been verified by simulations with up to four cascaded stages.

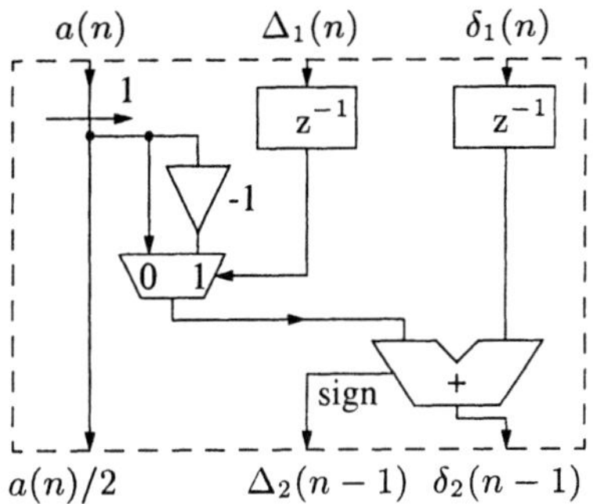

Figure 11.12.   Calculation of one additional bit of delay information

11.4.2.2   Implementation.   The algorithm was implemented calculating 3 bit of subsample delay information for a delay resolution of $T_s/8$. The calculation of the subsample delay information as shown in figure 11.11 only requires knowledge of the difference $B - A$. The wordlengths of A and B have been significantly reduced by an overflow correction of the accumulators.

To further reduce the power dissipation, the clock rate has been halved by choosing $T_u = 2T_s$. This reduces the delay information by 1 bit and thus requires one additional stage for the calculation of subsample delay information. Since the thresholds are evaluated only every second sampling period, the delay accuracy is slightly reduced. A power-optimized VLSI macro was realized following the proposed DPG-based design methodology (section 11.2).

While the adders were selected with regard to a minimized power dissipation, the registers were chosen to support low power standby when the delay calculation is inactive. The presented design was realized with only 8 adders and 8 registers per channel. Operating in a 40 MHz system at 3.3 V, it dissipates 2.1 mW using the described technology.

Three different algorithms have been implemented using the datapath generator. A comparison in figure 11.13 shows that only the linear power increase of the cascaded calculation offers a high delay resolution at reasonable power

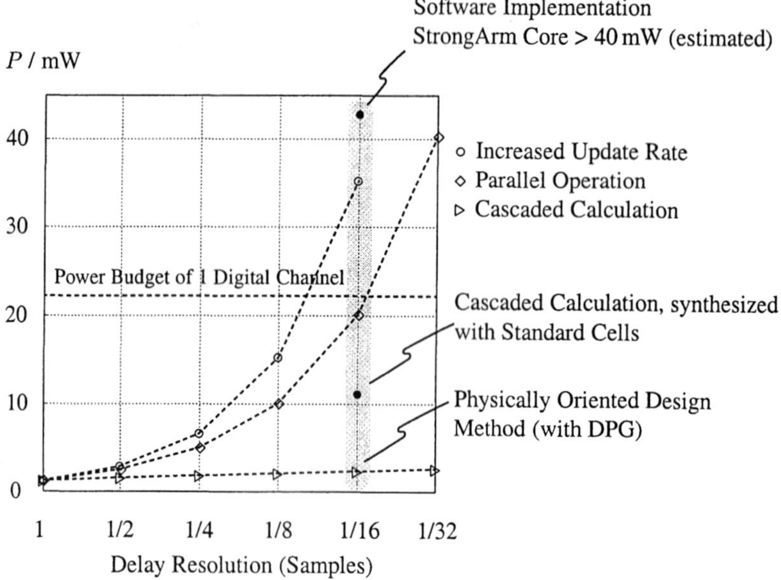

Figure 11.13.   Concepts for an increased delay resolution: power dissipation of delay calculation

dissipation. For a delay resolution of $T_s/16$, the DPG-based design of this algorithm is compared to a synthesized standard cell approach and to an on-chip processor-based software solution, and both would significantly affect the outlined total power budget for the digital part of the channel.

11.4.3   Fine Delay Interpolation Filter

For the applied interpolating beamforming scheme, the implementation of the required interpolation filter is presented in the following. The filter realizes the subsample delay adjustment in the digital part of each channel. Because the total power dissipation of the digital part is mainly caused by the interpolation filter, its optimization offers the highest potential to reduce to the overall power dissipation.

The required delay quantization of $1/L$ times the sampling period can be achieved by an L-fold interpolation filter. In a straightforward approach, this could be realized by padding $L - 1$ zeros between the input samples at an L times higher sampling rate followed by lowpass filtering. Because of the zero padding, most of the multiplier operations would not contribute to the filter result. Therefore, the use of L subfilters in polyphase structure, each operating at the input sampling rate, becomes favorable [11, 18]. In the conventional approach the interpolation filter is realized as a programmable filter using a small look-up table containing L different sets of coefficients.

An attractive alternative uses L hardwired subfilters providing the L interpolated timing phases which can be selected by a demultiplexer (figure 11.14a). For $L = 4$, three of the subfilters are designed to provide the delays in steps of quarters of the sampling period (i.e. 0.25, 0.5, and 0.75). As the fourth subfilter has to realize a delay by one sampling period it simplifies to a single delay.

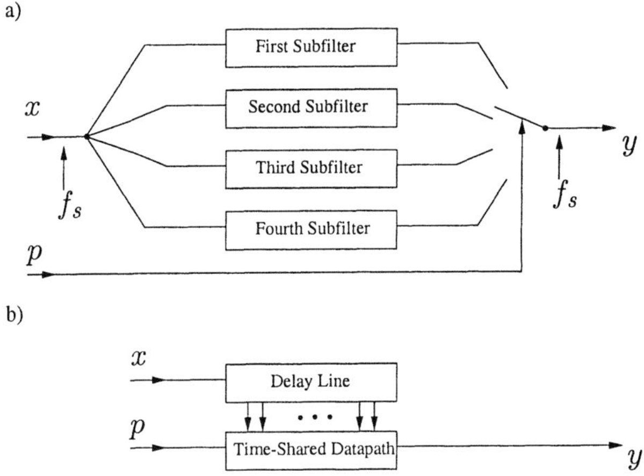

Figure 11.14.　Principle of a) interpolation filter and b) time-shared datapath with p denoting the timing phase selector

For the desktop application a lowpass filter consisting of four 12-tap subfilters was designed featuring an improved flatness of the passband frequency response, while for the hand-held application a lowpass filter consisting of four 6-tap subfilters with lower performance regarding image quality was chosen in favor of power reduction.

11.4.3.1　Architecture of the interpolation filter.　With respect to the hand-held application lowest power dissipation has been the primary concern during the optimization of the filter structure. Therefore, various power reduction strategies have been applied on all levels of CMOS design from system down to physical layout level to minimize the power dissipation.

Figure 11.14a shows the principle block diagram of the interpolation filter described above.

The input samples x are provided at sampling frequency f_s and this data rate is retained through the whole filter. According to the required timing phase the switch position p selects the proper output y from one of the subfilters. Because of the frequent switching between the timing phases during typical operation, the four subfilters would have to be continuously active in such a parallel implementation, thus leading to a high power dissipation. Since only one output of the four subfilters is used at a time, an implementation of the filter with a shared datapath for all subfilters becomes attractive (figure 11.14b). The sharing of the datapath allows to reduce the number of required adder stages by 33% and a corresponding power reduction is attained. However, now additional effort for generating the partial products becomes necessary.

The combined subfilter with the shared datapath is implemented in direct-form 1 which integrates the delay line. Additionally, here the direct-form 1 requires less delay elements than the transposed direct-form 1 because of the large intermediate filter result wordlengths.

The input samples enter the filter in a two's complement representation and the partial products are generated as two's complement numbers as well. This requires the extension of the sign bit of the partial products up to the most significant bit of the corresponding intermediate result which leads to a higher capacitive load at the partial product gates in the most significant weight. In addition, the time critical path increases which limits further power reduction techniques. To minimize this load to a small constant load in all weights, a biased number representation has been chosen for the input samples, similar to the technique proposed in [25]: By inverting the sign bit of each input sample, the input samples are converted to a binary offset representation and can be interpreted as strictly positive values. The resulting error of the filter output is corrected by subtraction of a constant value. In conclusion, a significant reduction of the power dissipation can be achieved. For example, the power dissipation in the 6-tap filter mode was reduced by 30%.

To reduce the number of partial products to be accumulated in the filter, the three subfilter coefficient sets can be encoded in canonical-signed-digit (CSD) representation [27] to minimize the number of non-zero bits.

Table 11.4.   Power dissipation of a moderately and a fully pipelined 12-tap filter

P @ 40 MHz	logic	clock	total
moderately pipelined (3.3 V)	24 mW	5 mW	29 mW
fully pipelined (1.8 V)	4 mW	7 mW	11 mW

The rather low throughput rate requirements could be met with a lower degree of pipelining, and based on carry-save techniques [17] the filter would dissipate 29 mW at 3.3 V. The higher throughput potential of a fully pipelined datapath structure can be traded by architectural voltage scaling for a further power reduction. The simulation results of the two alternative implementations are summarized in table 11.4.

11.4.3.2 Structure of the datapath. In general, CSD-encoding of the filter coefficients requires the generation of a positive, a negative or zero partial product pp. This is shown in table 11.5 on the example of three different coefficient values ($c_j = \sum_i c_j^i \cdot 2^i$) for one tap. Due to the sharing of the filter datapath among the subfilters, three different partial products might be required, as can be seen in the least significant bit c_j^0 (bold type column). This requires a slow partial product gate (figure 11.15a) with high power dissipation.

Table 11.5. Modified SD-encoded coefficient values

coefficient value c_j	initial CSD representation				modified SD representation			
4	0	1	0	**0**	0	1	0	**0**
7	1	0	0	**-1**	1	0	-1	**1**
5	0	1	0	**1**	0	1	0	**1**

However, recoding to a modified SD-representation enforcing the non-zero bits of all three coefficients in each weight to be either $c_j^i \in \{0, +1\}$ or $c_j^i \in \{0, -1\}$ allows the simplification of the partial product gate to a single NAND- or NOR-gate (figure 11.15b).

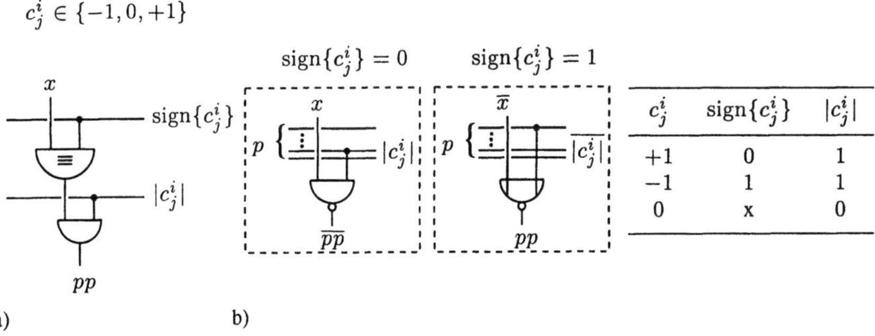

Figure 11.15. a) Partial product gate for CSD-coefficients, b) partial product gates for modified SD-coefficients

This optimization reduces the hardware effort by a factor of three and significantly reduces the power dissipation of the partial product generation. Due to the modified SD-representation additional adders in the datapath might be necessary, however in the presented filter application only 2 additional adders were required, which corresponds to an increase of just 4%.

A simple example of a signal flow graph resulting from the optimizations described above is shown in figure 11.16. The example has two non-zero bits for c_0, two for c_1 and one for c_2. It has been chosen to show the structure of the filter, independent from the actual coefficient values. Pipeline cut-sets are introduced after the carry-save adder stages and after the partial product generation. According to the control signal p, one of the four coefficient sets is selected and the proper partial product is generated. A further advantage of using direct-form 1 is that now the control signal p has to be synchronized according to the pipeline delay but not to the system delay. Each combination of the modified SD-values of the coefficient set is encoded to one of the m bits of the synchronized control signal p. Thus each partial product gate is wired to the according bit of p.

Because of the typical control signal switching activity characteristics in the given application (i. e. frequent switching in the beginning and less activity towards the end of the data acquisition within one scan line), the additional synchronization overhead for eight parallel control bits does not noticeably affect the power dissipation.

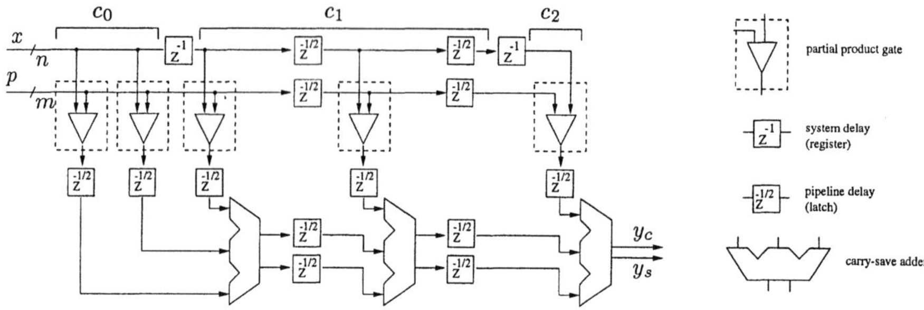

Figure 11.16.   Interpolation filter: Structure of the datapath

11.4.3.3   Comparisons.

The filter was implemented using the datapath generator (section 11.2). For comparison with other state-of-the-art low power filter implementations reported in the literature, power dissipation per tap was normalized according to

$$P_{\text{norm}} = \frac{P_{\text{tot}}}{\#\,\text{taps}} \cdot \frac{8}{w_{\text{coeff}}} \cdot \frac{10}{w_x} \cdot \left(\frac{1.8}{V_{DD}}\right)^2 \cdot \frac{0.5\mu\text{m}}{\text{technology}} \cdot \frac{40\,\text{MHz}}{f_s}. \qquad (11.7)$$

The filter implemented here has 10 bit input wordlength, 8 bit coefficient wordlength and an internal wordlength of up to 22 bit. The output wordlength of the filter is quantized to 12 bit.

Comparing just programmable filters without look-up table overhead required in a conventional polyphase-based approach with the optimized filter structure derived here leads to a power reduction factor of up to 8 (table 11.6).

Even compared to a highly low-power-optimized filter implementation [16], a power reduction by a factor of 2.6 is achieved [12].

Table 11.6. Comparisons with state-of-the-art implementations, scaled to a 0.5 μm technology

Reference	# taps	$w_x \times w_{\text{coeff}}$	Freq. MHz	Supply V	Techn. μm	P mW	P_{norm} mW
[23]	10	8×10	20	5	1.5	300	7.20
[19]	8	6×6	230	3.7	0.8	426	6.25
[26]	40	10×12	100	5	0.9	3100	4.13
[31]	32	8×8	10	2	1.2	14	6.56
[16]	128	10×12	80	3.3	0.5	415	2.34
this work	12	10×8	40	1.8	0.5	11	0.90

11.4.3.4   Filter redesign.   Due to increased image quality requirements, it was necessary to enhance the resolution of the fractional delay adjustment to 1/16 of the sampling period. As the previously described switchable 12-tap interpolation filter took up 1/3 of the overall power dissipation with a moderate delay resolution of 1/4 of the sampling period, a further optimization of the interpolation filter was inevitable. The transfer function of the new filter was designed as three cascaded transfer functions using a fixed coefficient prefilter with 4 non-zero bits, a switchable filter, and a fixed postfilter with 9 non-zero bits. With this separation it is possible to realize an accurate fractional delay with only 4 switchable filter taps using simple coefficient sets of 6 bit or less. Based on ultrasound simulations of wire phantom images, figure 11.17 shows

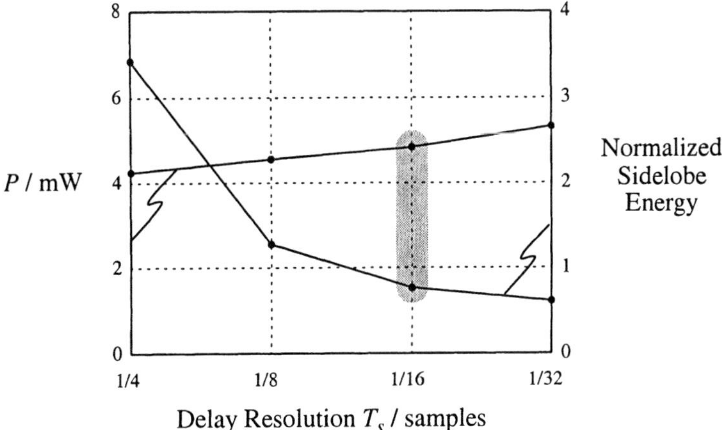

Figure 11.17.   Power dissipation versus image quality

the image quality in terms of sidelobe energy versus the power dissipation of 4-tap filters with different numbers of coefficient sets. The highlighted 16-fold interpolation offers the best tradeoff in this case and was chosen for implementation. Additionally, the image simulations showed that the postfilter can be implemented only once for all parallel channels after the channel summation without visible degradation of the image quality.

Since the implementation of the pre- and postfilter is straightforward, only the implementation and optimization of the switchable filter is discussed here. The chosen filter structure is a direct-form 1 structure with the delay line integrated in the datapath. The coefficient ROM is integrated into the filter datapath in a distributed manner, including pipelining of the delay information, where the local coefficient ROMs contain only the 16 coefficient bits of the 16 subfilters for the respective filter tap and coefficient weight. This way, locality in the SFG of the filter is preserved down to the physical implementation level leading to a power efficient implementation. Compared to a standard precharged ROM architecture, power savings by a factor of 6 are obtained with a dedicated ROM implementation using n- and p-channel devices and optimized for linear addressing and low switching activity of the coefficients. Maximum potential for voltage scaling is preserved by avoiding sign bit extension with a biased number representation [25].

As the switchable filter is based only on positive coefficients, it can be implemented using only one highly optimized carry-save (CS) filter leaf cell (figure 11.18). For the leaf-cell selection the possibility to exploit a quantitatively optimized clustered voltage selection was considered and several alternatives have been investigated with respect to their power dissipation at the minimum possible supply voltage. The complete optimized filter macro, including the pre- and switchable filter part as well as vector merge adders, is based on only a few leaf cells. It is automatically assembled using a datapath generator (section 11.2) and is fully parameterizable with respect to the number of taps and

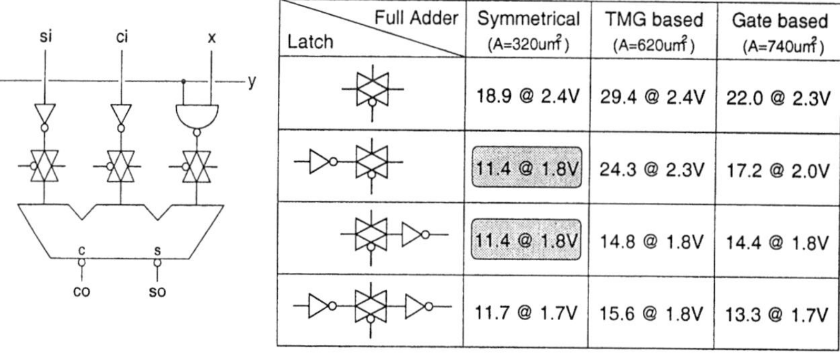

Latch \ Full Adder	Symmetrical (A=320um²)	TMG based (A=620um²)	Gate based (A=740um²)
	18.9 @ 2.4V	29.4 @ 2.4V	22.0 @ 2.3V
	11.4 @ 1.8V	24.3 @ 2.3V	17.2 @ 2.0V
	11.4 @ 1.8V	14.8 @ 1.8V	14.4 @ 1.8V
	11.7 @ 1.7V	15.6 @ 1.8V	13.3 @ 1.7V

Figure 11.18. Filter leaf cell power dissipation in μW at 40 MHz

coefficient sets, input data quantization, coefficient wordlength and values [9].
A comparison to the previous filter design is shown in table 11.7.

Table 11.7. Interpolation filter comparison

Reference	Delay resolution	A	P
4-tap filter	$T_s/16$	$0.75\,\mathrm{mm}^2$	$4.8\,\mathrm{mW}$
6-/12-tap filter	$T_s/4$	$2.35\,\mathrm{mm}^2$	$10.9\,\mathrm{mW}$

11.4.4 Apodization Multiplier

The apodization multiplier allows to improve the directivity of the transducer
aperture and thereby the lateral resolution by multiplying the delayed data with
a channel-dependent constant. For flexibility in transducer choice and depth of
transmit focus, the weighting factor of each channel has to be programmable.
Different types of 12×8 bit multipliers as required for the digital channel have
been evaluated for the apodization. Estimations of the total power dissipation
resulted in a booth-encoded carry-save array multiplier as best choice.

Using optimized pipelining with one latch per full addition and an optimized
carry-save basic cell [17], the maximum possible throughput rate is determined
only by the propagation delay of one full adder. The throughput potential
can be efficiently traded for further power reduction by lowering the supply
voltage [3], which would not be possible for a non-pipelined carry-ripple based
implementation.

11.4.4.1 Architectural level optimizations. Concerning the architectural level, tree structures are expected as being attractive implementations for
low power multipliers [24]. As the apodization multiplier needs a multiplicator
wordlength of only 8 bit, here a tree structure offers no advantage in the count
of required adder circuits. Because of the irregular and long wiring within tree
multipliers, an array structure has been chosen for the implementation of the
apodization multiplier.

A reduction of the number of partial products by an application of the second-
order modified booth algorithm has been evaluated and compared to a regular
carry-save array implementation. As an analysis revealed, the reduction of
adder stages and pipeline elements due to booth recoding reduces the overall
power dissipation, although more complex partial product gates are required.

An attractive alternative for the use of full-adders within the booth-encoded
multiplier is the application of 4:2 compressors [29]. Because of the multipli-

cand wordlength of 8 bit a maximum of only four partial products would have
to be added per result bit-position in two's complement. This would lead to
a very simple and regular design of the 4:2 compressor based approach. One
basic building block could consist of four booth-decoders, four latches and a
compressor cell. Only one of these cells would be necessary per digit position.
However, the multiplicand is an unsigned number which increases the number
of partial products per digit position to 5. Thus one more pipeline stage has to be
implemented which makes the multiplier inefficient compared to the full-adder
implementation.

11.4.4.2   Circuit level optimizations.   The main goal for optimizations
on circuit level was to find the optimum combination of adder-circuit and latch
type to obtain a multiplier basic cell with the lowest possible power consumption. Minimum sized transistors and circuits with low transistor count have been
chosen to reduce parasitic capacitances. At the same time a high throughput
rate has been aspired in order to attain a significant supply voltage reduction.
The full-adder has been implemented using the symmetric 24-transistor cell because of the low number of transistors and the compact design, resulting in very
low power dissipation. The adder has been combined with different types of
latches, each combination running with an appropriate supply voltage to meet
the 40 MHz constraint. The power optimal basic cell for a regular carry-save
array as well as for a booth-encoded carry-save implementation is achieved
using a latch consisting of a transmission-gate and an inverter.

Within the investigated full-adder based multipliers about 20% of the total
power is dissipated by the partial product gates. Therefore, different circuits for
partial product generation including booth-decoding have been evaluated with
respect to the small switching activity factor σ of the booth-encoded multiplicator. Gate based and different types of transmission gate based implementations
for the multiplexer as well as for the EXOR-gate have been combined.

Instead of broadcasting s^i and $\bar{s}^i$ (figure 11.19), a local inverter is used and
because of the low switching activity σ of the multiplicator, the additional
power dissipation can be neglected. Since the height of the cell is dominated
by horizontal wiring, this solution results in a smaller layout area and the wires
for the multiplicand with the higher switching activity can be kept shorter.

11.4.4.3   Implementation.   A power efficient voltage scaling technique
was applied. With only one partial product gate or one full-adder being placed
between two pipeline stages all critical paths are kept short. Thereby, the supply
voltage for the multiplier can be reduced towards the sum of the threshold voltages of a p-channel and an n-channel transistor, while preserving the required
throughput rate.

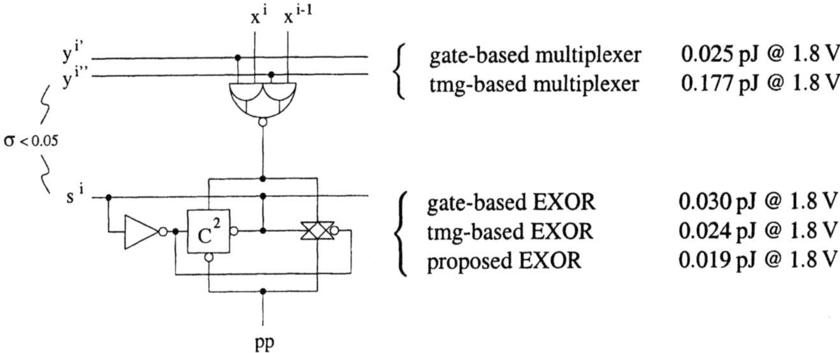

Figure 11.19. Partial product gate for booth-encoded multiplier

For low power dissipation, sign-extension within the multiplier is avoided by using the biased number representation which has been previously applied to low power accumulators [5]. The necessary addition of a correction number can be implemented at almost no extra cost in hardware, because the remaining free input of the first adder in the adder chain can be used.

The critical path through the partial product gates is kept independent from transitions of the booth-encoded multiplicator by implementing large buffers. Because of the very low switching activity of the multiplicator this can be done at almost no cost in power dissipation.

Booth-encoding is performed according to a scheme as proposed in [22] which differs from the conventional booth-encoding scheme. By avoiding a redundant representation of the number zero, the switching activity within the partial product summation circuitry is reduced.

The implementation offers a speed potential allowing a clock frequency three times higher than the required sampling rate of 40 MHz. Lowering the supply voltage down to a supply voltage of 1.8 V results in a power dissipation of 1.8 mW (6.0 mW @ 3.3 V) for the regular carry-save multiplier and 1.6 mW (5.4 mW @ 3.3 V) for the more efficient booth-encoded version, respectively. Comparing the presented apodization multiplier [9] to reported multipliers (figure 11.20) shows that the features of the multiplier elaborated here are well in the trend of low power multiplier design.

11.4.5   Accumulation Chain

The partial results from all channels are summed on-chip in the channel accumulation stage (figure 11.4). Scalability in the number of active channels is attained by cascading several chips. Therefore, an interface allows to add the output of an adjacent LUCS chip to the accumulated result of the on-chip channels.

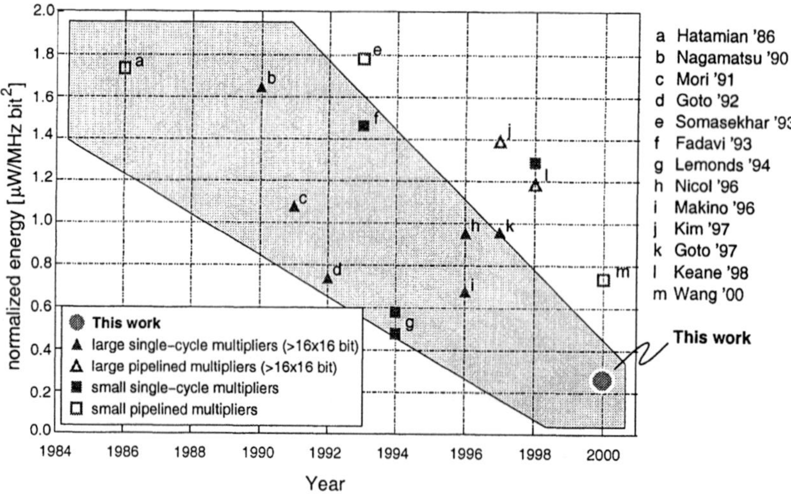

Figure 11.20.   Trend of normalized multiplier energy conversion

All power numbers are based on a layout-topology, where the physical place-ment of the datapaths for the channels represents a stack of 8 or 16 channels with an estimated height of 1 mm per channel. Two possibilities to implement the multioperand-adder, a pipelined tree- and a pipelined chain-structure, are shown in figure 11.21. A comparison of both structures shows that the chain-

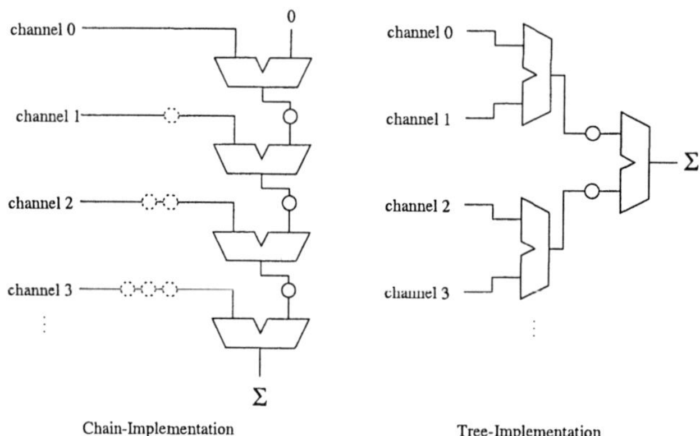

Figure 11.21.   Implementation of the channel summation

implementation results in lower estimated power dissipation. This is due to the larger wiring capacitances for the adder tree (increase of >50% for 16 active channels). For the chain-structure there is no skewing necessary as the individ-ual skewing-delays (in figure 11.21 shown dotted) can be shifted into the coarse delay operation of the beamformer. In this block, the additional delays can be realized without increasing the power dissipation. Again, using highly pipelined carry-save arithmetic yields a higher performance-potential than carry-ripple or carry-select based adders. This speed-advantage can be traded for power reduction by reducing the supply-voltage down to 1.8 V, resulting in a lower possible power dissipation. The relatively long wires connecting the adjacent channels of the beamformer have an estimated capacitive load of 200 fF. Before driving this large load, the carry-save number-representation is transformed into a modified non-redundant carry-save number (figure 11.22), thereby reducing the transition activity on these wires [14]. Although the overhead of the logic

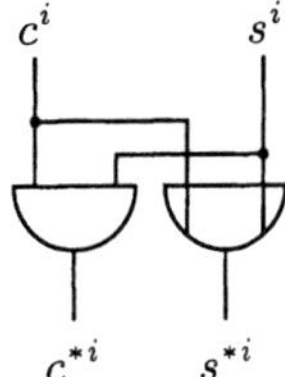

c^i	s^i		c^{*i}	s^{*i}
0	0		0	0
0	1		0	1
1	0		0	1
1	1		1	1

Figure 11.22. Non-redundant carry-save representation

to generate the non-redundant carry-save representation consumes some of the saved energy, a significant overall power reduction can be achieved. Table 11.8 shows the estimated power dissipation based on typical data of the proposed implementation for the multioperand adder. It can be seen that the chain structure consumes about 10% less power compared to the tree solution.

Table 11.8. Power dissipation of the channel summation at 40 MHz

Implementation	P @ 3.3 V	P @ 1.8 V
8 channels, tree	17 mW	5.0 mW
8 channels, chain	16 mW	4.7 mW
16 channels, tree	39 mW	11.6 mW
16 channels, chain	34 mW	10.1 mW

11.5 Conclusion

Figure 11.23 shows a die photograph of the realized test chip with the A/D converter on the left, and the digital beamformer on the right. It comprises the complete functionality of a single processing channel and is designed to be cascadable with up to 32 chips to provide the functionality of two cascaded 16-channel beamformer chips. Table 11.9 shows the power dissipation breakdown for the optimized digital processing channel. Compared to the estimated

Table 11.9.   Power dissipation breakdown per digital channel

Block	Proposal (20 MHz)	Test Chip (40 MHz)	Update (40 MHz)
Delay Memory	30.0 mW	6.2 mW	6.2 mW
Delay Generator	6.4 mW	2.8 mW	3.1 mW
Control Unit	2.4 mW	2.1 mW	2.1 mW
Interpolation	48.0 mW	10.9 mW	4.8 mW
Apodization	3.2 mW	1.3 mW	1.3 mW
Adder Chain	0.4 mW	2.1 mW	2.1 mW
Clock	–	5.8 mW	5.8 mW
Σ	90.4 mW	31.2 mW	25.4 mW

power dissipation of the project proposal, and taking into account the doubled sampling rate, a power reduction by a factor of 7 was achieved for the updated design. For minimum system power dissipation, clustered voltage scaling was reduced to two levels to avoid excessive power overhead for off-chip DC/DC conversion. The digital core dissipates only 25.4 mW @ 40 MHz from a dual 1.8 V / 3.3 V power supply, thus now shifting the optimization focus towards the A/D converter with an estimated power dissipation of 70 mW [9].

The design and quantitative optimization of a digital beamformer chip for use in hand-held ultrasound scanner systems has been presented. By consequently applying a quantitative optimization strategy, the implementation of a single chip beamformer comprising 16 parallel channels dissipates about 400 mW in the digital signal processing part of the beamformer. Thereby, high quality battery-powered ultrasound scanner systems become possible. Flexibility is given by incorporating additional features like beam steering capability together with a large delay memory, thus making the chip suitable for high-performance desktop ultrasound systems.

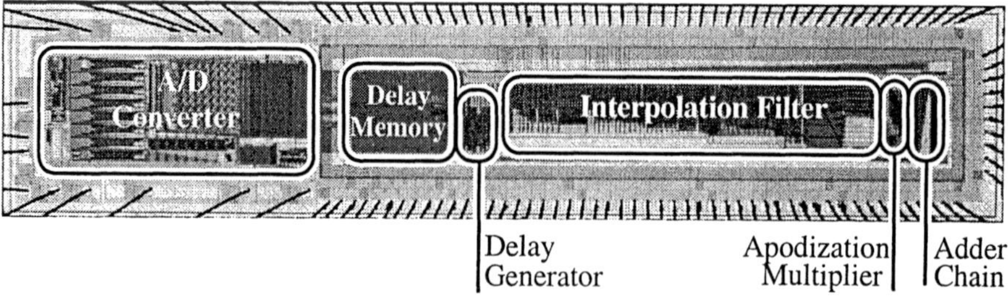

Figure 11.23.   Photograph of the realized single channel test chip

References

[1] J. R. Van Aken. An efficient ellipse-drawing algorithm. *IEEE Computer Graphics and Application Magazine*, 4(9):24–35, 1984.

[2] M.-H. Bae. *Focusing Delay Calculation Method for Real-Time Digital Focusing and Apparatus Adpoting the Same.* US Pat. No. 5,836,881, 1998.

[3] A. P. Chandrakasan, S. Sheng, and R. W. Brodersen. Low-power CMOS digital design. *IEEE JSSC*, 27(4):473–484, 1992.

[4] P. D. Corl. *Digital Ultrasound System with Dynamic Focus.* US Pat. No. 4,974,211, 1990.

[5] M. D. Ercegovac and T. Lang. Low-power accumulator (correlator). In *Proc. IEEE ISLPE*, pages 30–31, August 1995.

[6] H. T. Feldkaemper, R. Schwann, V. Gierenz, and T. G. Noll. Low power delay calculation for handheld ultrasound beamformers. In *Proc. IEEE UFFC Symposium*, pages 1763–66, October 2000.

[7] T. Gemmeke, V. S. Gierenz, and T. G. Noll. Scalable, power and area efficient high throughput viterbi decoder implementations. In *Proc. 27th ESSCIRC*, pages 476–479, September 2001.

[8] V. Gierenz, O. Weiss, T. G. Noll, I. Carew, J. Ashley, R. Karabed, and J. Rae. A 550 mb/s radix-4 bit-level pipelined 16-state 0.25-um cmos viterbi decoder. In *Int. Conf. on Application-specific Systems, Architectures and Processors*, pages 195–199, July 2000.

[9] V. S. Gierenz, R. Schwann, and T. G. Noll. A low power digital beamformer for handheld ultrasound systems. In *Proc. 27th ESSCIRC*, pages 276–279, September 2001.

[10] M. Hashimoto, M. Nomura, K. Sasaki, K. Komatsuzaki, H. Fujiware, T. Honzawa, K. Abe, T. Tachibana, and N. Kitagawa. A 20-ns 256k×4 FIFO memory. *IEEE JSSC*, 23(2):490–499, 1988.

[11] R. A. Hawley, B. C. Wong, T. Lin, J. Laskowski, and H. Samueli. Design techniques for silicon compiler implementations of high-speed FIR digital filters. *IEEE JSSC*, 31(5):656–667, 1996.

[12] C. Henning, R. Schwann, V. Gierenz, and T. G. Noll. A low power reconfigurable 12-tap FIR interpolation filter with fixed coefficient sets. In *Proc. 26th ESSCIRC*, pages 120–123, September 2000.

[13] K. Itoh, K. Sasaki, and Y. Nakagome. Trends in low-power ram circuit technologies. *Proceedings of the IEEE*, 83(4):524–543, April 1995.

[14] Ch. Lütkemeyer and T. G. Noll. An optimized update processor for high-throughput adaptive equalizers. In *Proc. 1997 ASAP, Zürich, Switzerland*, pages 519–528, July 1997.

[15] T. H. Meng, B. M. Gordon, and E. K. Tsern. Portable video-on-demand in wireless communication. In *Low Power Design Methodologies*. Kluwer Academic Publishers, Dordrecht, Netherlands, 1996.

[16] C. J. Nicol, P. Larsson, K. Azadet, and J. H. O'Neill. A low power 128-tap digital adaptive equalizer for broadband modems. In *IEEE ISSCC Digest of Technical Papers*, pages 620–627, February 1997.

[17] T. G. Noll. Carry-save architectures for high-speed digital signal processing. *Journal of VLSI Signal Processing*, 3(1–2):121–140, 1991.

[18] T. G. Noll. *VLSI Implementations for Image Communications*, pages 171–211. Elsevier, 1993.

[19] D. Pearson and et al. 250 MHz digital FIR filters for PRML disk read channels. In *IEEE ISSCC Digest of Technical Papers*, pages 80–81, February 1995.

[20] D. K. Petersen and G. S. Kino. Real-time digital image reconstruction: a description of imaging hardware and an analysis of quantization errors. *IEEE Transactions on Sonics and Ultrasonics*, 31(4):337–351, 1984.

[21] J. G. Petrofsky. *Method and Apparatus for Distributing Focus Control with Slope Tracking*. US Pat. No. 5,724,972, 1998.

[22] R. V. K. Pillai, D. Al Khalili, and A. J. Al Khalili. Energy delay analysis of partial product reduction methods for parallel multiplier implementation. In *Proc. ISLPED*, pages 201–204, August 1996.

[23] D. Reuver and H. Klar. A configurable convolution chip with programmable coefficients. *IEEE JSSC*, 27(7):1121–23, 1992.

[24] T. Sakuta, W. Lee, and P. Balsara. Delay balanced multipliers for low power / low voltage DSP core. In *Proc. IEEE Symposium on Low Power Electronics*, pages 36–37, October 1995.

[25] O. Salomon, J.-M. Green, and H. Klar. General algorithms for a simplified addition of 2's complement numbers. *IEEE JSSC*, 30(7):839–844, 1995.

[26] L. Thon and et al. A 240 MHz 8-tap FIR filter for disk-drive read channels. In *IEEE ISSCC Digest of Technical Papers*, pages 82–83, February 1995.

[27] W. Ulbrich. *Design of MOS VLSI Circuits for Telecommunications*, pages 236–271. Prentice-Hall, 1985.

[28] H. Wang and P. C. Liu. Double-edge-triggered address pointer for low-power high-speed FIFO memories. *Electronics Letters*, 33(5):387–389, February 1997.

[29] A. Weinberger. 4-2 carry-save adder module. *IBM Technical Disclosure Bulletin*, 23(8):3811–14, January 1981.

[30] O. Weiss, M. Gansen, and T. G. Noll. A flexible datapath generator for physical oriented design. In *Proc. 27th ESSCIRC*, pages 408–411, September 2001.

[31] C. Xu, C.-Y. Wang, and K. K. Parhi. Order-configurable programmable power-efficient FIR filters. In *Proc. 3rd Int. Conf. on High Performance Computing*, pages 357–361, December 1996.

Chapter 12

EPILOGUE

Christian Piguet

CSEM: Centre Suisse d'Electronique et de Microtechnique, Neuchâtel, Switzerland

LAP-EPFL, Lausanne, Switzerland

christian.piguet@csem.ch

> "My interest is in the future,
> because I am going to spend
> the rest of my life there"
> *Charles F. Kettering,*
> *American Inventor*

Future Perspectives

Power consumption reduction, after the well-known speed and area constraints, is certainly the new challenge in the design of integrated circuits. The design of such chips, in very deep submicron technologies down to 0.13 µm, with several hundreds of millions of transistors, supplied at less than 1 Volt, is a very challenging task if design, verification, debug and industrial test are considered.

The microelectronic revolution is fascinating; 55 years ago, in late 1947, the transistor was invented, and everybody knows that it was by William Shockley, John Bardeen and Walter H. Brattein, Bell Telephone Laboratories, which received the Nobel Prize in Physics in 1956. Probably, everybody thinks that it was recognized immediately as a major invention. Not at all!. When Bell Telephone Laboratories announced the invention of the transistor on June 30, 1948, six months later, the general press was almost indifferent. The New-York Times carried the news on the next to the last page of the paper, with four short paragraphs. Even technical journals were slow to appreciate the inherent possibilities of the transistor. To stir enthusiasm for the device, Bell Laboratories licensed it freely in the US and publicized it extensively in seminars and papers. A free license for the transistor! This means that nobody had understood what was such a device! Its direct competitor was the vacuum tube, a strongly established commercial product. Engineers did not like transistors; they pre-

D. Soudris et al. (eds.), Designing CMOS Circuits for Low Power, 271–274.
© 2002 *Kluwer Academic Publishers. Printed in the Netherlands.*

ferred vacuum tubes. The first market pull came from the hearing aids market, for which miniaturization was a must. The first transistorized hearing aid was announced by Sonotone in February 1953; it contained 5 transistors, but still required a pair of miniaturized tubes for the input and driver stages.

How the transistor was invented? Was such a device really needed? The answer is simple; the transistor had its origin in scientific theory rather than in technological developments. From the beginning of the last century, more and more studies have been performed on solid-state physics, metals and semiconductors. In 1935, a patent was issued to O. Heil for a field effect triode, although he was not able to explain how it worked. It is ironic that the concept of field effect transistors, so marvelously simple, provides practical implementations after the invention of the far more complex bipolar transistor. After World War II, which interrupted many studies in semiconductor materials, new research programs have been started. Bell Labs had such a program.

The Bell Labs group, including the three inventors of the transistor, decided to limit their research to germanium and silicium, the simplest semiconductors. The group believed that the only explanation for the continued lack of understanding of semiconductors - despite intensive worldwide research - was the diffuse experimentation on so many different complex materials. Silicon and germanium, on the other hand, were elemental, simple; moreover, their atomic structure had revealed the same strong covalent binding as a diamond, and thus their crystals tended to be strikingly free of defects. This is a main point to realize working devices; keep them as simple as possible.

In 1958, it was the invention of the integrated circuit. So pressing was the need for fabricating entire circuits in a single semiconductor block that had neither Jack Kilby nor Robert Noyce conceived the integrated circuit in 1958, someone else surely would have done! The device was qualified "as the most significant development by Texas Instrument since the commercial silicon transistor". Also in 1958, the first field-effect transistor was working. It was called "Tecnitron" by its creator, S. Teszner, working in France. In 1962, RCA was fabricating multipurpose logic block comprising 16 MOS FETs on a single chip. By 1963, RCA had fabricated large arrays of several hundreds MOS devices. They were however extremely sensitive to static charge, supply voltage was higher than those of bipolar transistors, and the speed was slower. Production was also plagued with oxide defects. In mid-1965, only two companies were producing MOS ICs - General Microelectronics and General Instruments - the other companies took a wait and see stance.

Everybody knows how the story continues: MOS technology, the first microprocessor in 1971, CISC machines, RISC microprocessors in 1981, superscalar and VLIW machines today, with a shift in CMOS technology in 1984-1985 with the 80386 and 68020. The 1999 SIA Roadmap predicts in 2014 a 0.035 mm CMOS process (probably SOI) with 19 billions of transistors for high-performance microprocessors, 0.6 Volt, 180 watts and more than 10 GHz as local frequency.

Intel, in late 2001, announced the smallest transistor ever built: 15 nanometers (0.015 µm) of gate length capable of switching at 2000 GHz! It could be used in 20 GHz microprocessors around 2009. This evolution is clear in the 2001 SIA Roadmap that predicts 0.032 µm technology in 2005 instead of 0.065 µm in the 1999 SIA edition.

As a result, there are today more transistors in the world (10^{17}) than ants (10^{16}).

However, what is the future of microelectronics? Are we close to the end of this marvelous story? Is the future belonging to nanotechnologies that could completely replace microelectronics (although 0.015 µm transistors have a 15 nanometers length)? Nano-devices have been constructed, capable of switching a current or single electrons with a ratio between the on/off current of 1 thousand to 1 million. Such elements could be promising as their sizes of some nanometers and extremely low or no power consumption are very attractive. Carbon nanotubes, quantum dots, single electron devices or molecular switches are the most promising nano-devices. For instance, a carbon nanotube has a diameter of 1 nanometer, and depending on its diameter, is a semiconductor device (otherwise, it is a conductor, not usable as a switch). However, if one over 10 nanotubes is semiconductor, how to select and interconnect the semiconductor ones to provide a useful logic function?

Quantum dots are based on the Coulomb blockade effect and electrons are moved one by one from dots to dots. They have been constructed atom by atom by atomic force microscopes. Due to noise, it is better to construct cellular automata with several dots, and to define a given state of the automat as the logic "0" and another state as "1". Majority gates have been demonstrated, as well as AND/OR gates. The main problem is still how to interconnect these gates to provide useful functions. Furthermore, it is hard to construct atom by atom a complete chip with several billions of elements.

Design methods could be completely different from today, as nano-devices could be constructed randomly, without any predefined schematic or layout. However, a useful function could emerge from this huge number of nano-devices, or some auto-organization could occur. It is a little bit similar to natural selection for which only the useful functions will survive. But it will be hard to design a predefined and very complex function like a Pentium microprocessor.

It is very hard to make predictions, especially for the future (Mark Twain). However, the most probable future is that microelectronics will be used until perhaps 2020. Then it will be not replaced by nanoelectronics, but both technologies, i.e. microelectronics and nanoelectronics will co-exist with probably different applications.

It means that there are many opportunities for new European programs in the context 6th IST Framework about very low power design in microelectronics and obviously, in nanoelectronics.

Index